Lecture Notes in Civil Engineering

Volume 82

Lecture Notes in Civil Engineering (LNCE) publishes the latest developments in Civil Engineering - quickly, informally and in top quality. Though original research reported in proceedings and post-proceedings represents the core of LNCE, edited volumes of exceptionally high quality and interest may also be considered for publication. Volumes published in LNCE embrace all aspects and subfields of, as well as new challenges in, Civil Engineering. Topics in the series include:

- Construction and Structural Mechanics
- Building Materials
- Concrete, Steel and Timber Structures
- Geotechnical Engineering
- Earthquake Engineering
- Coastal Engineering
- Ocean and Offshore Engineering; Ships and Floating Structures
- Hydraulics, Hydrology and Water Resources Engineering
- Environmental Engineering and Sustainability
- Structural Health and Monitoring
- Surveying and Geographical Information Systems
- Indoor Environments
- Transportation and Traffic
- Risk Analysis
- Safety and Security

To submit a proposal or request further information, please contact the appropriate Springer Editor:

- Mr. Pierpaolo Riva at pierpaolo.riva@springer.com (Europe and Americas);
- Ms. Swati Meherishi at swati.meherishi@springer.com (Asia - except China, and Australia, New Zealand);
- Dr. Mengchu Huang at mengchu.huang@springer.com (China).

All books in the series now indexed by Scopus and EI Compendex database!

More information about this series at http://www.springer.com/series/15087

Joseph Lim

Oil Rig and Superbarge Floating Settlements

Springer

Joseph Lim
School of Design and Environment
National University of Singapore
Singapore, Singapore

ISSN 2366-2557 ISSN 2366-2565 (electronic)
Lecture Notes in Civil Engineering
ISBN 978-981-15-5296-0 ISBN 978-981-15-5297-7 (eBook)
https://doi.org/10.1007/978-981-15-5297-7

This Springer imprint is published by the registered company Springer Nature Singapore Pte Ltd.
The registered company address is: 152 Beach Road, #21-01/04 Gateway East, Singapore 189721, Singapore

Preface

Floating cities have long been part of architectural and engineering history. Although contemporary versions have intensified in the last few decades, they are conceptually similar. They exist as structures built on modular platforms, which float directly on water and can be connected into larger surfaces. Unlike these typologies where the architecture exists above the raft and can be anything, this book explores the structures of oil rigs as elements of form to define space in new architecture.

An architecture of high-density configurations was made from megastructure vessels, which can be towed or sailed to locations with benign environments and anchored in a range of settlement forms to meet emergent need for human habitation in an age of rising sea levels. This book is an outcome of my design research studio with students in the Masters of Architecture Programme at the National University of Singapore.

I would like to thank my students—Christopher Wijatno, Davis Wong, Chen Qisen, Wang Yigeng, Sakinah Halim, Bek Tai Keng, Roy Tay, Nazirul Salleh, Judy Lee, Khairul Anwar, Cao Jinming, and Chen Shu Hua, without whose efforts this book could not have been written.

I would also like to thank Keppel Offshore and Marine consultants—Mr. YY Chow, Dr. Aziz Merchant, Mr. Gary Wong, and Ms. Wang Rong.

I am grateful to Prof. Wang Chien Ming for his comments and insight.

Last but not least, special thanks to Ismurnee Khayon for her critical input and technical support.

Singapore, Singapore | Joseph Lim

Introduction

This study began with the idea of repurposing oil rigs. To contextualize the task beyond isolated conversions, a wider platform for investigation was used to garner design research propositions responding to emergent need in the light of climate change. Vessels from the oil and gas industry were repurposed to offshore and nearshore settlements as counterpoints to megacities in an effort to create ecological footprints that are more sustainable. The continued depletion of natural resources leading to land and resource scarcity has resulted in frequent displaced communities worldwide.

Questions concerning alternative forms of settlement have led to explorations at sea. What if we floated on water instead of consuming land inefficiently? And could we use wave energy instead of nuclear energy? Could we replenish food supply and regenerate marine eco-diversity? How would our lives be shaped by new offshore settlements? What would we use as structures for shelter, farming scaffold, and recreation?

Floating cities emerged in the 1960s with Buckminster Fuller's Triton City and Kenzo Tange's A plan for Tokyo. Current manifestations include Vincent Callebaut's Lilypad, the Seasteading Institute's Floating Cities, and the mile-long Freedom Ship accommodating 60,000 people. As an alternative to these examples, three types of vessels in the oil and gas industry, namely, the jack-up platforms, the semi-submersibles, and the superbarges are repurposed as small footprint habitable propositions to accommodate 20 percent of a projected global population of 8.1 billion people in 2025 (United Nations).

Floating settlements are spatially conceived with food and energy estimates for housing, recreation, and education at sea, as well as post-disaster healthcare and resettlement for nearshore deployment. Population densities of oil rig floating settlements exceed those of current proposals using modular rafts and are closer to a viable proposition addressing the issue of population growth.

This book explores the idea of repurposing oil rigs and barges for human habitation in nearshore and offshore marine settlements. It suggests developmental options to cold stacking or decommissioning oil rigs with serviceable physical structures or the immediate deployment of recently constructed oil rigs in a depressed market when operational costs will not yield returns.

Chapter "Industry Challenges as Basis for Repurposing Oil Rigs and Barges" discusses the environmental threat of the oil and gas industries at all stages of exploration, operation, and decommissioning, to marine ecosystems. The scale of the technical complexities and financial implications is a challenge to both environmentalists and industry seeking to protect the environment. The reuse of the abandoned structures to minimize disruption to marine ecologies is one alternative. However, the periodic occurrence of unused good condition rigs due to oil price fluctuations and time lags in construction is costly to owners. Thus, they may explore viable solutions through repurposing. This study focuses on the repurposing of jack-ups and semi-submersibles as they have tall structures and/or a wide range of deck areas to explore additional built-up areas. The implications of adding more useable floor area and increasing structural load pose challenges to capsizing. Design strategies depending on the type of rig are discussed to overcome instability and to give an idea on cost implications.

Chapter "Intensifying Food and Housing at Sea" portrays emergent need in the context of climate change and rising global population and proposes as counterpoints to megacities, resilient offshore and nearshore solutions to food production and supply, energy and wastewater management. Jack-ups, semi-submersibles, and barges are repurposed into specialized vessels. Of key interest is a typology of a high-density housing on jack-up or semi-submersible to meet the rising population in capacities that surpass cities on modular floating rafts. The spatial quality of the architectural conversions capitalized on the use of derrick and deck structure, moonpool and pontoons, to create spaces for communal gathering and for recreation. Infrastructure became architecture. The density and living quality of these propositions are explored through their range of communal recreation spaces. These are benchmarked with those of landed housing blocks in Singapore and landmark international examples.

Chapter "Optimal Settlement Size and Masterplan Strategies" studies the optimal sizes of flotillas, which attempt a circular economy involving aquaponics production and fish farming to support population sizes adequately sustained by renewable sources of technology. Estimates of capacity are made from specialized vessels designed for accommodation, food production, waste management, and medical care. These are then grouped into flotillas for 40,000 persons to be agglomerated into entire settlements. The vessels will each form elements of water-townscapes where city square is interpreted as a water court between vessels docked in specific configurations; water foyers beneath semi-submersible decks used as outdoor living spaces; and wharfs for recreation and commerce. Combinations of flotilla forms generate variants of entire settlements using jack-ups for up to 40 m depth of water and semi-submersibles for deepwater locations. The population densities of these settlements are compared with those of current floating cities, township densities in Singapore, and other high-density cities. Using UNESCO guidelines for offshore settlements to estimate the spacing between settlements, these autonomous settlement fleets conceived as sea cities occupy 38 sq. km of sea space and are an alternative to land-based mega cities accommodating 100,000 persons per sq. km. 20% of the projected population in the year 2025 can be accommodated on 6,510 oil rig settlements spaced 240 sq. km apart over 54.25 M sq. km of sea. Related to high-density settlements is the concern with environmental quality through the provision of public space. One indicator is the rate of recreational area per person. The space standards of each marine inhabitant—at 50 sqm recreational area per person—exceed 27 sqm in London, which is equivalent to Amsterdam, with one settlement type reaching 115 sqm per person equivalent to Vienna. Singapore by comparison has 65 sqm of recreational area per person.

Chapter "Post-Disaster Applications for Displaced Populations" generates a separate fleet of floating settlements for post-disaster relief operations. Given an increase in the frequency of natural disasters caused by climate change, there is a corresponding increase with the number of displaced populations. For both climate and political refugees, the timeliness of aid for survival is not the only issue. Beyond the immediate need of restoring damaged infrastructure and providing medical aid for survivors is their long-term rehabilitation and recovery. The United Nations (UN) guidelines for post-disaster and humanitarian aid are adopted for response times in providing aid with a flotilla comprising semi-submersibles converted into a 1,500-bed hospital. UN space standards for post-disaster housing are fitted in superbarge accommodation units with additional units for cremation, parks, and farm plots to enable survivors to start up activities for self-subsistence. The idea is not only to create a fully deployable fleet for immediate disaster relief, but also to provide housing, production, and communal facilities for establishing and maintaining a settlement in the long term.

Chapter "Other Forms of Repurposing" explores six specific applications when repurposing oil rigs into specialized vessels. The proposals are premised on nearshore/offshore prototypes which improved by design, the challenging issues of inefficient land plot requirements (e.g. prisons, funeral facilities, resorts) or unsustainable business models (e.g. Olympic stadia) or which improved industrial workflow (e.g. LNG bunkering). This leads to a streamlining of processes saving time and land resources. A jack-up funeral complex integrates wake halls, chapels, funeral parlors, and crematorium in a single nearshore facility. With the exception of religious burial practices, cremation is a land-saving option with the advantage of resource recovery.

A prison on a semi-submersible is a solution to unsanitary conditions of many prisons worldwide that are operating in severe overcapacity. An offshore prison obviates the need for land set aside for incarceration of future convicts. It is designed for the rehabilitation of inmates involved in daily routines of farming production and acquiring skills for communal integration upon release. Design strategies provide higher cell capacity with fewer prison staff in a compact form. This facility uses waterscapes for recreational sporting activities and moats as escape prevention.

The construction of sports stadia for the Olympics and the FIFA World Cup series in separate cities is often a risky financial investment when stadia were underutilized in post-event scenarios. The proposal is for a fleet of semi-submersibles converted into floating stadia capable of staging international sporting competitions involving games, courts, and fields and which can be mobilized and reused at sea. The facility can be leased intermittently to organizations that may not have the means for large capital investment, but that can rent a number of semi-submersible stadia and recover costs through ticket sales.

Conventional port planning separates bunkering facilities from transshipment hubs, which sort out container cargo for re-exports. This proposition uses jack-up oil rigs arranged radially to integrate LNG bunkering with wharves in a nearshore facility to reduce time taken for loading and unloading cargo separately from time needed for bunkering. This facility is designed with the turning radii of tankers and cargo ships to maximize the number of berths in a single facility with a sizeable cargo TEU turnover. It can be conceived as a satellite facility to improve port efficiency.

With jack-up rigs, recreational parks requiring large amounts of water can be placed nearshore so as to free up coastal land for strategic and essential developments. Larger deck and airspace come from combining two jack-ups. A water treatment facility minimizes the high volume of water and energy usage by designing the layout of water rides in ways to minimize water pumping. Most significantly, the large ecological footprints can be avoided in the proposed jack-up waterpark.

Resort developments on coastal land risk damage to sensitive marine ecosystems. The proposition of offshore resort to mitigate environmental impact is explored with a semi-submersible rig and a jack-up rig of 5,000 sqm deck area, to accommodate the number of hotel rooms and a range of facilities recommended in the hospitality industry. Accessible by sea vessels, these studies have capacities larger than the Seaventures Dive Resort on a converted jack-up rig off Sipadan, Borneo in the Celebes Sea.

Contents

About the Author

Dr. Joseph Lim is Associate Professor and Director of the Masters of Architecture Programme at the Department of Architecture in National University of Singapore. He has a special interest in prototypical structures addressing emergent spatial and environmental need. His design projects have won SIA Design Awards and international awards including an Honorable Mention for Lee Treehouse at the Kenneth F. Brown Asia Pacific Culture and Architecture Design Award in 2003; and a Merit Win for Dragon Bridge in URA Southern Ridges Bridge International Design Competition in 2004. Joseph's focus in Industry and Infrastructure explores land intensification through architectural investigations, which have significant implications on the planning of future settlements, townships and infrastructure. In 2008, he pioneered corporation-funded design research studios, collaborating with Jurong Town Corporation in 2008–2010. Maritime and Port Authority of Singapore-funded research studio followed in 2014, which explored an urban design study for Tuas Port 2027 and was exhibited at the Singapore Maritime Week Exhibition and SG50 NUS Exhibition in 2015. The research studio was also featured in a Channel News Asia documentary—Futuropolis Episode 2: Keeping Afloat televised in 2017—on the effects of rising sea levels and the viability of floating settlements. Joseph is the author of Bio-structural Analogues in Architecture (2010) and Eccentric Structures in Architecture (2012); both publications are in their third reprint and have been translated in Korean and Chinese languages. His latest publication, Skybridge investigates forms of air rights structures built over Ayer Rajah Crescent, Singapore and the research conclusions were presented at the Jurong Town Corporation i3C Symposium on Industrial Infrastructure Research in 2017.

Design Authors

Christopher Wijatno Jack-up Prison
Davis Wong Jack-up Fish Production and Semi-submersible Fish Wharf
Chen Qisen Jack-up and semi-submersible Housing
Wang Yigeng Jack-up Funeral Facility and Barge crematorium
Sakinah Halim Superbarge post-disaster Housing and semi-submersible WWTP
Bek Tai Keng Semi-submersible Hospital
Roy Tay Jack-up Waterpark
Nazirul Salleh Semi-submersible Logistics Unit
Judy Lee Jack-up Luxury Resort
Khairul Anwar Semi-submersible Family Resort
Cao Jinming Semi-submersible LNG Bunkering Facility
Chen Shu Hua Semi-submersible Sports Facility

Industry Challenges as Basis for Repurposing Oil Rigs and Barges

Abstract

This chapter discusses the environmental threat of the oil and gas industry at all stages of exploration, operation and decommissioning, to marine ecosystems. The scale of the technical complexities and financial implications are challenges to both environmentalists and industry seeking to protect the environment. The reuse of the abandoned structures to minimize disruption to marine ecologies is one alternative. However the periodic occurrence of unused good condition rigs due to oil price fluctuations and time lags in construction is costly to owners. Thus they may explore viable solutions through repurposing. This study focuses on the repurposing of jack-ups and semi-submersibles as they have tall structures and/or a wide range of deck areas to explore additional built-up areas. The implications of adding more useable floor area and increasing structural load pose challenges to capsizing. Design strategies depending on the type of rig are discussed to overcome instability and to give an idea on cost implications.

1 Oil Exploration and Environment

The impact of oil and gas exploration on marine ecosystems is of concern to both industry and environmentalists. On-site pollution continues even after operations have ceased and the scenario of relying on non-fossil sources in future leads to the possibility of repurposing large numbers of vessels such as oil rigs to avoid obsolescence. Routine activities at stages of exploration, production and decommissioning have detrimental effects on the marine environment. Whereas sediment re-suspension, sea anchor and pipeline burial at installation stages are restricted to a 100 m radius on site, discharges of water-based and oil based drilling muds including produced water extend to a radius over 2 km. Impacts on marine ecologies at sea floor can be experienced as far as 300 m from source [3].

Operational accidents with major oil spills have been recorded in history. Of eleven cited worst disasters in operations, the majority of accidents were attributed to strong waves, blowouts and explosions [20].

(1) July 1988—gas leakage from a condensate pipe that ignited gases and a series of explosions (Piper Alpha platform, North Sea, United Kingdom)
(2) March 1980—capsizing caused by structural failure of bracing—undetected fatigue crack in weld—to one of five legs on semi-submersible due to 12 m high waves (Alexander L Kielland semi-submersible, North Sea, Ekofisk Field, Norway)
(3) November 1989—capsizing due to a 3.8 km drill-pipe in its derrick creating a high CG when toppled by 12 m high waves (Seacrest drillship, South China Sea, Thailand)
(4) February 1982—capsizing due to broken porthole window to ballast control room by 20 m high waves and 190 km/hr wind (Ocean Ranger semi-submersible, Hibernia Field, off Newfoundland, Canada)
(5) October 1983—capsizing by 138.9 km/hr winds (Glomar Java Sea drillship, South China Sea)
(6) 5 November 1979—capsizing due to punctured deck and flooding in fierce wind while under tow (Bohai 2 jack-up, Gulf of Bohai, off China)
(7) August 1984—blowout fire and explosion at central platform (Enchova Central Platform, Campos Basin, Rio de Janeiro, Brazil)
April 1988—blowout during conversion from oil to gas production (Enchova Central Platform, Campos Basin, Rio de Janeiro, Brazil)
(8) July 2005—fire from ignition of gas emitted from ruptured gas export riser due to collision with multi-purpose support vessel caused by strong waves (Mumbai High North, Arabian Sea)

J. Lim, *Oil Rig and Superbarge Floating Settlements*, Lecture Notes in Civil Engineering 82,
https://doi.org/10.1007/978-981-15-5297-7_1

(9) October 2007—ignited gas leak from collision between jack-up cantilever deck and production valve tree of platform caused by 130 km/hr wind and 8 m waves (Usumacinta jack-up, Campeche Bay, Gulf of Mexico)
(10) June 1964—blowout on drilling barge (C.P. Baker barge, Gulf of Mexico)
(11) April 2010—explosion from natural gas gush from concrete core of well led to capsized rig caused by ruptured riser and 87 day oil spill costing BP $65 bn (Deepwater Horizon semi-submersible, Gulf of Mexico).

Oil wells in operation are commonly abandoned temporarily when their yield potential is being re-evaluated or when a plan is being developed to overcome a drilling problem including damage from a storm. It is also likely that wells discontinue production until such time when oil prices rise to a level that make drilling operations viable. Temporarily abandoned wells will be sealed with fewer plugs and subject to less testing with a metal cap to stop corrosion from seawater [26].

The nature of the environmental threat posed by oil wells abandoned as early as the 1940s lies in their repressurizing as their structures gradually corrode and weaken over time. From 1989 to 1994, the General Accountability Office, warned U.S. Congress of environmental disaster caused by leaks from offshore abandoned wells. The Office cited Environmental Protection Agency data, which estimated then that up to 17% of the U.S. wells on land, were improperly sealed. 17% in the case of offshore wells would amount to 4,600 badly sealed wells in the Gulf of Mexico. According to a March 2018 data by Bureau of Ocean Energy Management and Enforcement, 52,964 wells were drilled in the Gulf of Mexico, of which 27,405 have been permanently abandoned.

In 2006 the Environmental Protection Agency reported the lack of proper planning and implementation in the overall abandonment of wells on land. Leakages continue from wells abandoned in recent decades despite regulatory plugging procedures. Repairs are routine in continued 're-plugging' or 're-abandonment'. 30–60 m long cement plugs in targeted zones of wells are used to block off oil or gas flow permanently. Heavy drilling fluid is added. In the offshore context, the piping is to be severed 5 m below the sea floor [26].

There were 12,000 offshore installations in operation globally after the first offshore drilling platform was commissioned in 1947. However, there were 2.5 million abandoned wells in the U.S and 20–30 million worldwide, which were not monitored or inspected for leakage [1].

According to a 2001 study commissioned by the U.S. Minerals Management Service, some abandoned oil wells in the Gulf of Mexico may be leaking crude oil. Abandoned wells that continue to leak oil, brine and greenhouse gases adversely affect the marine environment by altering marine habitats and polluting aquifers. Even fully depleted wells can flow again from two common causes. First is through the fracking process where pressurized fluid or gas injections are used to stimulate nearby wells, and this activates the abandoned well. The second cause is from pressure exerted by underlying aquifers. Some of abandoned wells are capable of explosive, toxin-spreading detonations due to primitive capping technologies. Although sealed with cement, each abandoned well can also be repressurized by accidents, earthquakes and natural erosion [12].

2 Environmental Risks of Decommissioning

Decommissioning oil platforms involve technical complexities, which place marine environments at risk. Oil platforms routinely use lubricants to protect their drills. The oil- or water-based lubricants contain metal additives, such as barium sulphate, which increase the viscosity and density of the lubricant and which help prevent well blowouts. Used lubricants are returned to the surface for reuse after extraction and separation from the mixture of lubricants and seabed fragments. Only from 2001 was this remaining mixture banned by OSPAR (an international treaty of 15 European nations that aims to protect the environment of the Northeast Atlantic) from being discharged onto the seabed. Previously discharged underwater slag has accumulated to building height mounds of several meters on the seabed, poisoning entire ecosystems. It is estimated that there are more than a million cubic meters of slag on 1,650 km^2 the North seabed [21].

OSPAR demands the massive superstructure of a platform to be removed and returned to shore for recycling. However it is unlikely that the piles in the seabed remain intact while most of the topside equipment is being removed from the water above. The topside includes the entire working core of the oil or gas rig, the drilling, production and processing modules, the helicopter deck and the accommodation for the crew [16].

When decommissioning oil rigs, the complete removal of a large platform while leaving in place its the drill cutting pile, would result in polluting a large area of sea. Leaving behind the toxic cutting piles after removing the steel rigs poses more environmental risk than benefit. This is because the structure, which has provided the habitat for thriving marine colonies, would be removed.

3 Environmental Benefits of Leaving Rigs in Place

In decommissioning, steel rig structures are severed just below the seabed, and either towed ashore in one piece or dismantled offshore. Early concrete structures which were massive and weighed as much as 400,000 tons would remain at sea as there was no way of moving them [21]. Rigs weighing less than 10,000 tonnes are required to completely remove their supporting structure. Oil and gas companies could however apply for exemption to remove the structures of heavier platforms built before 1999 and to leave a substantial part of the rig in place [16]. Where submerged structures are too heavy to be moved, there are four alternatives possible.

- leave the entire structure in place with markers for the safe navigation of sea going vessels
- partially remove the structure,
- move the structure to a new place (tow-and-place)
- topple the structure and lay it on its side.

Anne-Mette Jorgensen, founder of North Sea Futures (a non-profit organization that promotes ecosystem-supportive design and management of offshore structures) and Sam Collin, a marine scientist at the Scottish Wildlife Trust oppose the removal of oil rigs and advocate the 'rigs-to-reefs' program where obsolete rigs are repurposed into artificial reefs. The attempt to create new reef environments for the breeding of marine life species provides a strong case to leave a substantial part of oil rigs in place where locations are environmentally suitable. Over time the submerged part of the rig structures formed isolated clusters of biodiversity, some of which have merged and integrated with wider marine ecosystems.

Removal of rig structures risks the release of trapped chemicals creates disturbance to the seafloor and increases sedimentation. Furthermore the established marine colonies growing on the structure would be exterminated. Conversely, inactive rig structures when cleansed of pollutants can continue acting as reefs forming part of vital marine ecosystems. Thus the environmental benefits of non-removal outweigh those of removal of rig structures.

Platform ecosystems can evolve to mimic those of natural wild marine habitats. In the North Sea, the submerged structures of the rigs provide extensive rigid surfaces for invertebrates to colonize and to proliferate. These in turn attract predators such as fish, mammals and seabirds [7]. Early research into rig fouling communities reveals a symbiosis between the mussels' anemones and sea stars in a self-contained system. In nurturing marine benthic ecosystems, the rigs are excellent sites for ecological research in studying the pattern and interactions among the population of marine flora [29].

Together with pipelines linking the oil rigs and circulating undersea currents, the submerged elements form an infrastructure "undersea scaffolds" to repopulate species of Atlantic cold-water corals whose population had been diminished by destructive trawling [7]. The marine life in the partially decommissioned Murchison and Thistle A, produce Lophelia pertusa, larvae, a rare cold-water coral to populate the protected coral zones of nearby Aktivneset [21].

Decommissioned platforms can also be designated as marine conservation zones. These include a British owned Centrica platform nearby Markham's Triangle where sand eels thrive to provide the feeding ground of seals. ConocoPhillips platforms in its Viking gas field are refuge to harbor porpoises and white-beaked dolphins, which usually prefer coastal water, shelf edges and slopes to hunt for fish, crustacean and squid. Where dolphins are hunted for marine parks, aquaria and meat, the abandoned rigs provide protection from this threat, as fishing is restricted within 500 m of water around oil platforms. According to Collin, only one percent of the North Sea area provides important refuge in the form of these restricted areas. In this respect, a trust policy adopted in 2013 advocates the reconsideration of a complete removal of offshore infrastructure.

There were approximately 500 platforms in the Gulf of Mexico which have attracted marine life for over four decades in a rigs-to-reef program adopted in the. This figure is ten percent of the total number installed. While the Gulf has reportedly some success in encouraging marine life, it does not mean that all locations will benefit ecologically with the same strategy [21]. The oil platforms off the California coast create thriving habitats for large populations of demersal fish because of its substantial proportion of structural surface area in relation to seafloor surface area [15].

Offshore oil and gas platforms provide the substrate for invertebrates such as sponges and corals and also act as a refuge for fish, seals and whales. Offshore platforms provide hospitable substrate for invertebrates in their delicate larval stages. While a community of fish species may be formed within hours of platform installation, the formation of a habitat emulating a complex reef would take five to six years [27].

4 Decommissioning Costs

An average of 120 offshore rigs are decommissioned annually worldwide. The task at hand is a total of 2,000 aging offshore oil and gas platforms, subsea wells and related assets.

Offshore oil drilling at Gulf of Mexico is estimated at $26 billion. Future closures in the U.K. continental shelf alone, is estimated at £50 billion. IHS Markit predicts that decommissioning expenses will increase from $2.4 billion in 2015 to $13 billion a year by 2040 [18]. Europe alone accounts for half of global decommissioning spending, with its North Sea installations.

The 1998 Convention for the Protection of the Marine Environment of the Northeast Atlantic (OSPAR) prohibits the dumping of oil installations and requires operators to restore the healthy condition of the marine environment on vacating the site. In this respect, the total removal of 470 platforms, 5,000 wells, 10,000 km of pipelines and 40,000 concrete blocks will cost £17.6 billion between 2016 and 2025 [18].

Of concern is the financing of decommissioning costs. While major oil companies may have previously set aside funds from the revenue of new projects to defray the decommissioning costs of older rigs, the low oil prices today do not allow much flexibility. At the same time, governments need to protect their budgets from increasing costs of decommissioning and it is necessary to structure alternative sources of funding and legislation to protect the environment and the North Sea economy [21].

The financial structure to support decommissioning costs is important and these have been considered by energy companies and governments acting as stakeholders in safeguarding the marine environment. Major companies, such as Shell, have budgeted for the cost of decommissioning. It was the practice of companies to bear the costs of decommissioning rigs from old projects with the revenue from new projects. This was possible when oil prices were high but is not the case today. Currently low oil prices pose challenges to the sufficient accumulation of decommissioning funds therefore the funding mechanism is from tax reliefs offered by governments who profited in periods of high oil prices.

Governments, regulators and energy companies seek alternative funding solutions to offset increasing decommissioning costs. Private sectors establish a form of pension fund for oil and gas fields. In this scheme, an oil company would make annual contributions to a Special Purpose Trust (SPT) protected from insolvency, to provide for the decommissioning and reinstatement costs of installations. Public funds are insulated from decommissioning liabilities with the recovery of the oil and gas economy, as in the case of the North Sea. Insurance policies to protect against environmental liabilities during and after decommissioning allow asset protection for operators [18]. It becomes worthwhile to explore viable alternatives to decommissioning given the scale of financing resources to be deployed for the sake of the environment.

4.1 Disassembly Techniques

The high decommissioning costs are attributed to the complex processes in removing very heavy large-scale structures with specially fabricated large-scale vessels operating complex equipment in longer time frames [8].

To illustrate the weight of these structures, the total weight of steel in the nine Norwegian rigs at Ekofisk field is 113,500 tonnes—"equivalent to the weight of 54 London Eyes", says the Royal Academy of Engineering. Shell's concrete platforms in Brent field weigh 300,000 tonnes, equivalent to the Empire State Building in New York, according to Duncan Manning, decommissioning manager for Royal Dutch Shell's Brent field [16].

There are three methods of removing the topsides where the main component of weight reside.

- 'piece small' uses the rig's crane to dismantle the superstructure into 'small' pieces before transporting them to a coastal recycling center by barge;
- 'reverse install', is in effect the reverse of a rig building process. The structure is dismantled in large modules by floating cranes with large lifting capacities moored next to the platform. These entire modules are then transported large barge and shipped to recycling centers.
- removes the entire 24,000 tonne topside in a single lift operation using a series of bespoke cranes supported between the bows of two oil tankers welded together. This method is used by Shell on its Brent Delta platform [16].

Apart from removing the topside, concrete supporting legs are left behind, but if the supports are made of steel, they may be cut below the seafloor and lifted in one piece to the surface. Robot submarines or remotely operated drones equipped with abrasive water jets, hydraulic shears and diamond saws, enable cutting [18]. It is possible that lasers may be used in future to speed up the cutting process. Cutting technology has since progressed with lasers attached to flexible arms, used in dismantling nuclear reactors.

Platform topsides with structures in serviceable may be refurbished but much of the other components are not reusable. Their main support columns in concrete are difficult to move and are likely to be left in place. Also left in place would be oil contaminated drill cuttings, which are difficult and expensive to remove.

Complete removal is unfeasible with no solution to disposing of the recovered material. Only the metals, cabling and plastics worth recycling are salvaged, while the rest are landfilled after decontamination.

As an alternative to dismantling, the reuse of such structures is possible. Off the coast of Malaysia, is the Sea Adventures Center, which consists of a 25-room hotel and diving school converted from an oil rig. Where the scale of the investment limits isolated conversions, it is possible that larger scale infrastructural applications such as wind turbines, data storage centers, marine research stations, military and coast guard seaposts may be viable. Even larger scale applications are explored in later sections of the book where oil rigs and barges are used as alternatives to mega-float or large raft solutions for floating settlements.

4.2 Shipbreaking Yards and Scrap Value

Jack-ups and semi-submersibles can be towed or sailed to shipbreaking yards offering scrap value for material. The recovery of scrap value reduces the cost of decommissioning. The scrap value of a secondhand vessel is established by the price of reclaimed steel in the local market. In Turkey, it is about $190 per tonne while in China it is $210 per tonne. Shipbreaking yards in Alang, India, Chittagong, Bangladesh and Gadani, Pakistan, will offer higher prices at $280 per tonne. By contrast, rates are less competitive at EU-approved shipbreaking sites, which are regulated by environmental waste laws. These regulations protect water quality by requiring vessels to be dismantled in partitioned quays or dry docks.

In this respect European yards do not purchase scrap material but instead charge a collection fee. Of the 864 vessels that were dismantled worldwide in 2016, only nine were completed in Europe, according to Patrizia Heidegger of Shipbreaking Platform. A dozen were dismantled in Mexico, the Philippines, Russia and South Korea while 668 were dismantled in Turkey, China, India, Bangladesh and Pakistan, where their shipbreaking yards offer the highest price per tonne. An estimated $80 or $90 per tonne in value, amounting to one million dollars in total had been forfeited by Transocean for their oil rig Winner scrapped at Turkey in 2016, compared with what a South Asian yard would have offered [14]. NGO Shipbreaking Platform in February 2020 reported that scrapyards worldwide received 674 vessels and offshore units in 2019 alone.

Out of this, 469 large floating platforms, tankers passenger liners and cargo ships were dismantled in Bangladesh, India and Pakistan. These three locations account for 90% of annual scrap tonnage and therefore, the accumulation of shipbreaking pollution is of environmental concern.

Shipbreaking Pollution A confidential 2015 report by the shipping consultancy Litehauz, obtained by journalists from the investigative agency Danwatch detailed the potential impact of ships that were torch-cut on tidal water. Approximately 120 tonnes of molten steel and two tonnes of debris paint are discharged into the sea for every 10,000 tonnes of steel that is torch-cut from a vessel [14]. The Basel Action Network in Mar 2019 estimates that scrapped ships which weigh 10,000 to 40,000 tons of which 95% of is steel is coated with between 10 and 100 tons of marine paint which has lead, cadmium,, arsenic, zinc and chromium. Each vessel has hazardous waste of up to 7.5 tons of asbestos, several thousand liters of hydraulic fluid, engine oil and lubricant. The shipbreaking industry produces different types of hazardous waste and yards are required by license to have oil-water separators incinerators and temporary storage facilities for non-hazardous waste material. Yet many waste treatment systems are not adequately maintained or licensed to be operationally effective in preventing damage to adjacent fisheries, agriculture and to the surrounding environments of local communities. It becomes necessary to consider the reuse of inactive oil rigs and vessels in good condition to avoid extended environmental damage through shipbreaking practices.

5 Economic Cycles, Backlogs and Costs

An oversupply of oil rigs is created by a combination of reasons. In sustained periods of falling oil prices, drilling operations are curtailed when production costs become unviable. The costs and risks to parties involved are affected by the amount of time between the date ordered and the final delivery of the rig. While it takes about 18–36 months to complete the construction of an oil rig, the time between contract finalization and rig delivery can take much longer due to shipyard backlogs. The risk for both sellers and buyers change with market conditions. Rig utilization and day rates may decline by the time the rig is constructed and this will render unfeasible, the rig buyer's investment. Similarly, the seller's profit margin is diminished when material and labor costs increase over time with delays and backlogs. A cost escalation clause is often included to protect parties in the event of long delay between contract finalization and commencement of construction [10].

Time lags between supply and demand also create large oil rig surpluses. Oil prices in the market can fall within the three-year construction period of oil rigs. In 2010–2011, the year-long waiting time for available rigs trebled worldwide orders. Yet $65 billion worth of offshore rigs were still being built in 2016 when idling rigs were in existence. By 2016, oil prices fell with economic downturns in China, Russia, India. At the same time shale oil was extracted in North America following the 2000–2008 oil price spikes whilst the Middle East sustained its supply volume. Rigs without commissions were either "cold-stacked" (without crew and anchored) until the market recovered, warm-stacked with motors running or sold for scrap [14]. Cold Stacking costs $15,000 a day while warm stacking costs $40,000 a day. Because rigs are not designed for idling, operators prefer bearing the costs of warm stacking and not risk machines requiring total replacement over indeterminate periods of disuse.

With North Sea hydrocarbon reserves depleting, several hundreds of oil and gas rigs are about to expire their productive life cycles and are costing more to operate. The low oil price compounding the economic downturn results in one-third of oil fields operating at a loss [16].

Political uncertainty on the matter of Britain and the EU also had an impact on North Sea Operations, which in 2000 was one of the world's largest sources of oil, yielding six million barrels a day. At only 1.5 million barrels now, an estimated 600 production platforms in the North Sea are due to be decommissioned. The British sector alone comprises 470 platforms and as many other offshore installations, 10,000 km of pipelines and 5,000 wells [21]. The British industry anticipates more than 200 decommissions until 2025.

The decommissioning of offshore oil and gas installations is only in pioneering stage in the North Sea, Australia and New Zealand where technological, environmental, regulatory, political and financial considerations are evolving. Decommissioning both steel and early concrete rigs are complex and costly operations.

It is estimated that over 60% of the jack-up rigs available will be over 30 years old by 2025 [19]. Given the immense decommissioning costs, the possibility of optimal measures is not straightforward. The impetus for rig owners' to renew their fleets is affected by overall industry performance and projection. In an effort to reduce the gap between supply and demand, rig owners are expected to scrap the excessive supply of rigs. Thus, it would increase the day rates of operators relieving the economic pressure on rig owners running at a loss. Rigs recently constructed but not deployed due to cancelled orders run the risks of mechanical failure when being "cold-stacked" over protracted periods of disuse. When rig demand is interrupted or orders cancelled due to oil pricing fluctuations, it becomes feasible to produce newbuilds with civilian applications and/or to refurbish rigs with serviceable structures for business opportunities outside the oil industry.

6 Offshore Oil Rig Types

Offshore drilling rigs are broadly divided into bottom-supported rigs and floaters (Fig. 1). Bottom-supported rigs are anchored to the bottom of the sea floor and their operations are limited to shallow water depths of 150 m. These are typically platform rigs and jack-up rigs. Platform rigs consist of steel or concrete platforms supported by fixed columns that are made of tubular steel members and driven into the seabed.

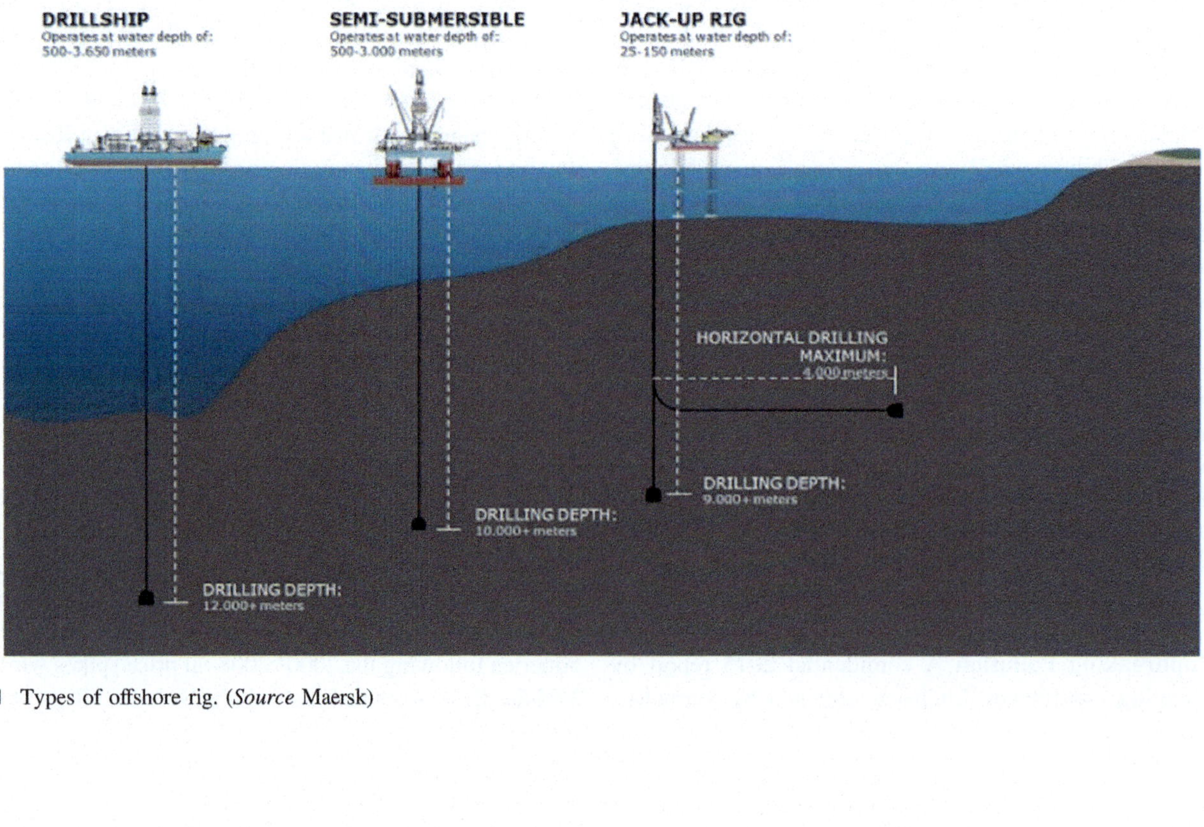

Fig. 1 Types of offshore rig. (*Source* Maersk)

Jack-up rigs are floated out to the drilling location where their retractable legs can be lowered down to the seafloor. They therefore operate in water depths, which are limited by the length of their legs that are fitted with a mechanism to elevate the topside above water. When drilling is completed, the legs are raised out of the water, and the rig becomes a floating barge that can be towed away 'wet tow' or placed on a large transport ship 'dry tow'.

Jack-up rigs are classified by their specifications and water depth ratings. The three common rig classifications are standard, high specification and harsh environment. Older standard rigs with low hook-load capacities have little automation in their systems and their drilling equipment is often mechanically operated. These rigs operate at water depths of less than 92 m. However, standard rigs can also complete the tasks of high-specification rigs but at a much lower rate.

Although high-specification rigs are widely used in Southeast Asia, they are also used in Mexico and the Middle East. The robust rigs are kitted with modern automation systems and drilling equipment to operate at water depths of up to 121 m. A special class of oil rigs is designed to withstand extreme weather conditions at sea and water depths of up to 150 m in the North Sea.

There are two types of floaters: semi-submersible rigs and drillships both of which are self-propelled and which have drilling equipment on-board. Floating rigs are anchored with the help of anchor handling tug supply vessels (AHTSV) in a complex, lengthy process. Unlike jack-ups, floaters move up and down with the tide, but the fixed wellbore does not move as drilling position is stabilized by a system of hydraulic wave-motion and heave compensators. Semi-submersible rigs float on large pontoons providing the buoyancy to sail to different locations. The pontoons are half submerged when in operation hence their name, which in short form are 'semis'.

Drillships and floaters, which use mooring lines to connect to the anchors on the sea floor, are limited to midwater operating depths of up to 800 m. Drillships and semi-submersibles with rotating thrusters to hold the rig in the exact location without mooring lines are dynamically positioned. The dynamic positioning systems (DPS) extend the operating depths of these floaters to deepwater ranges of 800−1500 m.

Drillships also have high load-carrying capacities and are ideal for deepwater and remote locations. Moreover they are mobile and maneuverable. Like semi-submersibles, drillships also use dynamic positioning systems to fix their location in the sea at an exact position. Floaters with advanced DPS and multiple thrusters can operate in ultra deepwater range of 1500 −2000 m [19]. Seventh generation semi-submersibles are capable of operating in 3000 m depth of water.

7 Jack-ups and Repurposing Considerations

The KFELS N Class rig has an overall height of 173 m and can operate in harsh weather conditions in water depths ranging from 122–152 m. This is 40% deeper than traditional units in benign waters. Its drilling depth of 10,600 m is 15% deeper than current jack-up rigs designed for harsh environments [11].

Its technical specifications meet the design requirements for the Norwegian sector of the North Sea, which is one of the world's harshest operating environments for offshore exploration and production. This state-of-the-art jack-up drilling rig is designed to accommodate process modules for production activities. Thus the flexibility of concurrent processing, production and drilling activities improves the efficiency and output.

The configuration of these robust structures to withstand harsh conditions and support a range of operational functions makes it suitable for their repurposing into a wide range of civilian applications in our study. The tall masted supports are ideal for additional built up volumes in providing high-density spaces. Their yield to investment ratio becomes viable when the capital expenditure may be recovered within acceptable investment periods of time. The positioning of large spaces spanning between structural masts and of cellular spaces cantilevered off the masts accommodates several typologies for marine settlements (see chapter "Optimal Settlement Size and Masterplan Strategies", Fig. 3). These include aquaponics production, markets and wharves, public conservatories and housing. Special typologies such as prisons, crematoria and hospitals are also easily accommodated. Larger deck plates to suit recreational functions are possible by combining a number of rigs.

Jack-ups are first built in 1954 and remain in demand today. With subsea high-pressure riser systems, they are able to operate uninterrupted in bad weather conditions up to and exceeding 50-year extreme storms with 30 m waves. A semi-submersible however will have to stop operations when wave heights exceed 2–3 m. As a jack-up rig is fixed above the wells, the offset of waves and vessel responses are less than that for a semi-submersible.

Tensile stresses on the riser system, subsea and surface equipment will also be lesser while reducing the impact on fatigue lifetime. Rental, operational costs and CO_2 emissions are also lower for jack-ups than with semi-submersibles. Jack-ups can float and when not deployed their legs can be raised. To save time, both topside modules and legs can be towed to distant locations by barges.

Once the jack-up is in location the legs are lowered to the seabed and its deck structure is raised above water surface and to a level above wave heights. Jack-ups can only drill in waters up to 175 m deep while semi-submersibles can operate in depths of up to 3000 m. In repurposing jack-ups to habitable functions, the open-truss legs made of huge steel tubes have vertically continuous voids within them to sufficiently accommodate passenger elevators and to attach escape staircases on their outer periphery. In supporting additional load, these legs would need to be reinforced with additional steel struts to increase the overall girth the masts and resist torsion, compression and lateral and bending forces caused by wind and wave action.

Foundation supports need to be upgraded for new load There are two types of supports: independent-legged and mat-supported jack-ups [22]. Independent legged jack-ups have cylindrical shoes with tapered ends called spud cans attached to each leg. The pointed ends are driven into the seabed to provide support and stability. In soft seabed, independent leg jack-ups can penetrate the seabed by more than 30 m to reach soil substrata with sufficient bearing capacity to resist vessel dead loads and environmental imposed loads of wind and waves. As penetration depth of the independent legs increase, the water depth capability decreases and for this reason, the mat support was developed. The mat is in fact a giant single pad footing connecting all the legs of the jack up. The mats were designed for soft flat seabed having low shear capacity of only 4.8 kN/km^2 and could transmit a compressive bearing load of up to 24 kN/km^2 on the soil [13]. This compressive bearing load includes the vessel weight, its consumables and environmental loads.

As the topside of the rig is jacked up, the mat penetrates the seabed until the ultimate bearing capacity of the soil is equal to the pressure exerted by the mat. At the same time, undersea currents, wind and wave exert forces onto the rig making it slide. Rig sliding resistance depends on mat area, mat skirt height, spacing, orientation, and the soil strength of the seabed. The stresses on the underside of the mat are the sum of pressures exerted by deadweight, redistributed forces from lateral loads, overturning moments and dynamic waves.

The mat bearing pressure must be supported by the seabed soil bearing capacity and so, soil tests are imperative in foundation design. The stability of the rig depends on the eccentricity limits between the mat center and both the transverse and longitudinal centers of gravity of the jack-up. To minimize eccentric loading, the CG of the structure has to be monitored and to avoid overstressing the seabed; a process of preloading the mats is used. Here, seawater may be pumped into ballasts within the mat to keep mat bearing pressure within the bearing capacity of the seabed substrata. To support increase in deadweight and imposed loads on oil rigs repurposed into applications requiring additional built-up areas, preloaded and enlarged mat foundations would be more likely than spud cans on individual masted legs (Fig. 2). The mat would be able to spread the load of the oil rig over a larger seabed area and keep within limited soil bearing capacities without risk of sliding or overturning.

Jack-ups have an elevation device to adjust the height of their topside decks according to the water depth. One type uses hydraulic cylinders with moving and stationary pins. These cylinders extend and retract to move vertically in either direction along the legs of the jack-up. Maintaining the same pressure for all elevating units moving along one leg is a challenge with piping bends and lengths.

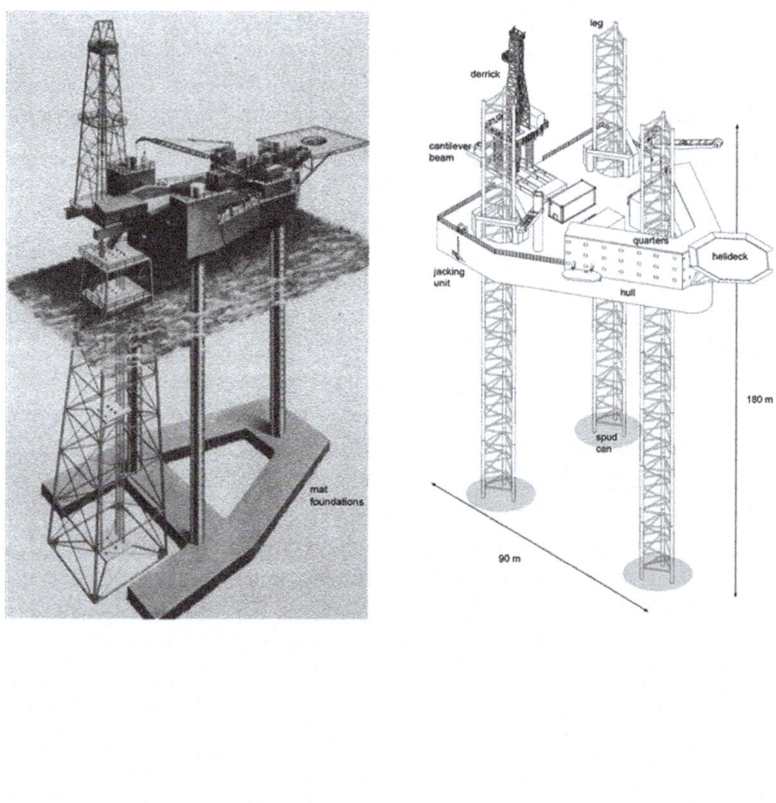

Fig. 2 Enlarged mat foundations instead of spud cans for stability in supporting increased deadweight and imposed loads on oil rigs

Another type of elevating device used is a rack with two pinion gears, which can turn to move the legs in vertical directions. Electrical motors can change motor speed with a change in pinion loads and speed-load characteristics to equalize pressure on all structural chords. This device is useful in constructing additional topside decks at lower level before hoisting them into positions along the height of the mast. Jack-ups with cantilevered derricks have a greater deck area intact compared to semi-submersibles, which have a moonpool and derrick in the center of their decks.

New topside decks are conceived as deep transfer structures with three-way girders spanning between the masted legs. Each transfer structure deck can support six to ten storeys structural frames to accommodate a range of shorter spans and useful headrooms. Here, each transfer deck 7.5 m deep can carry ten storeys of floor plates. Therefore a 180 m mast in 20 m depth of water has 120 m height for 40 storeys of accommodation at 3.3 m headroom per storey.

Transfer decks can accommodate different usages of space and therefore are ideal for offshore mixed developments common in larger scale settlements and individual complexes.

8 Semi-submersibles and Repurposing Considerations

In semi-submersibles, horizontal buoyant members called pontoons interconnect the legs supporting the large topside decks. The legs of early semi-submersibles were cross-braced with diagonal members to resist wave-induced prying and racking loads on the floating structure [2].

Column position affect the ease with which pontoons can be accessed to accommodate unloading/loading, docking operations for rescue missions, farming operations and supervision of recreational application, while twin hull pontoons are preferred to allow access to operational craft.

Doughnut arrangements of pontoons when partially submerged can be used to cordon off an inner pool. In this respect, pontoon configurations may enable flow-through water courtyards or create gated courtyards for security purposes controlling access. The base of a semi-submersible is a large-scale entrance harbor for work vessels or commuter craft or it can be used as a private water court for recreation.

Configurations that did not have diagonal bracings between legs are preferred for the same operational reasons of avoiding structural encumbrance. This is important for docking applications for rescue and transport vehicles. Multiple columns reduce the clear span of useful pontoon surface above water level. However they are good for larger displacements with a capacity to accommodate higher occupancy loads when repurposing. For example, Heerema Marine Contractors has a semi-submersible crane vessel with two units of 10,000-tonne revolving cranes supported on eight columns to lift larger integrated structures loads up to 20,000 tonnes in tandem. It is named 'Sleipnir', after Odin's eight-legged horse in Nordic mythology [23].

A major issue in semi-submersible design criteria is the rig lifespan, inspection capability, future class survey and repair. For permanently sited semi-submersibles, the environments, the fatigue requirements, and the difficulties in structural maintenance, repair and inspections are important design considerations.

9 Physical Limits to Additional Structure

In repurposing, the capacity of rig types to accommodate new floor area depend on the footprint size of the hull area and the extent to which the new structural volume can be safely supported off the enlarged masts of the jack-up without toppling, and on the upper hulls of semi-submersibles without risk of capsizing. The variable load comprises live load (occupants) and machinery, equipment and cargo. The motors and propulsion systems of the vessel are retained for mobility in deployment. The semi-submersible has a limit to the quantum of new built-up area it can carry in repurposing. The weight of the additional structures and the number of floors affects displacement and toppling.

The extent to which an oil rig can support an additional number of floors affects the viability of a specialized vessel to accommodate its intended usage with usable floor area. For a jack-up rig, the platform will be lowered to maximize the airspace between mast lengths above deck while allowing for waves height clearance below deck level.

In the example of a repurposed jack-up prison (see chapter "Other Forms of Repurposing", Fig. 37), estimates are made with a prefabricated prefinished volumetric construction (PPVC) system of hybrid steel-concrete cells each weighing 16.5 tons. If a three-legged jack-up rig can carry up to 15,000 tons and the additional load is 81,770 tons then additional columns welded to the masts will be required, including additional foundations into the seabed. Additional structural reinforcements are necessary in cases where new loading exceeds structural limits of existing rig structure. The use of lightweight material for modular construction of habitable units would reduce new deadweight and save construction time with the possibility of disassembly and relocation for deployability. This feature is crucial for cost recovery and flexibility in usage over the lifecycle of the repurposed structure.

For a semi-submersible prison, the limit to the number of storeys is constrained by flotation level and toppling in the

Table 1 Construction cost distribution for jack-ups and floaters

	Jack-ups		Floaters	
	Cost (million $)	Proportion (%)	Cost (million $)	Proportion (%)
Steel	15–40	10–20	25–60	<10
Labor	15–55	10–30	50–120	10–15
Drilling equipment	20–70	10–30	100–200	20–30
All other equipment	35–50	20–30	100–200	20–30
Profits	10–25	<10	50–75	<10
Total	175–225	100	500–700	100

transverse axis of the twin pontoons. Rigs designed for operating in harsh environments are heavier than those built for moderate environments. Harsh environment semi-submersibles have longer legs to maintain the air gap, and which increases the distance between legs and the pontoon size. Semi-submersibles built for harsh environments have columns, which are more substantial in length and in girth than those of moderate environment units [9].

In the example of a semi-submersible prison (see chapter "Other Forms of Repurposing", Figs. 30 and 31), its structure can carry 28,173.2 tonnes of live load; and dead load is subject to another 28,000 tons equivalent to twelve storeys of PPVC cells. But since the metacenter is still higher than the center of gravity, the semi-submersible rig will not topple but approximately half the height of the rig will be underwater. Two countermeasures are possible; one is to significantly reduce the weight of the PPVC cell by using lightweight reinforced plastics. The second countermeasure is to enlarge the pontoon dimensions of newbuilds to increase its displacement. Excluding the engine rooms of semi-submersible propulsion systems, which are retained, the tanks in pontoon hulls and legs may be freed up for use.

10 Rig Construction Costs

The primary factors affecting the construction costs of oil rigs are determined by market demand, its design specifications for its type and class, and the shipyard in which it was constructed. The impact of contract type, shipyard productivity, and scale economies on cost are difficult to ascertain, although they do influence cost.

Construction pricing is influenced by the global number of shipyards specializing in construction services and the market demand for newbuilds. At the same time, demand for newbuilds rise when investment returns to drilling contractors are viable with day rates and utilization rates. There are few shipyards worldwide capable of building rigs during periods of high demand. Price increases result from backlogs and sustained demand for rig construction services. These shipyards are in Singapore, China, Japan, India, South Korea, Russia, the United States, and the United Arab Emirates.

Where steel is the main material used in rigs, its material prices directly affect newbuild costs. There is proportionately larger steel component cost in the construction of jack-ups than in floaters—10–20% for jack-ups versus less than 10% for floaters (Table 1). Steel prices vary regionally with quality and the sectional profile of members specified for construction. As rigs are designed and constructed using different steel strengths, no single steel price per ton can be used to estimate rig construction cost. Where the majority of rigs are built in Asia, the Asian steel price index is an indicative measure of the rig construction market. The rate of increase in rig prices correlated with the increase in steel index over a decade. Variations in average jack-up rig prices as high as 70% can be attributed to high steel price indices. There is no observable relationship between floater prices and steel prices [10].

Equipment Costs The main equipment costs of an oil rig lie in engines, cranes, generators, drilling equipment, and dynamic positioning systems. OEM equipment is purchased by the rig builder and assembled at shipyard. The drilling equipment package is costliest at $20–$70 million for jack-ups and $100–$200 million for floaters. This ranges from 10 to 30% of total costs. Non-drilling related equipment together with drilling equipment range from 30 to 60% for jack-ups and floaters.

Where drilling and equipment costs are made of steel and influenced by steel prices, it is the demand from the oil, gas and commercial shipping industries, which influence manufacturing costs. Specialized labor also contributes significantly to equipment manufacturing costs.

While contracts for rig construction are in U.S. dollars, international shipyards may use national currencies of shipyards in euros, Chinese yuan, South Korean won, or

Singaporean dollars. A strong U.S. dollar lowers newbuild costs at international shipyards for U.S. buyers. The converse is true when the dollar declines relative to a local currency [10].

Labor Shipyard costs are determined by labor costs and productivity that can be boosted by mechanized processes. The costs of shipbuilding labor in the U.S. and Korea are similar. In Singapore, a rig can be constructed at 30% the cost in U.S. and South Korean yards Singaporean yards. The high labor costs in South Korea compensates for advantages in productivity over Singaporean and U.S. yards. Recently, Singapore and South Korea shipyards have experienced an increase in both labor costs and productivity. In the U.S., labor costs account for thirty percent of revenue generated. In Singapore and South Korea, labor costs comprise 10–15% of total rig construction costs [10].

Shipyard Characteristics Shipyards with lengthy rig building experience, effective supply chain management and advanced construction technologies with integrated and automated processes are sought after. South Asian yards are competitive in pricing as they have labour and production costs lower than elsewhere. Tax structures and government subsidies sustain heavy industries with large-scale operating budgets. This enables major rig building yards to specialize in design of proprietary technology.

Keppel-FELS and LeTourneau build customized designs on demand at a cost lower than other yards and have their own product lines acquired from industry experience (Tables 2 and 3). Rig operators prefer the advantages of lower pricing with economies of scale on orders of the same rig and engage with such builders in long-term contractual relationships. The same design is predictable when subject to similar environments and operating conditions in the same oilfield.

Design class Newbuild jack-ups designed for water depths between 61 and 150 m and variable deck load (VDL) of 3,750–7,000 tons cost $159–$530 million. The KFELS B Class, Letourneau 116E, and F&G JU-2000E are the most common designs. While operating depths and rig type affect costs, rigs of the same design do not vary much in costs, perhaps due to sensibility of market pricing between competitor builders.

Semi-submersible newbuilds with operating water depths from 500 m to 3000 m and VDLs between 5,000 and 22,000 tons cost $460–$771 million. Operational displacement varies between 42,000 and 62,000 tons. As semi-submersibles are heavily customized, and equipment costs are substantial, no two of the same design are in a similar cost range than in the case of jack-ups.

The costs of drillships and semi-submersibles are not related to their operational water depths because "floating units with the exception of their drilling risers and anchor handling systems, do not have elements that pass through the water column. Increased water depth is associated with

Table 2 Jack-up design class properties and newbuild cost (2011)

Design	Number	Cost (million $)	Water depth (ft)	Harsh	VDL (tons)
Friede and Goldman JU-2000E	11	190–220	400	Y	7,000
LeTourneau super 116E class	12	159–210	200–375	N	3,750
KFELS B class	20	180–210	350–400	N	4,500
PPL Shipyard Pacific class 400	3	190	400	N	3,750
Friede and goldman JU-2000A	4	220–229	350	Y	4,500
Friede and goldman JU-3000 N	6	220–245	400	N	7,000
KFELS super a class	5	230–260	400	Y	7,000
LeTourneau 240-C workhorse	3	194–257	400	N	3,000
GustoMSC CJ70	3	500–530	492	Y	7,000

(*Source* Bailey and Sullivan 2011; Industry Press) [10]

Table 3 Semi-submersible design class properties and newbuild cost (2011)

Design	Number	Cost (million $)	Water depth (ft)	Harsh	VDL (tons)	Displacement (tons)
GVA 7500-N	2	526–709	10,000	Y	8,250	62,000
F&G ExD	3	599–771	7,500–10,000	N	10,000	58,000
Ensco 8500	2	537–560	8,500	N	8,000	
CS-50 MkII (N)	2	510–526	9,843–10,000	Y	6,800	47,000
Sevan drilling 650	3	526–685	10,000	N	22,000	61,000
GM 4000	2	460–560	1,640–4,000	Y		60,000

(*Source* Bailey and Sullivan 2011; Industry Press) [10]

increased costs for risers, riser tensioners, mud pumps, anchor winches, drill strings, and mud storage facilities" [10].

Operating Environment Rigs built for harsh environments are built with higher structural performance specifications and therefore costlier than moderate environment rigs. Harsh environment jack-ups are designed with longer legs to increase operating levels above waves and waterlines. Increase in leg length adds to material usage with sturdier structural reinforcement and enlarged hull dimensions to acquire stable centers of gravity.

Similarly, semi-submersibles built for harsh environments must have longer and thicker columns than those designed for moderate environments [10].

11 Rig Repurposing Costs

Reinforcing the rig structure The cost determinants for repurposed rigs also include the additional structures needed to reinforce the existing rig structure and its foundations, to carry new floors/walls or volumetric modules adding to the VDL. For jack-ups, additional decks are needed as well as additional rings of columns around the existing masts, to support additional beam networks and floor structures. Enlarged spud cans to form connected mats are necessary to spread load over a larger area on the seabed. For semi-submersibles, the propulsion, ballasts and dynamic positioning systems are left in the pontoons but drilling equipment, workshops, storage areas, crew accommodation will be removed together with selected steel bulkheads to free up useable space. Power generation and water plants will have to be upgraded, separating those for the vessel from those for the repurposed applications such as farming and housing.

Merging shipyard and building construction Shipyards differ in construction methods, depending on the extent of automation involved in the construction process, the scope of work subcontracted to specialist parties, and the degree of serial production line usage. Methods of fabrication and assembly are unique to each shipyard and rig type. Also the available space and equipment type affect the productive capacity of the shipyard. Although Singapore shipyards are not larger than Chinese and South Korean shipyards, they are highly productive and able to construct state-of-the-art oil rigs at competitive prices to win significant international contracts. Larger South Korea shipyards use an advanced "mega-block" process of ship construction. Here large ship sections are prefabricated before assembly in a floating dock [10].

Similarly, prefabricated plants allow the specialization of structures built for dwelling, workspace and food production before transportation to the shipyard for installation and assembly onboard the oil rig. It is possible to conceive of integrating the finishing works of the habitable interior with the fabrication of the oil rig module. This would further reduce production costs, and provide quality standards instead of separating interior works from secondary structure and mechanical systems

Large-scale steel construction There is greater opportunity for the repurposing of oil rigs in countries in cost effective ways particularly where advanced shipbuilding and building construction technologies are both available.

One such example is the Sendai Mediatheque Library in Tokyo. The building completed in 2001 is a seven-storey building with two basements, 36.5 m in height and has a built-up area of 22,000 m^2 of library, informational systems art galleries, and visual image media center. Like a four-legged jack-up, four of its lattice steel tubes are located at the corners of its 52 by 52 m floor plan. These four largest tubes—20 m in diameter—which acted as light wells, accommodated elevators in them. The remaining nine lattice tubes were 5 m in diameter and located in positions on plan, between these four primary corner columns (Fig. 3). Thus, the special outcome of this structure is a seven storey building supported by lattice column/tubes where the edge condition of each floor is left free of perimeter columns. This allows the cladding of the building to be self-supported and to be framed by slender elements thus enabling a visually transparent façade.

Its construction of 400 mm deep steel-ribbed floor slabs made of I-sections sandwiched and stiffened between steel floor plates, which spanned up to 17 m between supports are made possible with shipyard fabrication methods.

The grid pattern of beams comprised three zones of density (Fig. 4): one pattern dense orthogonal connecting columns in line, one radial pattern in the zone around column heads, and one less dense orthogonal pattern between the first two. These patterns connected the rectilinear floor plate to the circular tube lattice columns of 5–10 m diameter. This structural configuration withstood an earthquake of magnitude 9.0 Mw in 2011. Translated into a marine context of harsh deep-sea environments, extreme lateral, flexural and torsional loads may be resisted by such structures. In the construction process, lattice column tubes were prefabricated to preset the radial connections to floor beams for site welding (Fig. 5).

While this system of regular spaced lattice tubes worked within the plan requirements of a digital information resource facility, a different structure is needed for floor-to-floor flexibility accommodating mixed use developments and production infrastructure in marine settlements. In this respect, a transfer structure, such as the one used in the upper hulls of jack-ups and semi-submersibles, would span between the masted legs of the jack-ups and more than one transfer structure may be used to stiffen the entire tall configuration and provide lateral

Fig. 3 Construction of Sendai Mediatheque. like a four-legged jack-up, four largest primary supports are located at the corners of its floor plan. (*Source* Naoya Hatakeyama—sendai mediatheque, from "Under Construction" (1998–2001))

Fig. 4 **a** Grid pattern of beams comprising three zones of density **b** detail of plate organization at tube

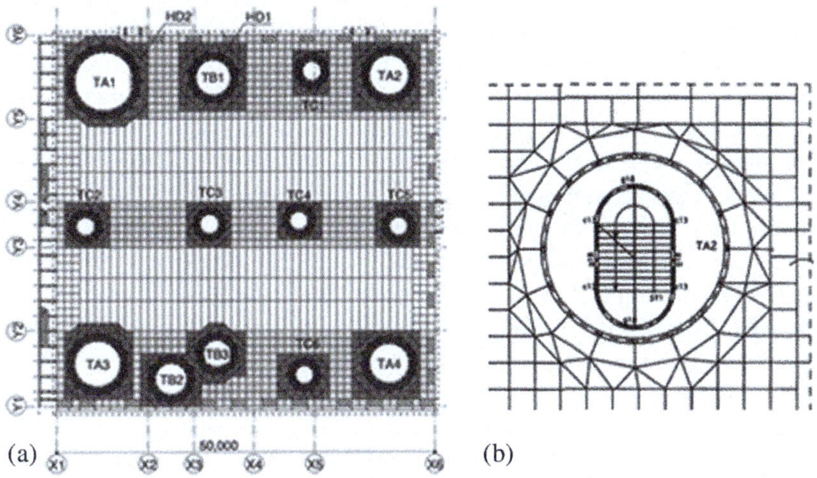

Fig. 5 Lattice column tubes to floor beam junctions. (*Source* Naoya Hatakeyama—sendai mediatheque, from "Under Construction" (1998–2001))

stability. This system is preferred to accommodate column-grid variations with multiple programs in the additional floors of repurposed oil rigs.

12 Exploration of Oil Rigs as Floating Settlement

It is estimated that 60% of global population will reside in cities. It is also anticipated that 90% of the largest cities are coastal and under threat of sea level rise. Historically, cities have extended their coastlines seaward in need of land to accommodate population growth. In Singapore, 25% of the city is built on reclaimed land, while 20% of Tokyo is built on artificial islands. The authorities in Dubai have built entire luxury complexes on artificial islands while huge tracts of Holland have been reclaimed from the North Sea. An intricate system of levees and dykes has protected Dutch urban areas from flooding.

Floating settlements are now re-explored as solutions to global housing shortage. Koen Olthuis, founder of the firm, Water Studio have designed floating platforms that can act as the foundations to support buildings. Initially conceived single villas or offices, Olthuis has upscaled the idea to plan for entire low-rise floating cities [4].

While most architectural proposals have traditionally focused on floating rafts or piled structures, for example, Triton City, Russian Oil Settlements, Lily Pad and Seasteading, the idea of using oil rigs on a large-scale is the basis of design research in this book. It has the advantage of small structural footprint compared to raft solutions.

The notion of an autonomous settlement with sovereign rights is possible when floating entities are located in international waters to create 'start-up' nations with their own legal systems. Floating settlements are believed to be a plausible response to sea level rise and an alternative to megacities in accommodating population growth and limited land resources for food water and in obviating the complexities of dense city cohabitations with nature reserves and hinterlands. Also, settlements for climate and political refugees in post disaster scenarios are worthwhile explorations in the medium- and long-term time frames.

Much of these explorations depend on issues of administrative structures, capital and operating expenses to comply with maritime, marine and environmental legislations. The costs of construction related to repurposing oil rigs are also dependent on several factors.

12.1 Conversion Costs

The viability of converting a jack up into an offshore installation depends on [1] the conversion and installation cost of the jack-up versus the construction and installation of a newbuild; and [2] the cost of a refurbished jack-up structure with a new or refurbished topside. In both cases, the capital expenditure is significantly reduced when the refurbishing party owns the oil rig structure. Upon refurbishment, the life cycle of the vessel can be extended for an additional 30 years.

The areas of costs for conversion are hull, marine systems, legs (structural and jacking system), drilling system comprising derrick, drill floor, and substructure, below deck systems, and piping and electrical systems. In addition, costs are influenced by accommodations, safety and lifesaving equipment, helideck, electrical power and distribution, and communications and electronic systems.

The hull costs are incurred from steel repairs and renewal, additional strengthening or reconfiguration for repurposing, and painting and corrosion prevention of its hull steel. Costs associated with the latter two can be predicted, while the accuracy of budgeting for the costs associated with the first item is dependent on the quality of the pre-budget inspections of the hull and associated steel. The marine systems with cost implications are fuel, ballast, cooling water, potable water, service air and black and greywater systems [6].

Table 4 shows a breakdown of the cost components of a semi-submersible rig. The largest component of 27.5–37.5% for drilling equipment can be omitted for repurposed newbuilds.

12.2 Geometry

The geometry of the frame design for additional structures is constrained by the geometry of jack-up masts of the oil rigs. While primary beams are set in a tri-axial or a three-way grid, it is possible to transit grids into rectangle ones to support regular floor plates cantilevered off the triangle masts. However, additional frame geometry should have a spanning strategy, which eventually is supported by the primary masts of the jack-up rigs. In the case of semi-submersibles, grid axes may be orthogonal, and connected to the structural grid of the semi-submersible topside structure. This is necessary to transmit all new loads resting on the topside to the legs of the semi-submersible to the pontoon structures floating on water.

Both jack-ups and semi-submersibles require additional reinforcements to their primary structures viz. In the jack-ups, the triangulated masted columns and spud cans need to be enlarged and reinforced to accommodate additional load. In the semi-submersibles, the legs and the pontoons need to be enlarged to resist toppling and to provide the required buoyancy. Superstructure loads on the topside of the semi-submersibles will have to be symmetrically positioned.

Table 4 Construction cost breakdown of a semi-submersible rig [24]

	Drilling (%)	Structure (%)	Machinery (%)	Hull outfitting (%)	Electrical/ Fitting (%)	Painting (%)	Others (%)
Material cost	27.5–32.5	4–6	13–17	8–11	5.5–7	0.5–2	
Labor cost	2–3	7–8	2–3	1.5–3.5	2–4.5	2–3	5–7

Frame design also has an impact on the cost of the sub-structure cladding and on the speed of the construction. The tallest structure of the jack-ups will cost more than the lower ones on the semi-submersibles.

12.3 Speculative Usage

For a structural shell to be useful in the long term, it should be designed to accommodate change of use. Commercial sector speculation often excludes tenants and purchasers in planning stages. Real estate research however, would collate for example, popular sizes of tenanted spaces and tenant preference for shapes of units and frontages to guide developers in project inception stages.

New structures added onto the oil rigs would need to be flexible in meeting the needs of a range of tenants in order to provide lettable space in demand. Such features include open plan floor areas (where columns in the interior space are avoided or minimized), maximized floor to ceiling heights and service zones laid out efficiently to serve remote locations on plan. Most importantly, the layout of the structure should enable multiple tenancies on any floor. This will affect not only column positioning but also those of elevator cores.

In the case of jack-up rigs, elevator cores are located within or around the triangulated masts positions. These configuration factors will impact on the cost of repurposing and its design. The speed of construction to generate timely returns on investment is crucial in development. For this reason, the techniques of oil rig construction and the jack-up systems of self-elevating decks can be exploited for ease of erection when repurposing individual oil rigs/vessels and when assembling entire flotillas or settlements.

Open plan lettable floor space and spans are typically in the range between 6 m and 16 m. Spans in excess of 12 m are common in commercial office schemes. With longer spans, steel has advantages over other construction materials. In terms of cost (kg/sq. km of steel), longer spans attract a cost premium, as the weight of steel frame per unit of floor area would be higher than for shorter spans. In the case of repurposing jack-up rigs, the use of short span frames sitting on transfer structures is one strategy to reduce cost. Although long span layouts may have a higher overall dead load and cost, there may be efficiencies in other areas of construction that can compensate for this. Optimizing cost efficiency of structure will be affected by key priorities in providing spaces suitable for a mixed range of uses which in turn change grid layouts between floors requiring the use of transfer structures and carrying floor to floor heights.

Non-standard sections In maximizing the number of floors within the full height of the jack-up oil rigs, shallow floor steel systems such as the Sendai Mediatheque is advantageous. Although this may have a higher cost per unit area than the standard floor construction, the gain in total floor area and numbers justifies investment. Long span fabricated cellular beams allow for the integration of services through opening in the beams (castellated beams). This feature increases the flexibility in fitting out in the repurposing work and reduces the overall depth of the floor/ceiling zone. It is costlier per tonne of steel to fabricate a cellular beam than for a standard I-beam, but this premium is offset by its reduced deadweight compared with other horizontal structures of the same span.

Fire protection requirements Fire protection accounts for 10–15% of the overall steel frame construction in multi-storey structures, so the requirements and design strategies have a significant impact on construction cost. The typical fire rating for elements of structure can range from one to four hours depending on sprinklers and the height of the structure affecting fire-fighting strategies. With heavier steel sections, the volume and cost of fire protection materials can be reduced. Specialists in fire engineering may be consulted to perform simulation studies on actual risks, to further reduce the costs of structural elements.

Frame pattern In repurposing both jack-up rigs and submersibles, the use of symmetrical and regular patterns in the layout of the structure can result in cost efficiency through the repetition of members feasible for fabrication and erection. Factors that reduce the level of repetition need to be studied in order to assess the impact of construction program and cost planning.

Construction at sea/shipyard The logistics and access to construction site would vary on the location on the oil rig construction/repurposing. The speed of erection and storage requirements are affected by construction type, technique and the capability of the shipyard. Erection of steel structures account for 10–20% of the total cost of the structure. Structural frame construction is a critical path activity and impacts on project cost through the construction program.

12.4 Fast and Lightweight Systems

The use of prefabricated prefinished volumetric construction (PPVC) and mass engineered timber (MET) options are possible when integrated with the repurposing of oil rigs or in newbuilds for habitable and civilian applications. The spatial accommodation supported off the superstructure frames may be fabricated on shore in modular units and brought to the site of the jack-up or the semi-submersible for assembly by cranes. In this respect temporary living quarters (TLQ) offshore accommodation units may firm the basis on which more permanent and varied forms of modular construction are conceived. While the TLQ are sized according to container box dimensions, a construction method using PPVC may allow for greater variety in configuration of space to meet mixed use applications. Free standing volumetric modules with wall, floor and ceiling finishes fitted at prefabrication yards can reduce offshore assembly time on location.

While in Singapore the PPVC concept began with bathroom modules and then household shelters and now entire housing units are precast in concrete, the idea of applying PPVC for oil rig repurposing may be extended to entire accommodation units of floors or the equivalent. Moreover, these accommodation units or modules of production space facilities may be conceived in fiberglass, expanded polystyrene (EPS) or MET in order to reduce deadweight. A combination of PPVC or individual panel system (IPS) may be adopted depending on configuration demands to minimize construction time.

A lightweight PPVC system may be configured in chassis form [17]. However, these are typically in dimensions of 3 by 4 by 6 m, integrated with a building structural frame. Such a chassis can be made of composite steel construction where small hollow steel sections may be infilled with concrete for fireproofing. To further reduce the weight of the unit, the flooring system is constructed in cold form floor joists/beams with a variety of floor boarding such as timber-based, gypsum or cement-based material. A more solid lightweight floor can be made from cold floor joists with troughed metal decks and lightweight screed, giving a 120–140 kg per km^2 floor system with acoustic properties and underfloor heating system integration.

While HVAC systems are integrated in TLQ modules, these would be connected to central or decentralized systems used in the new spaces for the repurposed oil rigs. Explosion proof air-cooled systems with both heating and cooling functions can be considered. Spaces onboard would need ventilation and pressurization fans, electrical preheater for air supply systems, fire dampers, extraction system for sanitary spaces, pressure control dampers and pressure loss alarm system.

It is interesting to note that a design strategy would be to optimize variable permutations by minimizing the variations of PPVC modules with standardized interfacing. Accommodation units range from two persons to 20-, 36- and 56-persons blocks (20 m length by 50 m wide beam). Accommodation blocks may be made from linked and stacked modules (GRS Offshore). Beyond the accommodation units are ancillary and work space modules which include restaurant, galley, recreation, changing rooms, laundry, gym, medical suite, freezer and store, lab, workshop, data rack and cable transits.

These production modules would need to be rethought at a scale of habitable water settlements. Special plant modules would be needed for sewage treatment, freshwater supply and non-potable grade water for maintenance. Close loop systems would need to be integrated on-board, individual vessels/oil rigs and planned with dedicated treatment plants for wastewater management and aquaponics/hydroponic production. It must be noted that the concept of accommodation and ancillary facilities on-board a vessel for oil and gas production with limited crew would need to be upscaled into requirements for marine settlements of higher occupancy and a wider usage of space. Mess and galleys are now expanded into restaurants and kitchens fronting promenades creating public spaces with water bodies. Instead of confining crew to minimum proportion of rig area to maximize production space, habitable space is now re-proportioned to occupy most of the vessel/rig area to prioritize quality space at sea. Accommodation is now planned as entire apartment blocks cantilevered off superstructure, with unit terraces, balconies, large-scale communal sky gardens, all with sea view frontages.

Height Structural frame costs rise with the increase in its height. There is a 10–15% per unit cost premium for a 20-storey structure rising to 20% for a 30-storey building, and up to 70–90% for structures between 45–60 storeys. The ability of a structure to resist wind load is greater with a high aspect ratio. Lower structures with deep plates are more wind resistant and therefore more economic in construction costs than a taller structure with smaller floor plates. The taller structure with a much smaller floor plates in proportion to its height is inherently less wind resistant and would require more steel for rigid configurations. Thus the structural design strategy in adding new floor plates on jack-up rigs is different from those in semi-submersibles. In jack-ups, additional transfer structure decks (equivalent to the topside footprint of the rig) serve four main functions:

- act as intermediate ties to the tall masts, making this a rigid primary frame resistant to wind load;
- support ten-storey short span frames which can vary in grid configuration for mixed use applications requiring different proportions of space;
- free up airspace in the mid-height of the jack-ups from structural bracing requirements for creation of sky parks and courtyards for ventilation and communal space; and
- allow transfers for mechanical and plant equipment.

In semi-submersibles the height of the additional floor plates is constrained by the limit to which the vessel stays afloat without toppling. The limit to additional dead load is related to the volume of water displaced and, the sizes and configurations of the pontoons providing the buoyancy. Thus the additional superstructures on the semi-submersible need no transfer structures but only to be adequately braced and stiffened for wind resistance and harsh sea environments.

13 Scale of Repurposing Oil Rigs

Owing to costs involved in stabilizing additional loads of new structure, private investors are conservative in refurbishing individual oil rigs as in the example of Seaventures Dive Resort, a 25—room hotel and diving school built onto the structure of an oil rig near Sipadan, Borneo coast. Globally, there are plans of converting old rigs into platforms for other usages such as wind turbines, data storage centers and luxury homes. Most of these plans are scoped as individual single owner/operator entities with viable returns and operational costs where risks are manageable.

Our design study, however, argues that the mitigation of wastage and environmental risk in the oil rig construction industry cannot be sustained by individual installations and isolated investments. Instead they have to be deployed as a collection of specialized vessels operating integrated processes in a cluster advantage, capable of combination at a larger scale. The wider agenda of accommodating global population growth on autonomous marine settlements demands a larger scale of investigation and can be facilitated by the planning of settlement fleets with flotilla components. Most importantly the use of wave energy and waste to resource recovery systems is necessary to reduce the operating expenses required in systems maintenance and production of resilient resources at sea for long term survival.

Design investigations into vessel and flotilla components are based on strategies to replenish and regenerate marine biodiversity with many settlements. Each settlement to be planned in ways that would not deplete food supply or natural resources at sea in pursuit of recreation and knowledge production with increasing population. Flotillas of repurposed oil rigs which can be deployed to sites with climates and environmental conditions optimal for food production and dwelling in order to form entire marine settlements. This study provides the basis for prototypical investigation in order to verify actual performance at sea.

References

1. Ars F, Rios R (2017) Decommissioning: a call for a new approach. Offshore technology conference. Houston
2. Chakrabarti S et al (2005) Historical development of offshore structures. Handbook of Offshore Engineering. Chapter 1.7.5.1. Elsevier
3. Cordes EE et al (2016) Environmental Impacts of the deepwater oil and gas industry: a review to guide management strategies. Front Environ Sci
4. Cosgrave E (2017) The future of floating cities—and the realities. BBC Future
5. GRS offshore. Global Renewable Shipbrokers. https://www.grs-offshore.com/en/any-vessel-any-time/equipment/accommodation-modules/
6. Hale L (2019) GC energy services, budget considerations for converting a jackup drilling rig into an offshore installation. GC Energy Services
7. Henry LA et al (2017) Historic scale and persistence of drill cutting impacts on North Sea benthos. Mar Environ Res 129. Elsevier
8. Kaiser M, Liu M (2015) Decommissioning cost estimation for deepwater floating structures in the U.S. Gulf of Mexico. J Ships Offshore Struct 10(4)
9. Kaiser M, Snyder B, Pulsipher A (2013) Offshore drilling industry and rig construction market in the Gulf of Mexico. Coastal Marine Institute Report
10. Kaiser MJ, Synder BF (2012) Reviewing rig construction cost factors. Center for Energy Studies, Louisiana State University. Offshore Magazine
11. KFELS N Class. Keppel Offshore and Marine. http://www.keppelom.com/en/content.aspx?sid=3642
12. Kotler S (2011) Planet sludge: millions of abandoned, leaking oil wells and natural-gas Wells destined to foul our future. EcoHearth
13. Kudsk G (2018) Design of jack-ups. Encyclopedia of Maritime and Offshore Engineering
14. Lamont T (2017) Where oil rigs go to die. The Guardian
15. Love M et al (2014) Oil platforms off California are among the most productive marine fish habitats globally. Proc Nat Acad Sci
16. Marks P (2016) What it takes to dismantle oil rig. BBC Future
17. Mills S et al (2015) Breaking the pre-fabricated ceiling: challenging the limits for modular high-rise. CTBUH 2015 New York Conference
18. Newman N (2017) The sharp rise of platforms decommissioning. SPARKS
19. Offshore Jack-up rigs. Daleel oil and gas supply chain portal
20. Offshore Technology (2019) The world's worst offshore oil rig disasters. [Online]
21. Pearce F (2018) As North Sea Oil Wanes, Removing Abandoned Rigs Stirs Controversy. YaleEnvironment360
22. Rigzone. How do Jackups Work? https://www.rigzone.com/training/insight.asp?insight_id=339&c_id=

23. Sembcorp Marine (2019) Sembcorp marine completes world's biggest and strongest semi-submersible crane vessel for Heerema
24. Sharma R et al (2010) Issues in offshore platform research—Part 1: semi-submersibles. Int J Nav Archit Oc Eng 2
25. Technical guidelines for biosphere reserves. https://en.unesco.org/op-wnbr
26. The Associated Press (2010) 27,000 abandoned oil and gas wells in Gulf of Mexico ignored by government, industry
27. Todd V et al (2016) Meals on wheels? A decade of megafaunal visual and acoustic observations from offshore oil and gas rigs and platforms in the North and Irish Seas. PLoS ONE
28. United Nations. Department of economic and social affairs, population division (2019) World population ageing 2019
29. Wolfson A et al (1979) The Marine life of an offshore oil platform. Mar Ecol Prog Ser 1:81–89

Intensifying Food and Housing at Sea

Abstract

This chapter portrays emergent need in the context of climate change and rising global population, and proposes as counterpoints to megacities, resilient offshore and nearshore solutions to food production and supply, energy and wastewater management. Jack-ups, semi-submersibles and barges are repurposed into specialized vessels. Of key interest is a typology of a high-density housing on jack-up or semi-submersible to meet the rising population in capacities that surpass cities on modular floating rafts. The spatial quality of the architectural conversions capitalized on the use of derrick and deck structure, moonpool and pontoons to create spaces for communal gathering and for recreation. Infrastructure became architecture. The density and living quality of these propositions is explored through their range of communal recreation spaces. Theses are benchmarked with those of landed housing blocks in Singapore and landmark international examples.

1 Global Challenges to Fish Production

By 2050, 68% of the world population is projected to live in urban areas [20], posing immense challenge to social and environmental sustainability. While cities remain as engines of economic production, their capacity to accommodate population growth sustainably is questionable when inhabitants have unequal access to resources and use them inefficiently. The relentless extraction and commodification of natural resources will result in large ecological footprints. While planners have relied on smart technologies as a way forward, the World Economic Forum prioritizes the development of inhabitant-centric cities, which attempt to minimize their environmental impact. Liveable cities are envisioned with quality environments, which can produce and consume resources without depleting or causing irreversible damage to natural reserves and remaining ecosystems.

World Health Organization cites fish as the primary source of protein for more than one billion people. Currently, the global annual fish consumption for a population of 7.4 billion is 148 million tonnes and the projected population of 8.1 billion in the year 2025 is 162 million tonnes [3]. Yet, global fish catches are declining at rates which suggest overfishing. Such phenomenon occurs when fishes from the sea are caught at faster rates than they can be replenished by natural reproduction. As such, the sole reliance on fishes caught from the sea is not sufficient to feed the growing world population; hence a shift towards an increasing aquaculture production is imperative to the global food security of the future.

By 2025, global aquaculture production is projected at 102 million tonnes and this will surpass current catches from the sea at 78 million tonnes [3]. Moreover, the decrease in reliance on fish catches from the sea allows for the gradual regeneration of the heavily depleted marine wildlife.

1.1 Singapore's Fish Production

The land-based and coastal fish farms in Singapore (Fig. 1) are used as case studies for design considerations in optimizing farm plot areas and production capacity. Harvest-ready fishes from the land-based and coastal farms are distributed daily to Senoko Fishery Port and Jurong Fishery Port where they will be sorted, packed and sold to various fish suppliers. These fishes are then delivered to the wet markets and seafood restaurants island-wide.

Statistics from Singapore's Agri-Food and Veterinary Authority (AVA) states that on average, Singaporeans consume 21 kg of fish per year. At this rate, Singapore would need 119,700 tonnes to feed her current population of 5.7 million and 153,000 tonnes of fish for a projected population of 7.3 million [18] in the year 2050. AVA targets to increase its fish production from 8 to 15% of Singapore's total consumption by 2050.

J. Lim, *Oil Rig and Superbarge Floating Settlements*, Lecture Notes in Civil Engineering 82,
https://doi.org/10.1007/978-981-15-5297-7_2

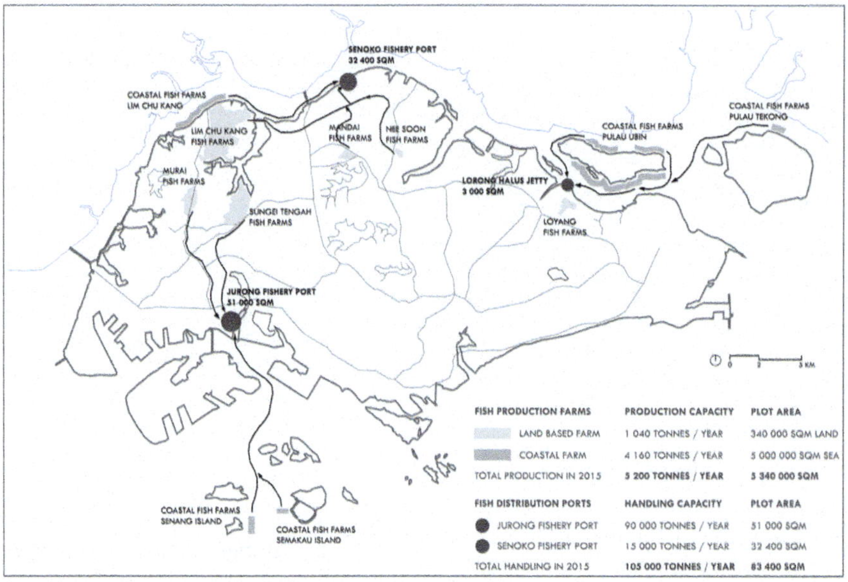

Fig. 1 Fish production farms and distribution ports in Singapore. (*Source* AVA)

Fish Reproduction Cycle Specialized technologies and equipment; spatial layout and distribution workflow for a sustainable fish production and farming, are studied via site visits to local coastal fish farms, Marine Life Aquaculture (MLA) (Fig. 3) near Pulau Ketam and AVA's fish research facility, Marine Aquaculture Centre (MAC) at St. John's Island.

MAC's fish research facility focuses on farming Asian seabass (a tropical fish species). Its sustainable farming technology uses matured adult fishes as brood stocks to spawn millions of eggs that hatch into larvae and grow into adult fishes. One reproduction cycle from an egg to an adult fish takes up to 11 months. The entire process is repeated for subsequent batches. The key to ensuring a long-term sustainable reproduction cycle is to keep the parent fishes in good and healthy

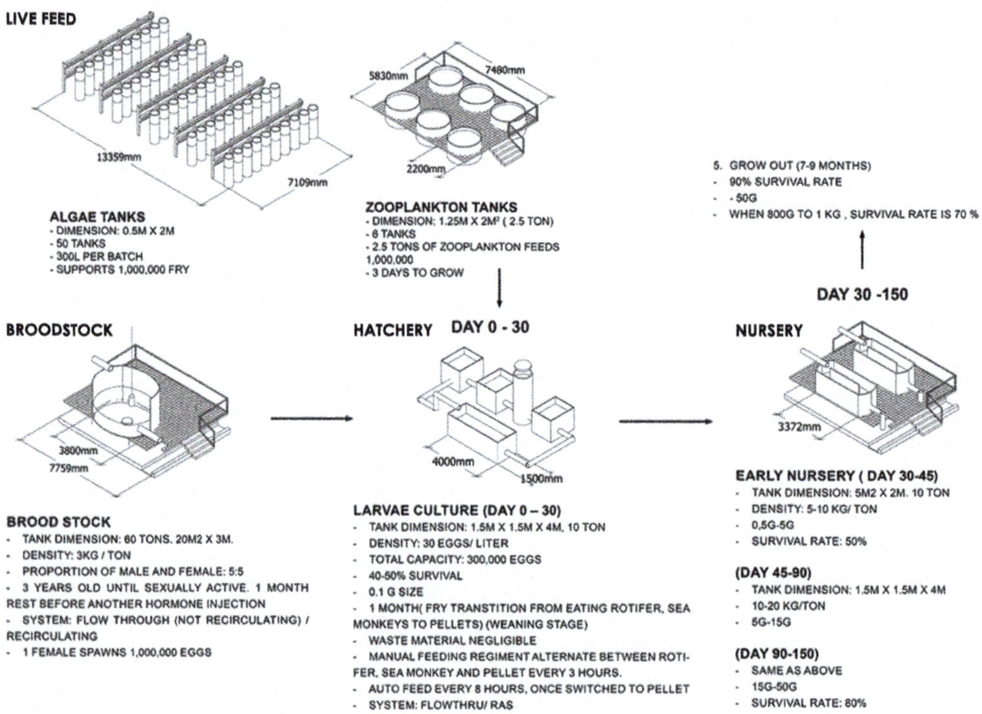

Fig. 2 Fish farming guidelines for optimal growth. (*Source* AVA)

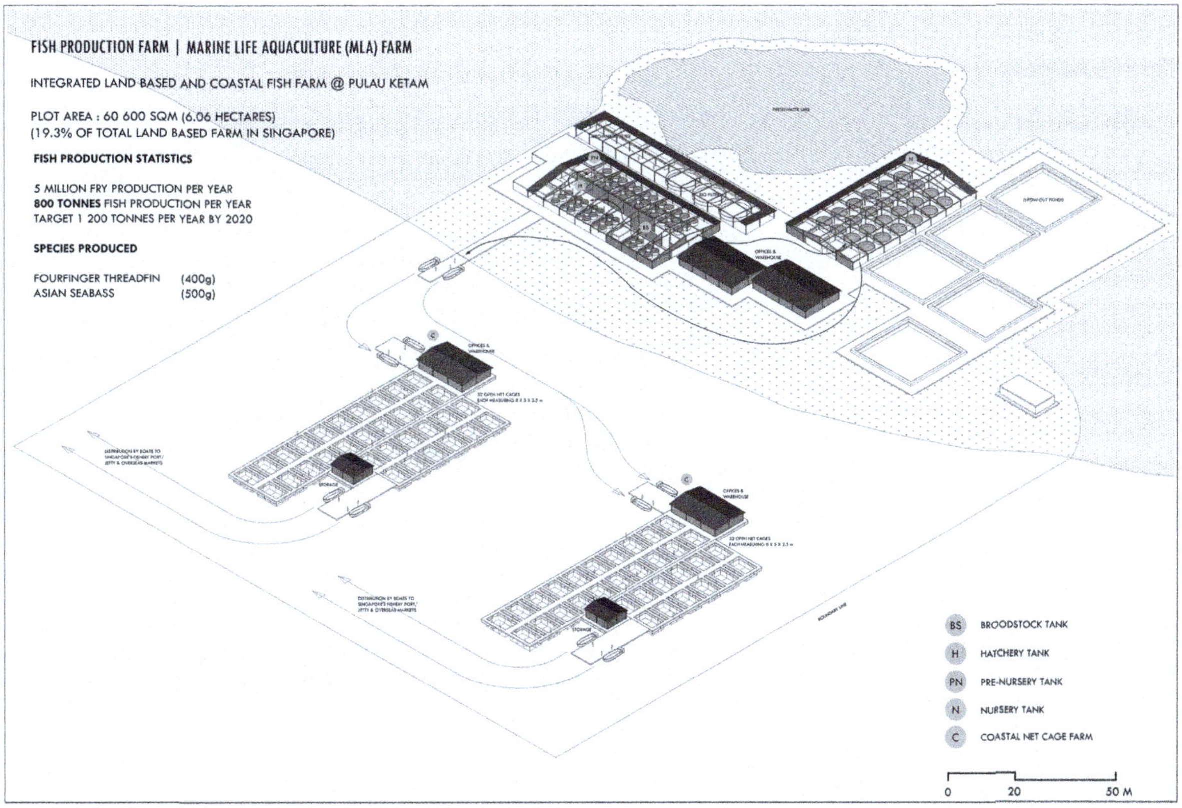

Fig. 3 Spatial layout and distribution workflow at Marine Life Pte Ltd (S) Aquaculture Farm

living conditions. This farming practice is opposed to the unsustainable harvesting of wild fishes directly from the sea that ultimately leads to overfishing in the long term.

Figure 2 illustrates the types of tanks required at different growth stages of the fish. The indoor and controlled environment is essential during early phase—first to the fourth month—of the fish life cycle. The fish research scientists at AVA provide guidelines on optimal tank dimensions and stock densities.

1.2 Repurposed Oil Rigs as Fish Production and Distribution Facility

In order that floating settlements does not diminish the natural resources of the sea, their food production must reverse the current effects of overfishing, by actively repopulating fish numbers and marine ecosystems. To do this, settlement production must aim beyond outputs in excess of consumption rates in order to replenish the marine supply chain. Settlement food production incorporates 'fish nurseries' to augment natural growth rates at sea. Floating fish production facilities are thus an integral component of an offshore settlement and are conceived as nodes of urban activity.

For an offshore settlement to be sustainable, it has to lower its ecological footprint in energy and food production [10]. Energy can be produced by renewable sources of wave, wind or solar energy while part of its energy may be recovered from waste treatment processes. Food production processes would be symbiotic (e.g. aquaponics) so that the provision of one necessity produces by-products for another with every precious unit of energy expended. The species farmed would be converted to many uses in addition to protein supply.

Oil rigs can be strategically repurposed as fish production and distribution facility with waste treatment, energy production and farm crew accommodation. The facility comprises a jack-up fish farm, a 'grow-out farming area' and a semi-submersible fish wharf.

2 Jack-up Versus Semi-submersible Fish Farm

The vertical fish farm uses the structures of the main platform decks in both semi-submersible and three-legged jack-up to accommodate nurseries, hatcheries while the 'grow-out farming area', and distribution wharf are separately planned. In comparison, a three-legged jack-up oil rig has a greater annual production capacity of 33,600 tonnes per year against 28,080 tonnes of fish per year from a semi-submersible (Figs. 5 and 4).

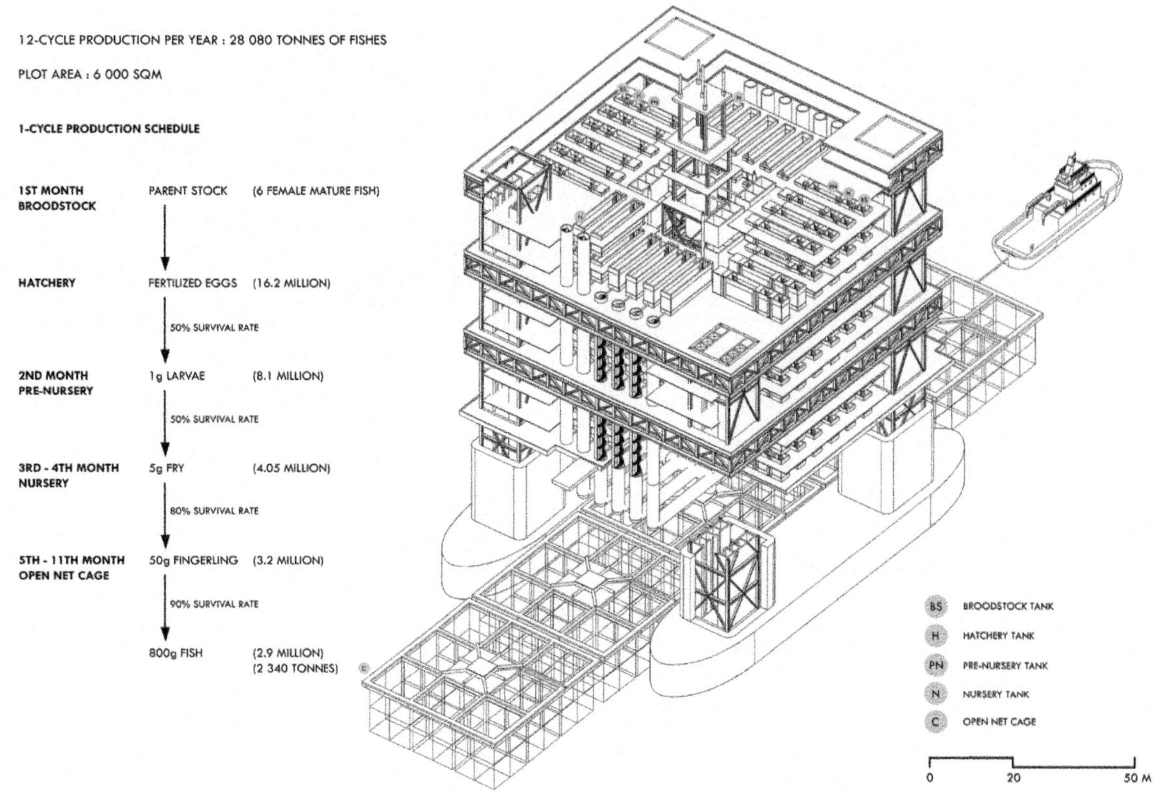

Fig. 4 A semi-submersible fish farm produces 28,080 tonnes of fish per year

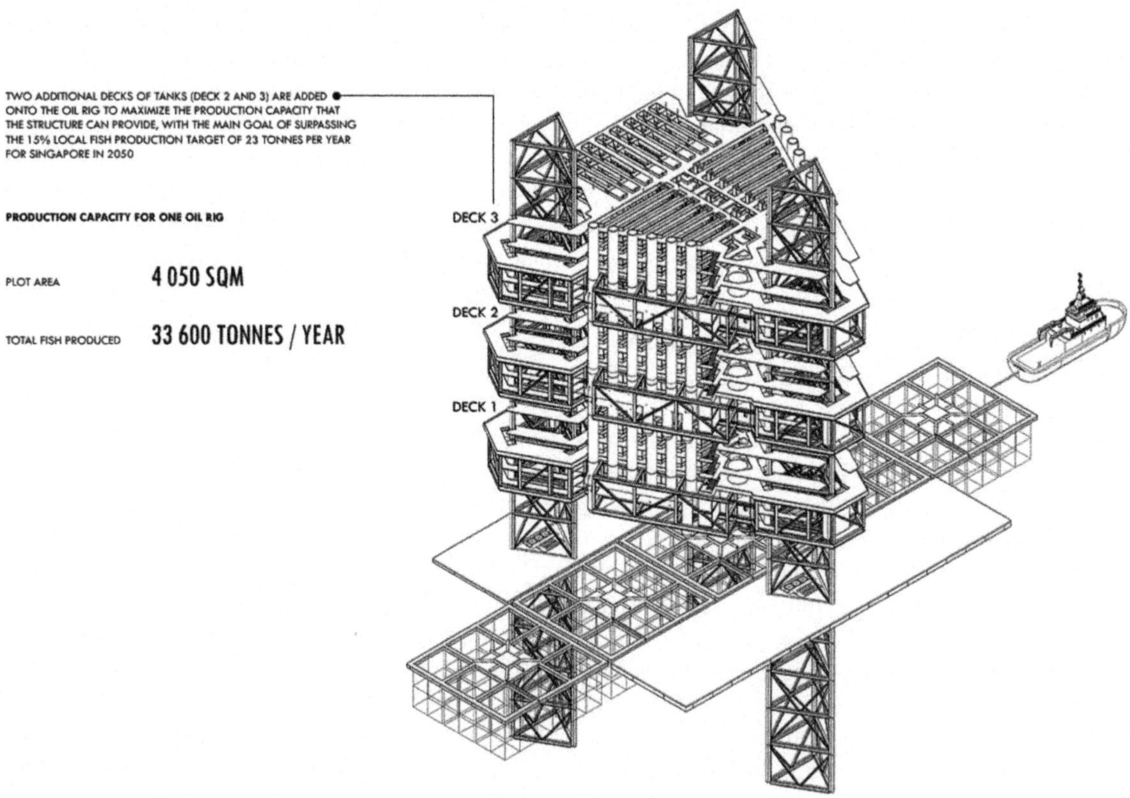

Fig. 5 A jack-up fish farm produces 33,600 tonnes of fish per year

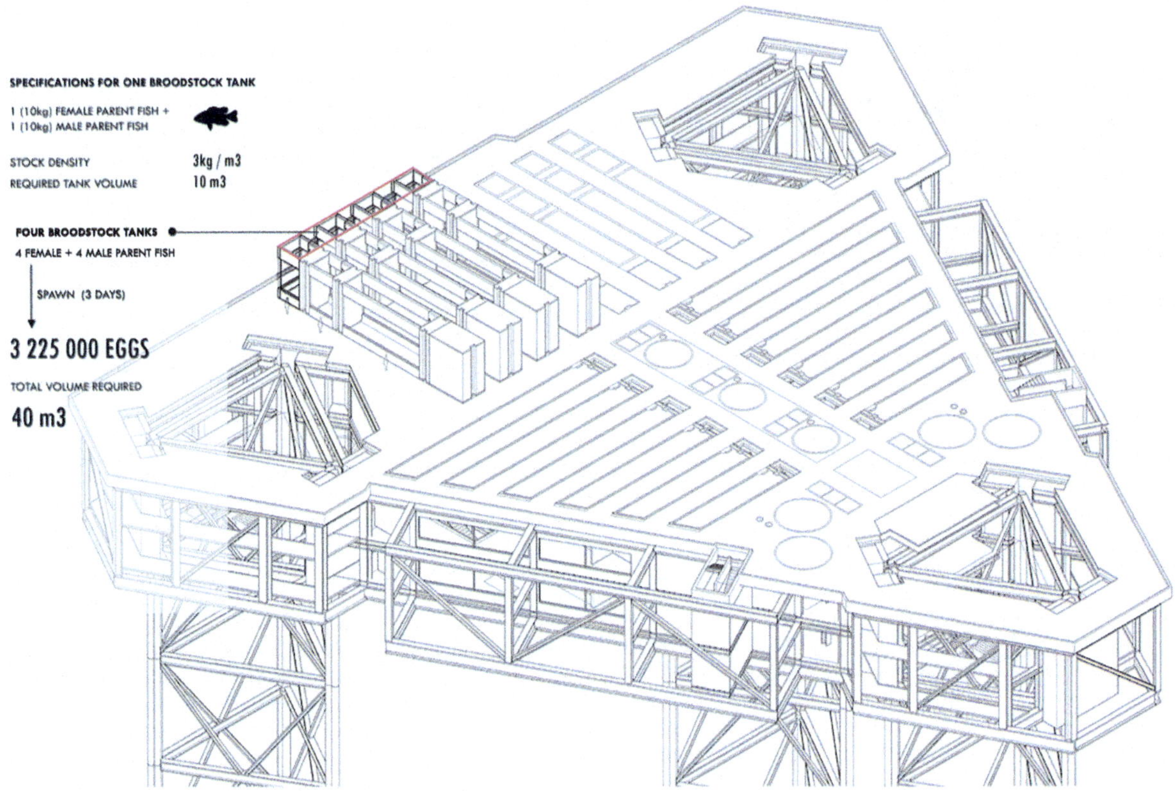

Fig. 6 Total of eight broodstock tanks on main deck

The annual fish production by one jack-up fish farm would exceed Singapore's target of 15% as set by AVA in the year 2050. Five jack-up fish farms can produce 168,000 tonnes of fish, which meets 100% of Singapore's projected fish consumption in 2050. The global annual fish consumption of 162 million tonnes by 8.1 billion people in 2025, can be met by 4,822 of such facility, occupying 642,290 ha of sea space.

Figures 6, 7, 8 and 9 illustrate the locations of various tanks on the jack-up fish farm, with specifications and production capabilities. To maximize fish production, additional steel decking can be added onto the main deck, to house more tanks (Fig. 10). From the first to fourth month (eggs to fingerling stage), production is indoors and in a controlled environment. From the fifth to eleventh month, fingerlings are transferred from deck to sea in net cages to the 'grow-out' farming area (Figs. 11 and 12). Figures 13, 14 and 15 detail the design of a jack-up fish farm for production efficiency.

2.1 Fish 'Grow-Out' Farming Area

Figure 16 illustrates the workflow from fish production to the 'grow-out' farming area. After the first four months of in-house fish farming, the fingerlings are transferred to open net cages where they will 'grow-out' to their adult stage. An offshore fish farm has vast space and at this growth stage, the fishes can withstand harsher waves of the open sea. The dimension of the net cage is based on the optimal stock density of 20 kg per cubic meter of sea (as provided in AVA's guidelines) to receive batches of fingerlings from the jack-up rig fish farm monthly. The fishes in the net cages are fed with solid feed pellets supplied through an automated feeding system on feed barges. These barges would be kitted with advanced aquaculture technology such as AKVA's feed systems, which specialize in deep sea fish farming.

The annual feed requirement for three commonly farmed fishes—salmon, seabass and grouper—is calculated according to Apparent Feed Conversion Ratio (AFCR) [3]. AFCR is the actual number of units of 'dry' aquafeed required to produce a unit of 'wet' fish. The AFCR for salmon, Asian seabass and grouper are 1.0, 1.5 and 1.8 respectively, and translates to 33,600; 50,400; and 60,480 tonnes respectively. The fish feed pellets may be produced at the fish wharf (Fig. 24) using fishmeal processed from fish waste byproducts [7]. Fishmeal is one of the ingredients in feed pellets.

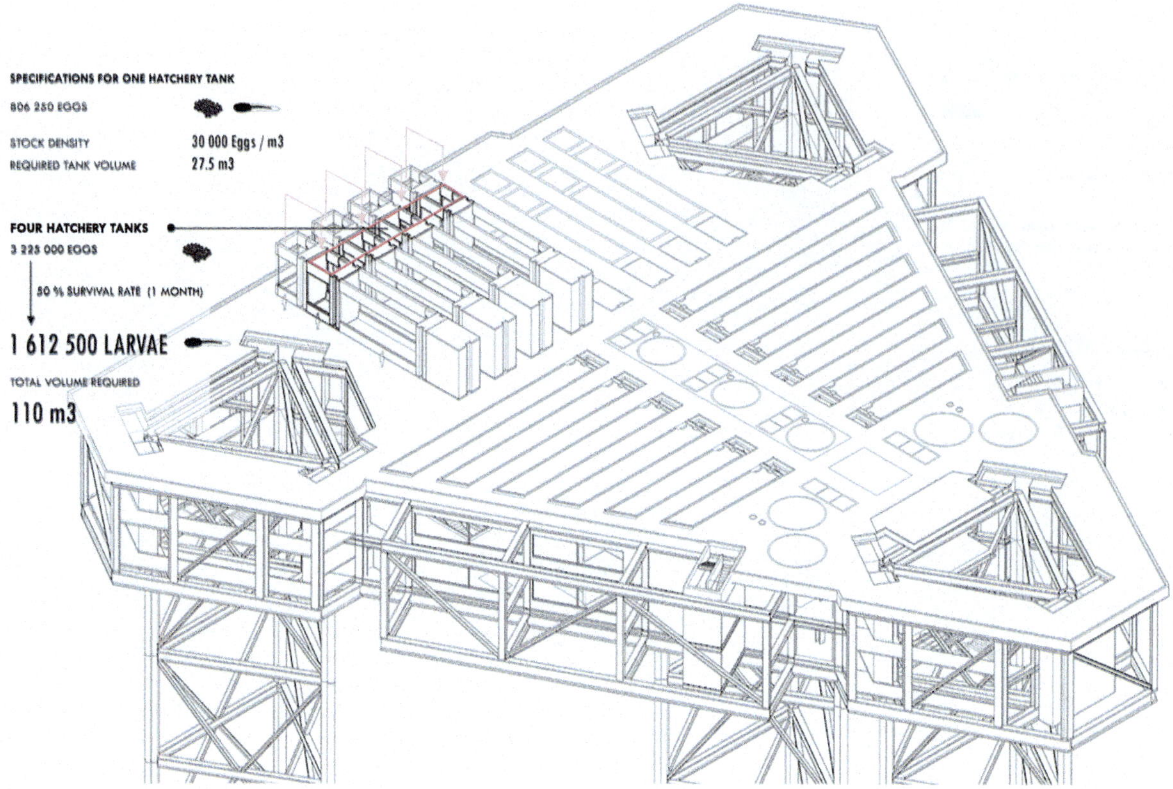

Fig. 7 Total of eight hatchery tanks on main deck

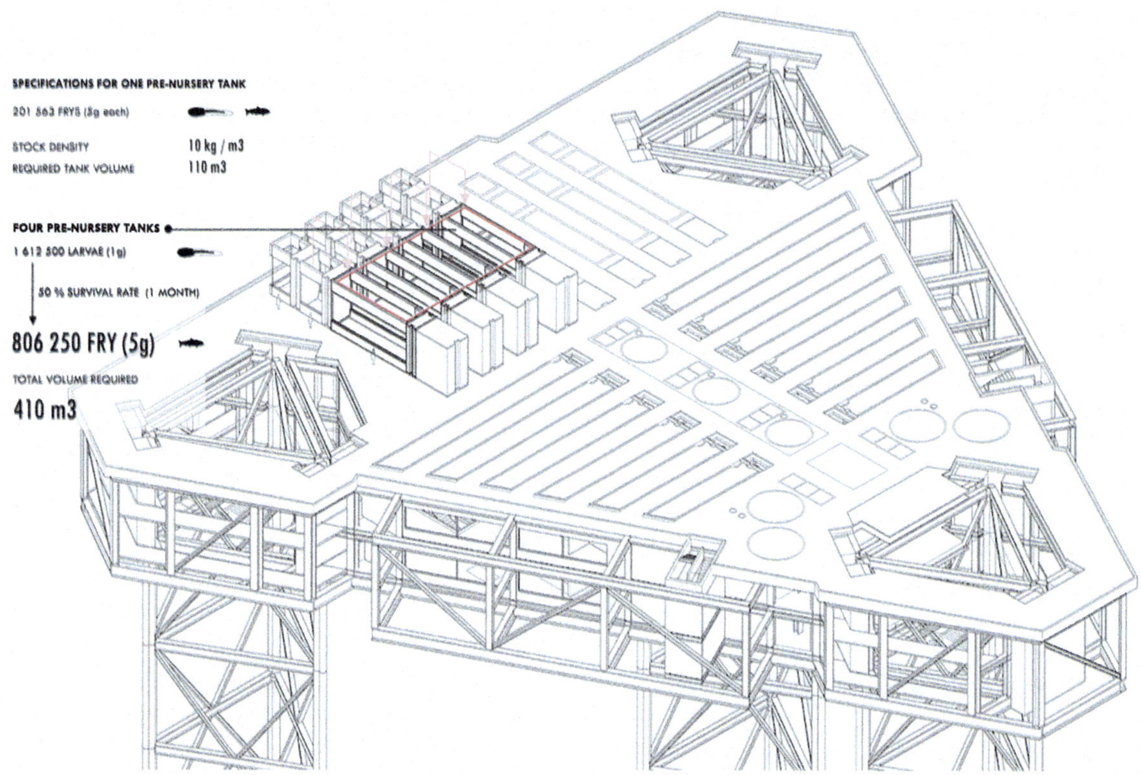

Fig. 8 Total of eight pre-nursery tanks on main deck

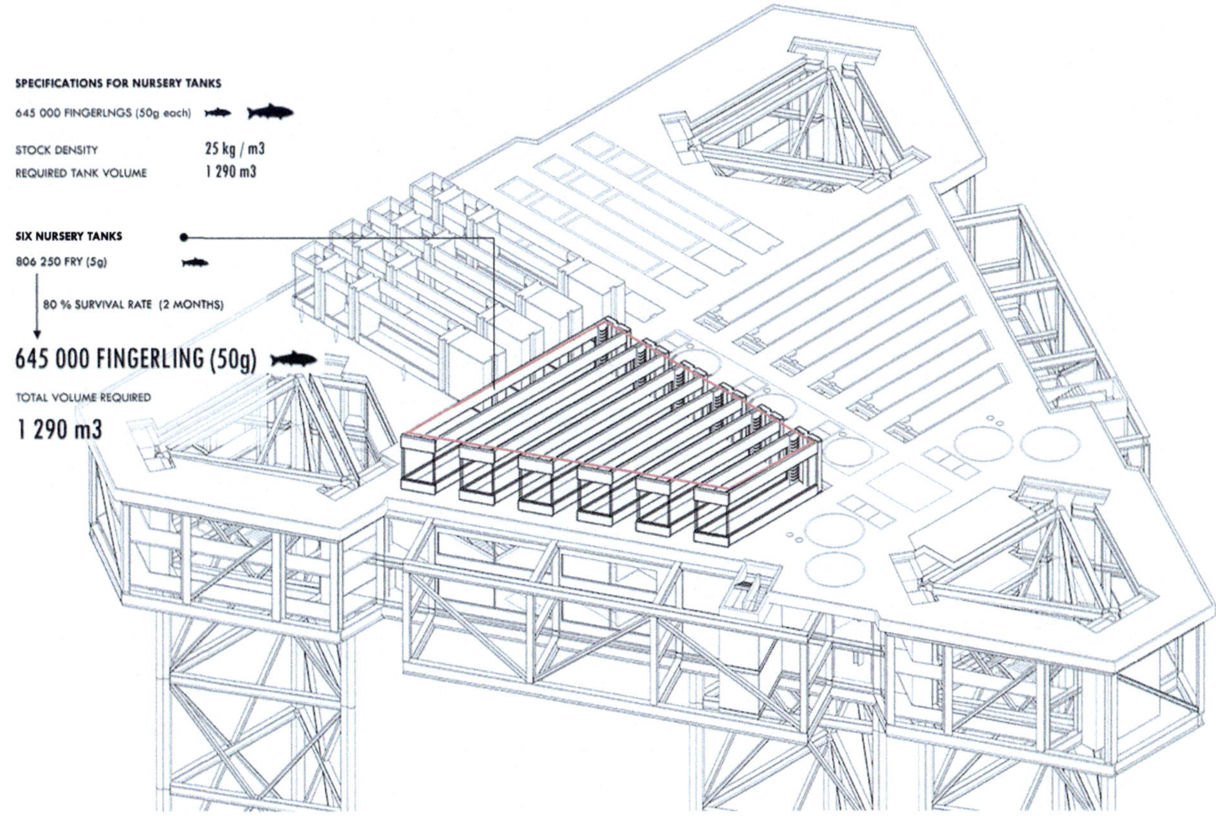

Fig. 9 Total of twelve nursery tanks on main deck

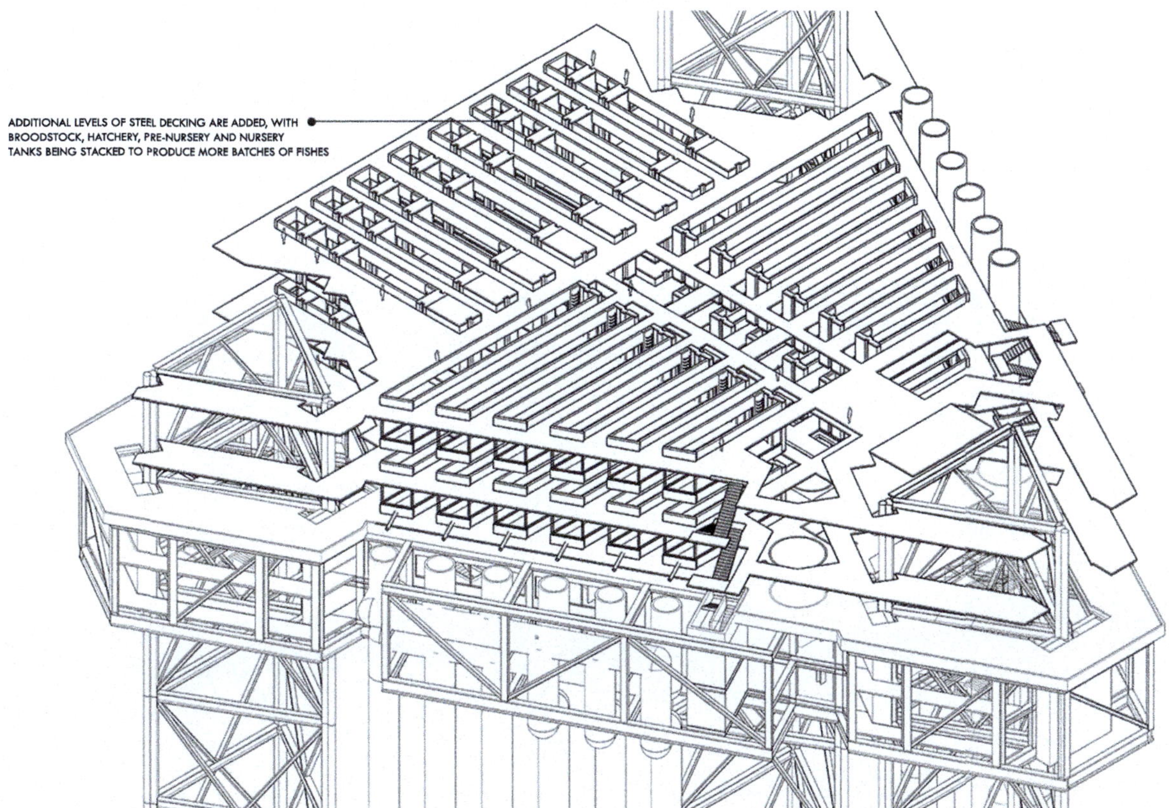

Fig. 10 Maximizing fish production with additional levels of steel decking to house more tanks

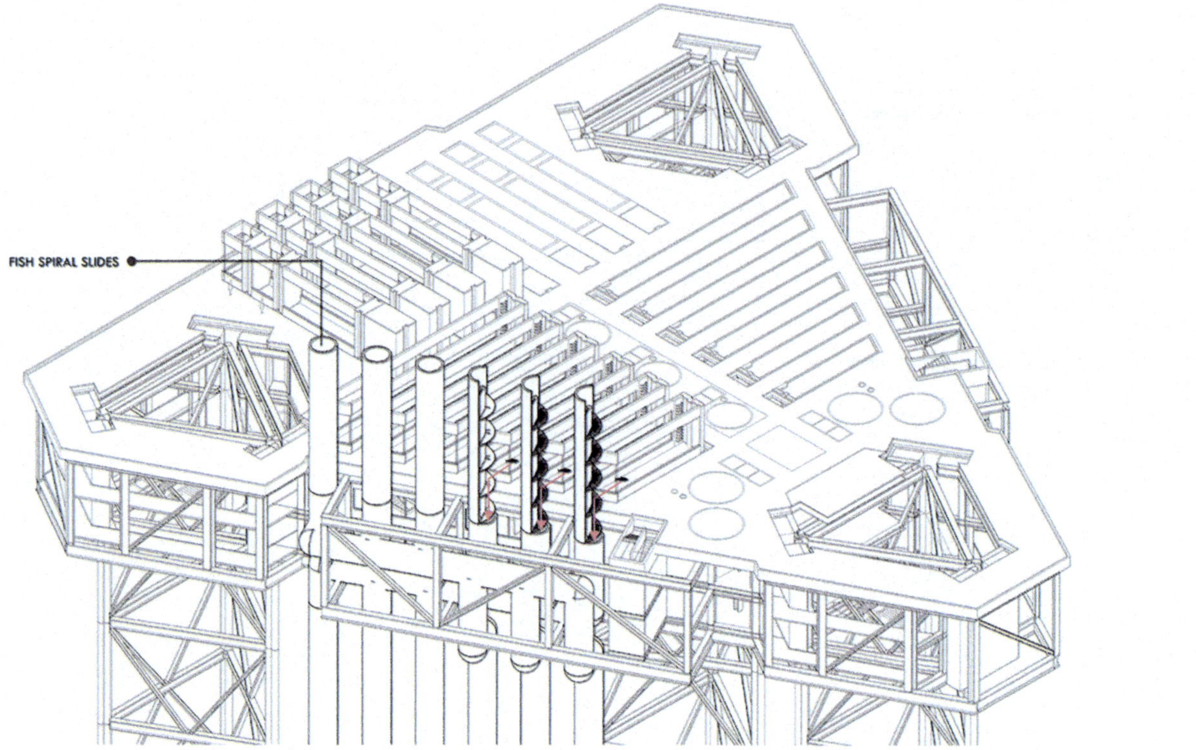

Fig. 11 Vertical automated pipes to transfer fingerlings down to net cages

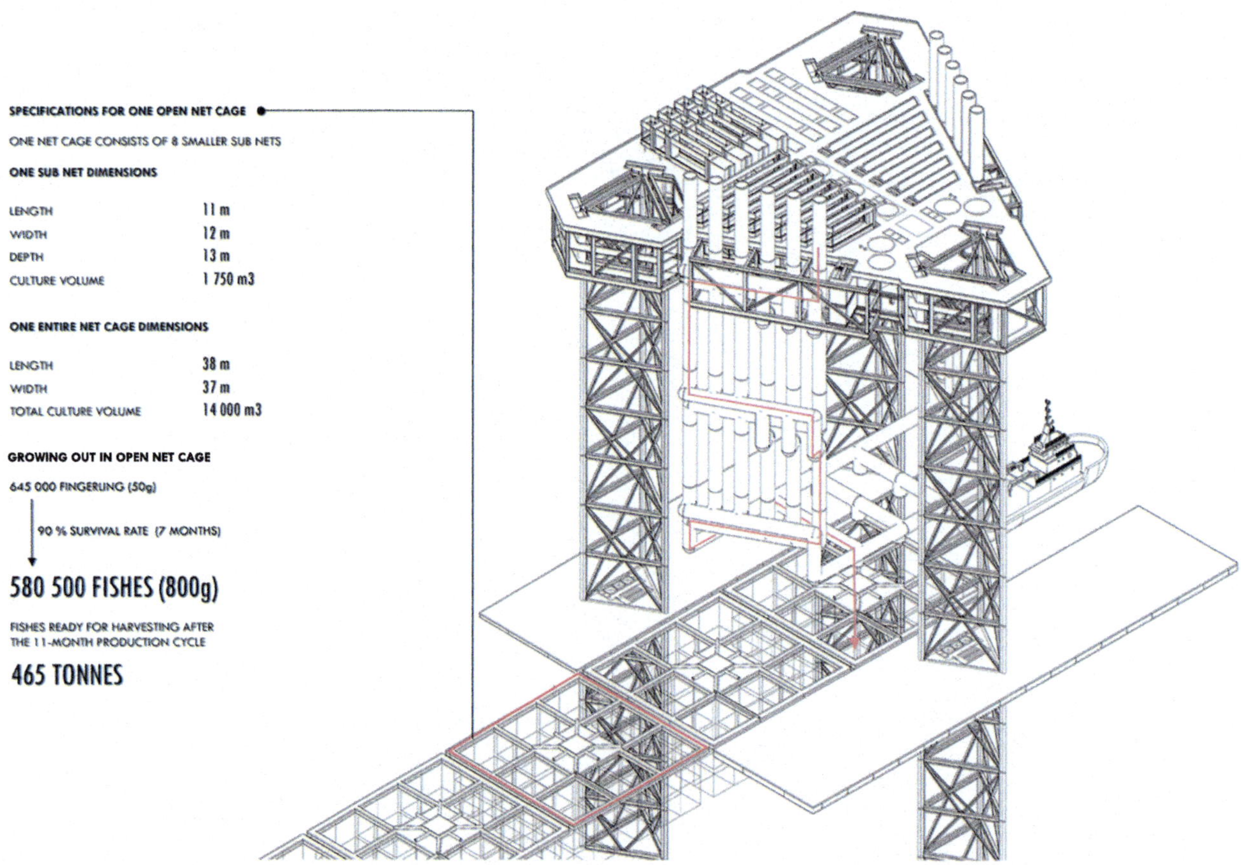

Fig. 12 A net cage of 38 m long with approximately 645,000 fingerlings will be transported to the 'grow-out' farming area from the fifth until eleventh month of their production cycle. Allowing for 10% attrition, about 580,500 fishes would grow to their adult stage, ready for harvest

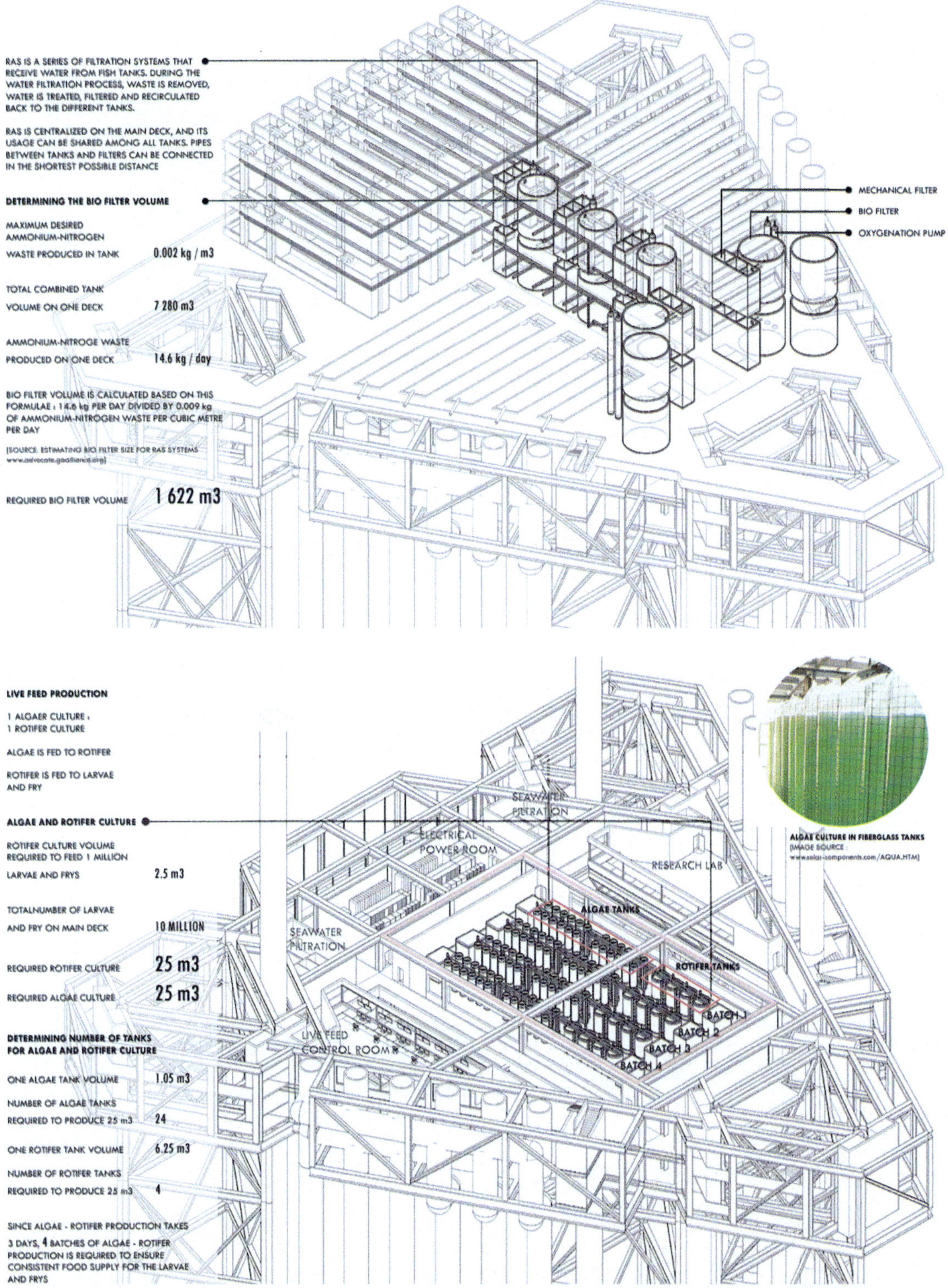

Fig. 13 Specialized technologies and automated systems are integrated into the jack-up fish farm for efficiency in fish production at a large scale. Placed on the main deck, the recirculating aquaculture system (RAS) is a series of filtration systems that treats water from fish tanks. During the water filtration process, waste is removed and treated water is recirculated back to the different tanks. RAS is centralized on the main deck to shorten pipe connections between tanks. The use of fiberglass tanks also accelerates the growth of algae as feed for the rotifers, which are then fed to the larvae and frys

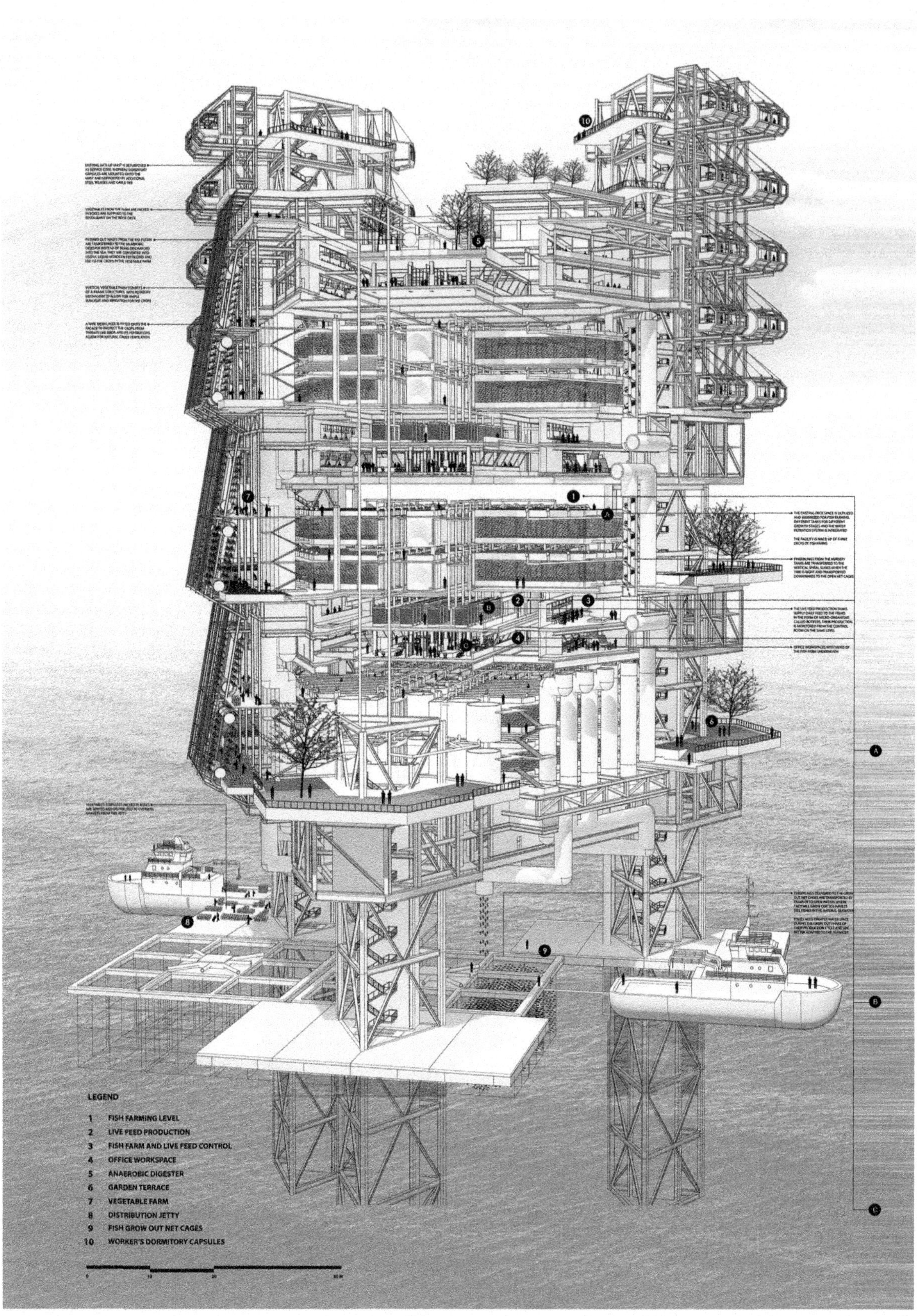

Fig. 14 Fish tanks are located on decks above rotifer production levels on transfer deck structures. Automated spiral pipes located on the inner periphery of the decks transfer fingerlings to net cages. Dormitory capsules for production crew and aeroponic scaffolds are located on the outer periphery of the jack-up structure

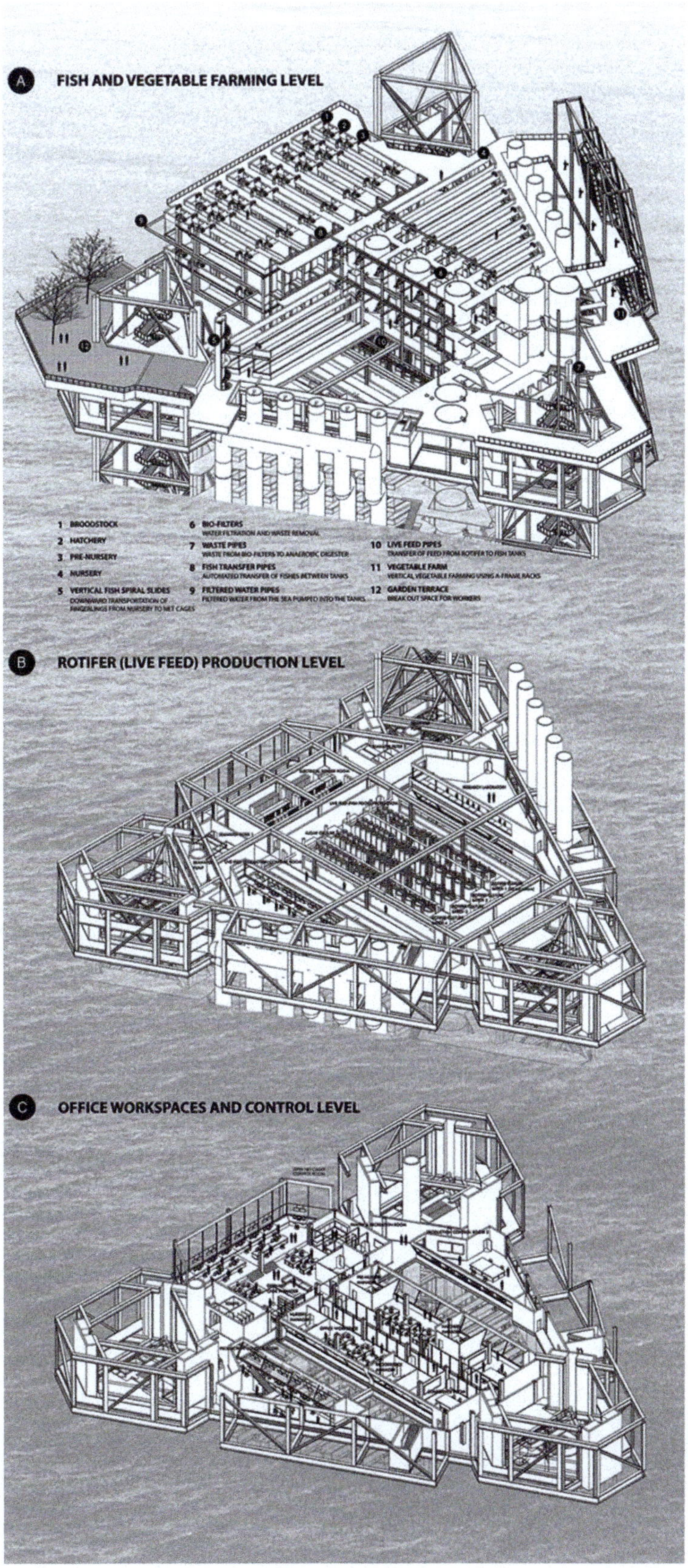

Fig. 14 (continued)

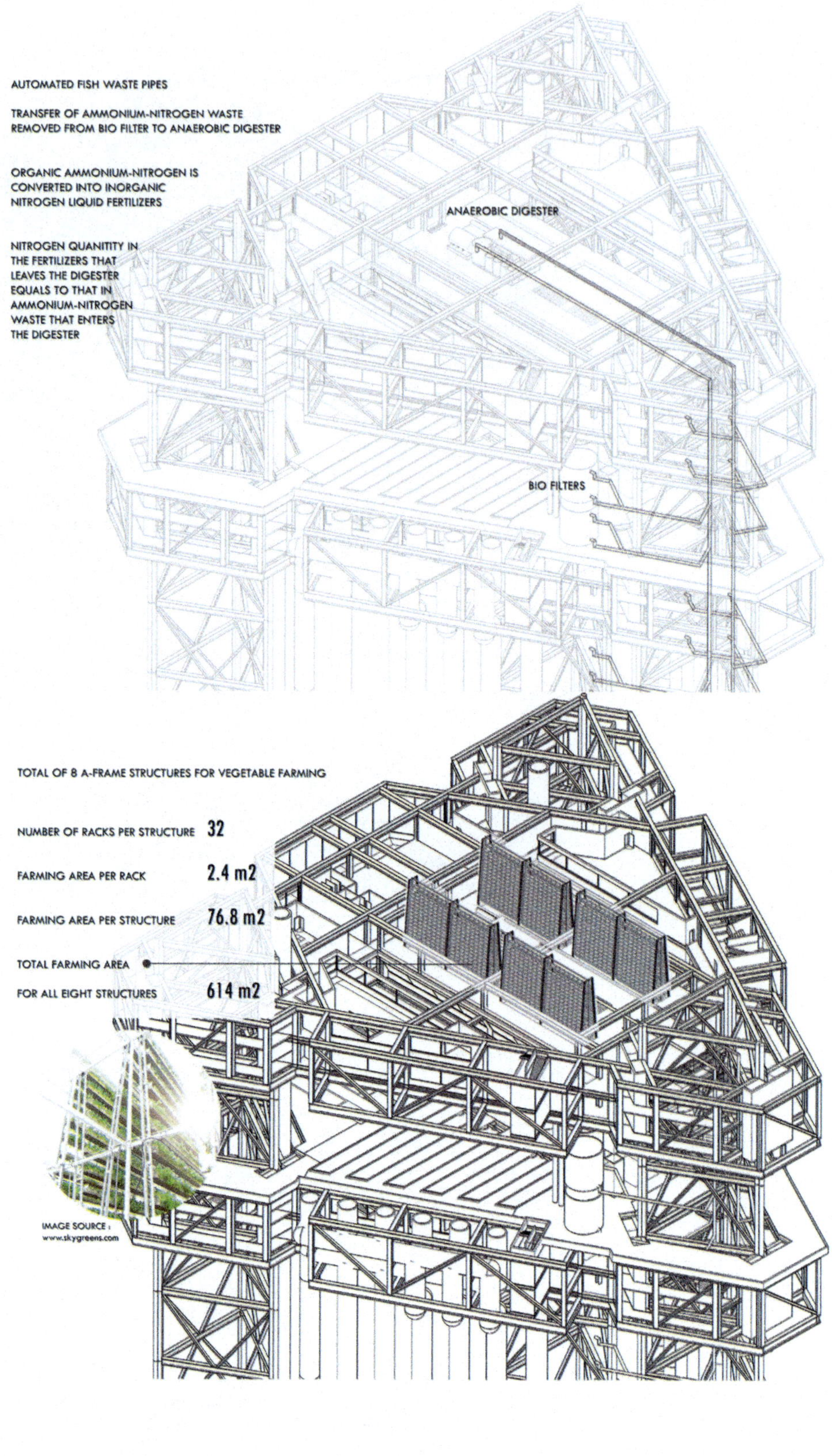

Fig. 15 Waste loop strategy. Automated fish waste pipes transfer the ammonium-nitrogen waste from the biofilter to the anaerobic digester, which is then converted into liquid fertilizers for vegetable farming. The vegetable farm platform is proposed above deck 3 of the jack-up rig. Crops such as spinach, tomato and cabbage can be grown onboard for the consumption of production crew

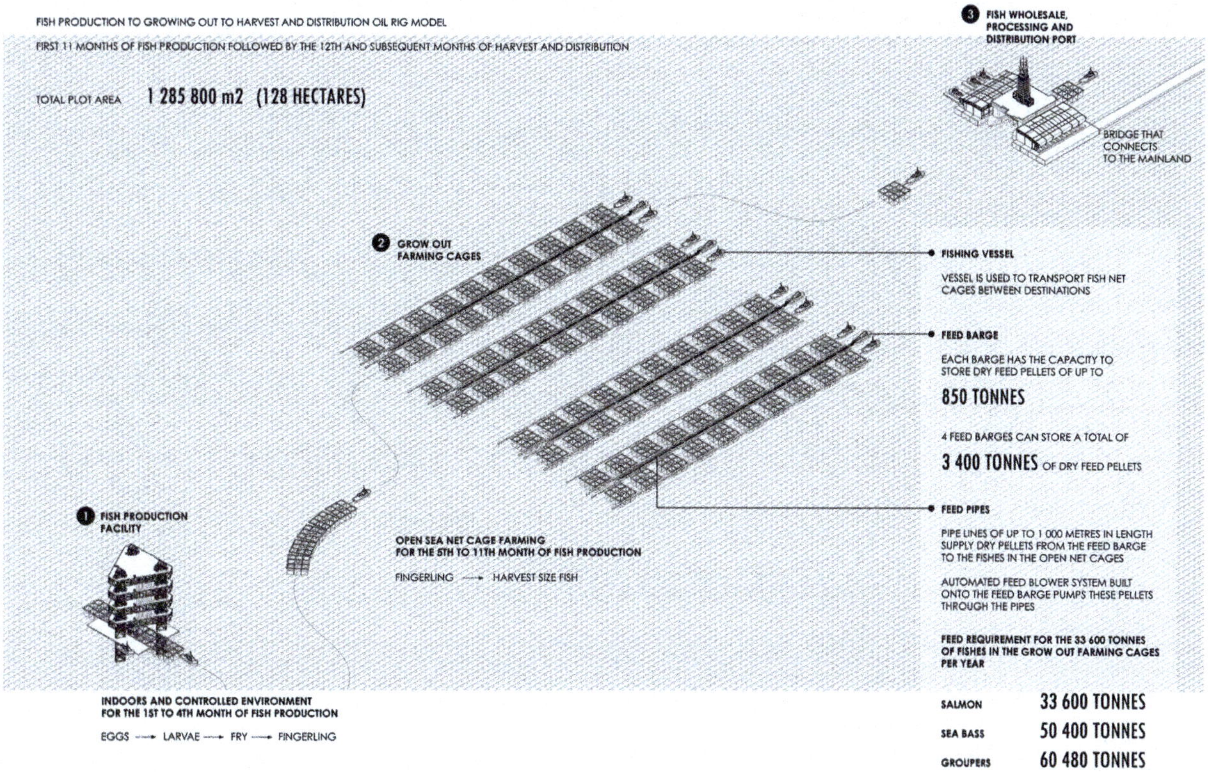

Fig. 16 Tugboats tow the nets out to a 128 ha growth area at sea. Automated feed pipes transfer food pellets through a feed blower system on board a feed barge. Net cages are reinforced against predators. Annual feed requirement is substantial depending on AFCR rates

2.2 Polluted Water Containment Strategy

Offshore fish farming can be vulnerable to threats of oil spill from collision accidents at sea. The sudden drastic change in water quality can cause loss to fish production and revenue. Hence, a containment strategy is imperative to counter such threats.

Containment booms can be deployed to contain and control oil spills on water. These booms are inflatable devices that remove oil on water surfaces via absorbent pads, skimmers and vacuums. A proposed containment strategy for the fish 'grow-out farming area' is to form a ring fence defensive perimeter as illustrated in Fig. 17. The ring comprises a series of oil spill storage containers on floating pontoons. Security outposts are included to monitor and control ships entering this protected zone and also safeguard valuable assets to other threats like piracy.

Figure 18 demonstrates an oil spill containment operation. Each oil storage container can store up to 15,000 tonnes of oil. The combined storage capacity for all 40 containers amounts to 600,000 tonnes. This amount is based on estimates reported in past accidents. For example, Singapore's worst oil spill in May 2010 from a vessel collision resulted in 2,500 tonnes of crude oil spilling into the Straits of Singapore [16]. The biggest oil spill in the United States was in the Gulf of Mexico in April 2010, where 585,000 tonnes of crude oil spilled into the sea after an explosion on BP's oil rig 'Deepwater Horizon' [14].

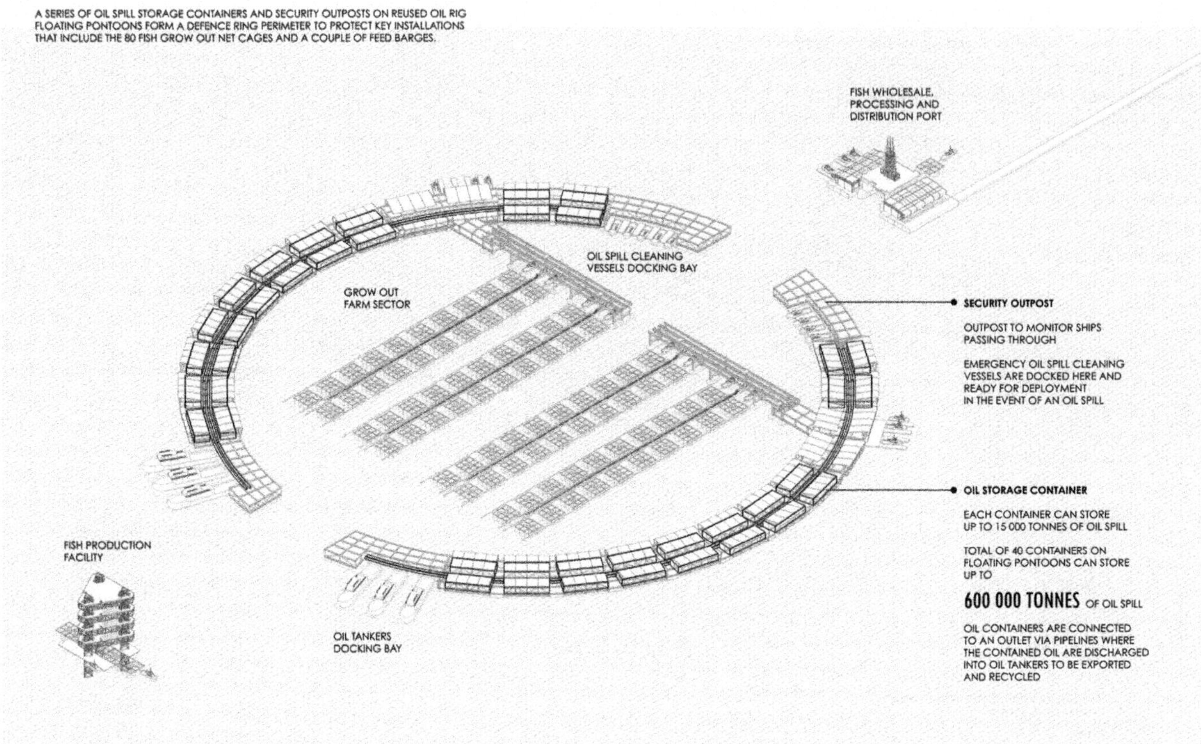

Fig. 17 Oil spill storage containers form a protective ring around the 128 ha growth area. Each container stores up to 15,000 tonnes of spillage

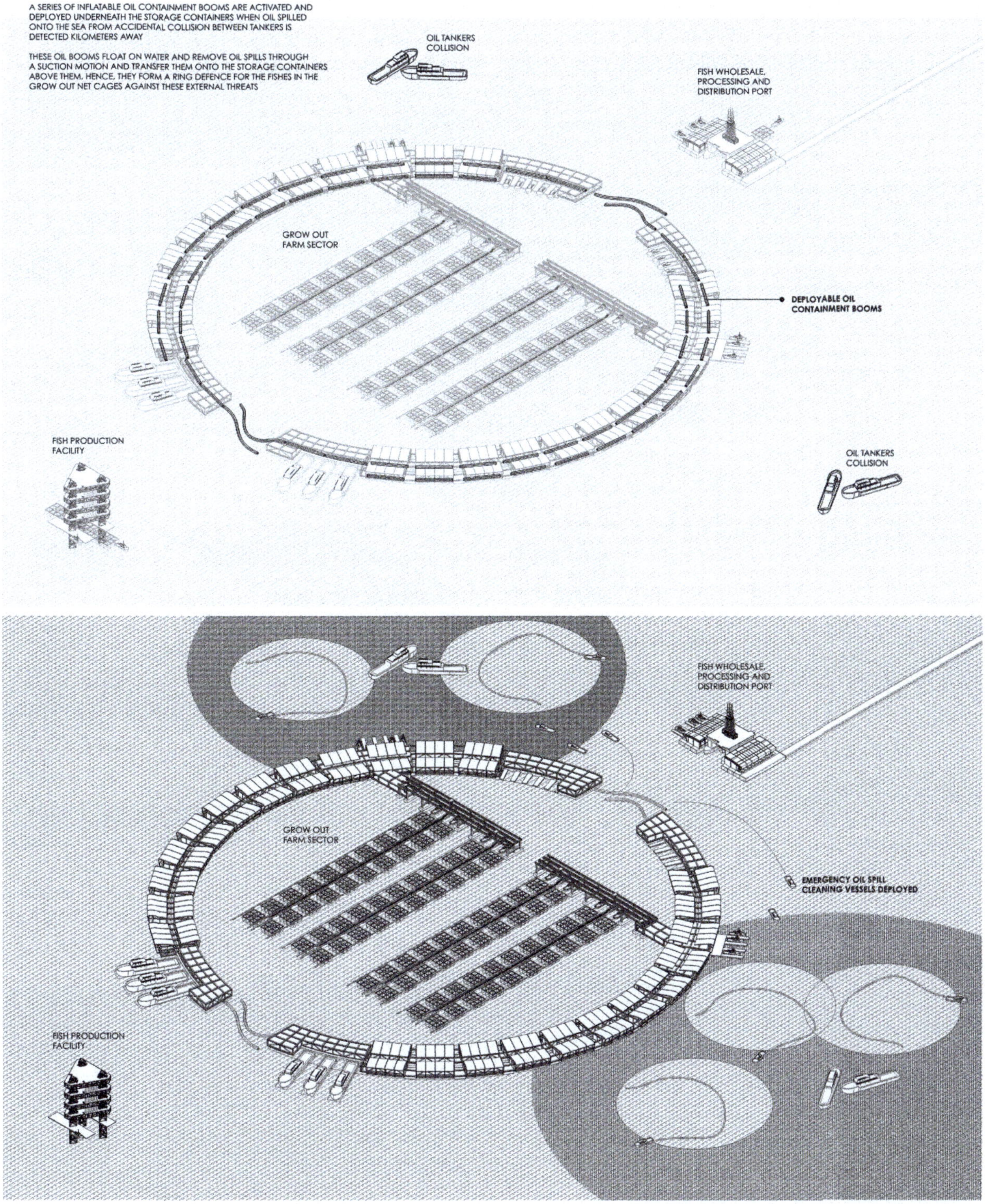

Fig. 18 In an oil spill accident, a series of inflatable oil containment booms are deployed. The booms remove oil on water surfaces via absorbent pads, skimmers and vacuums

2.3 Semi-submersible Fish Wharf

After the eleventh month, harvest-ready fishes are delivered to a repurposed semi-submersible wharf for processing, sorting, packing and distribution. Figure 19 shows that the structure of a semi-submersible rig allows its underside to be used as harvesting area while its adjacent pontoons become the loading bay for the catches. Floating pontoons are added on either side to provide stability to the semi-submersible rig's structure. The value proposition to integrate an offshore fish-processing wharf is to streamline workflow and to reduce the footprint as compared with current practice of transporting harvested fishes from distribution ports to the various processing plants on mainland. Figures 20, 21, 22, and 23 detail the design of a semi-submersible fish wharf for workflow efficiency and as a recreational public space.

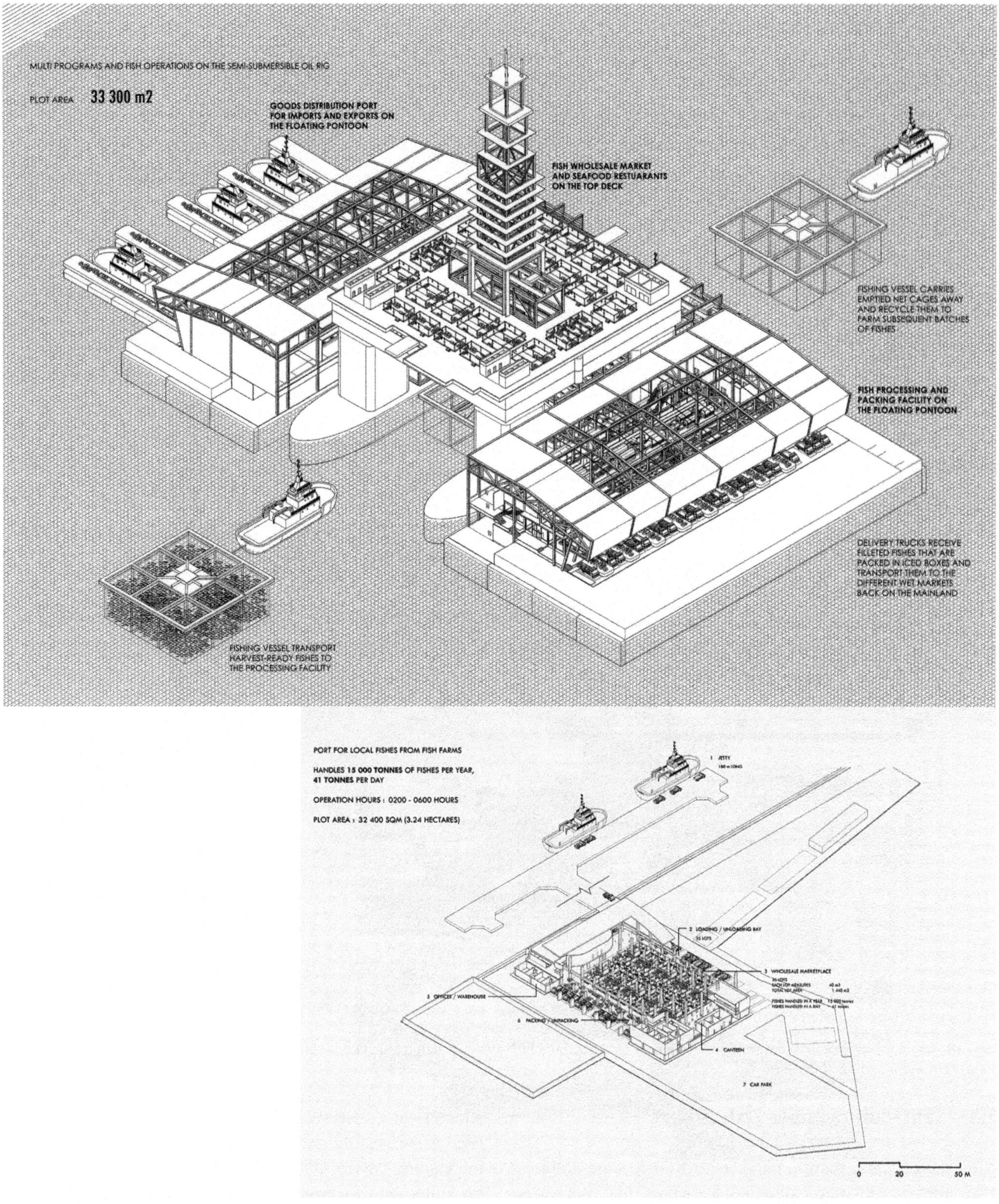

Fig. 19 Semi-submersible wharf for processing, packing and distributing harvested fishes compared to Senoko Fishery Port of Singapore, of equivalent footprint at 33,300 m^2. The semi-submersible wharf processes double the tonnage of fish that Senoko processes annually at 33,600 tonnes and 15,000 tonnes respectively

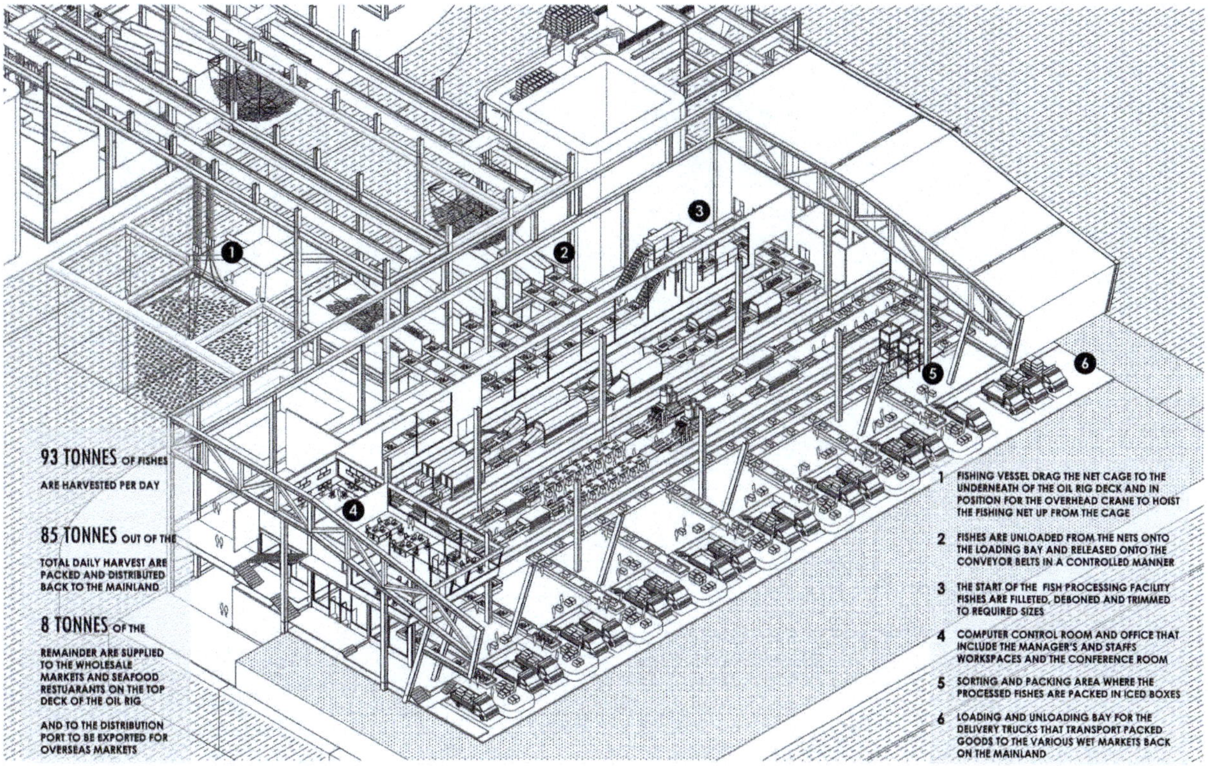

Fig. 20 Loading/unloading, sorting and distributing 93 tonnes of fishes daily

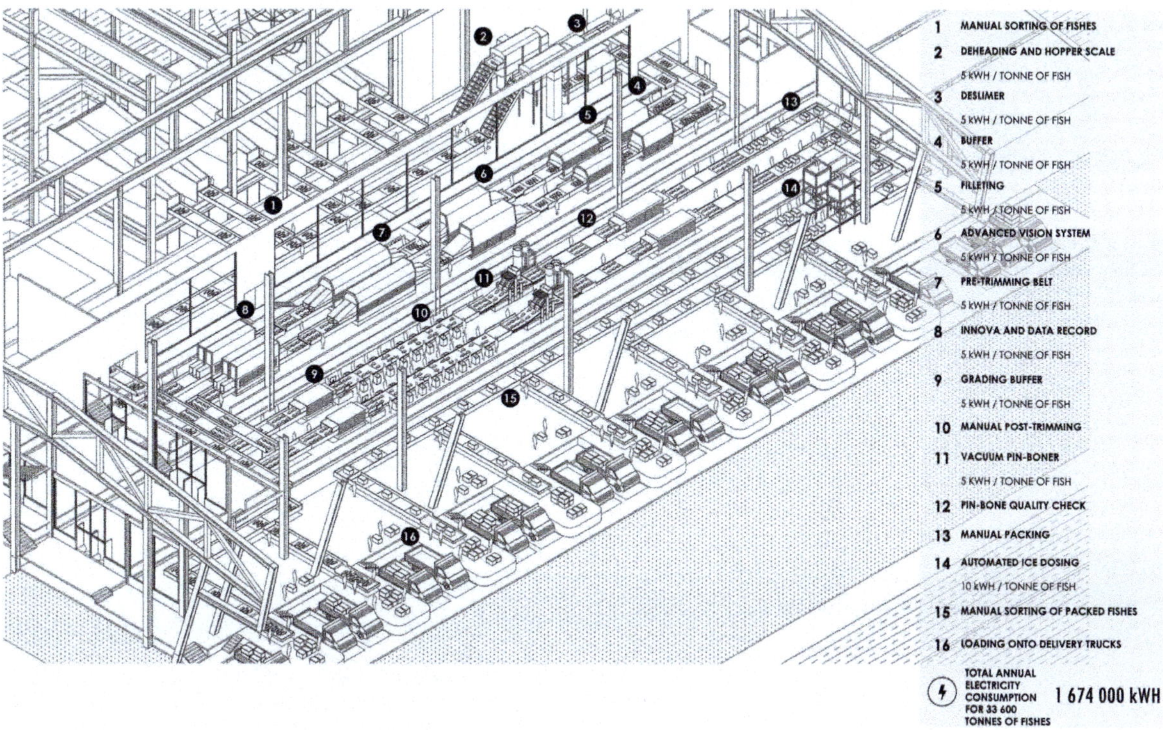

Fig. 21 Fish processing workflow with annual electrical energy consumption is estimated at 1,647,000 kWh

Fig. 22 At the fish wharf, attention is paid to the quality of both the work spaces and user experiences. Fish harvesting, processing facility, waste to fish feed pellet production, and office workspaces are integrated with public spaces comprising wholesale fish markets, open terrace dining and restaurants, to create a vibrant wharf atmosphere

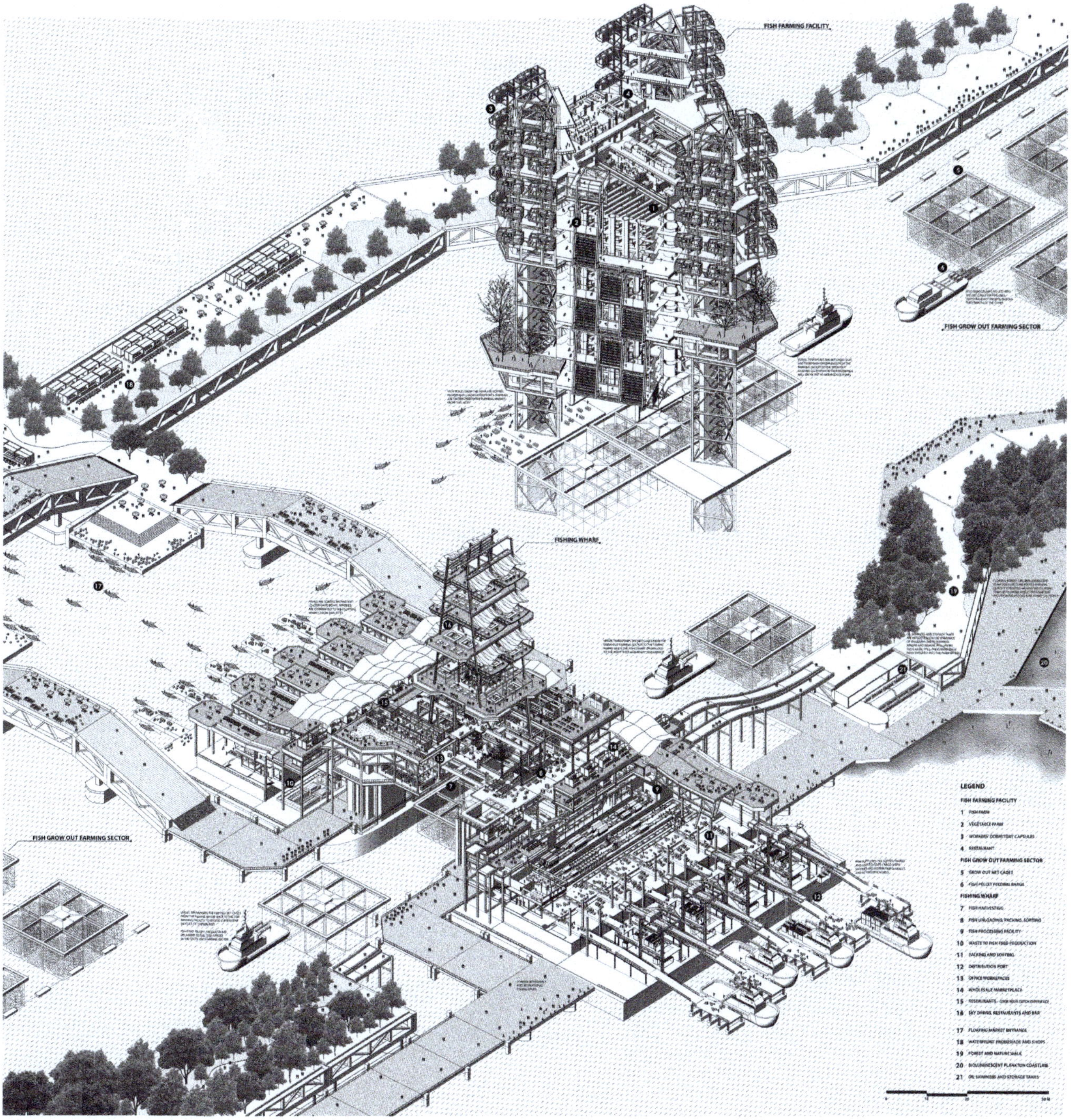

Fig. 23 An offshore fish production and distribution facility aims to intensify its oil rig footprint for annual fish production compared to coastal fish farms in Singapore. Public spaces such as wholesale markets and restaurants are incorporated above the processing and distribution wharf. The offshore fish farm facility can also serve as internship, research and educational reresource for a self-reliant community

3 Resource Loop Strategies

Table 1 shows the operational capacities of an offshore fish production and distribution facility, which can be sustained by renewable wave energy generated by wave turbines. 44 rows of 16 modular wave turbines installed at floating pontoons circumventing the 'grow-out area' can generate an estimated 308 M kWh all year round against an energy consumption of about 52 M kWh per year (Fig. 25c).

Both waste and energy resources are looped to integrate non-polluting and self-sustained processes. At the jack-up fish farm and semi-submersible fish wharf, fish waste is converted into useful byproducts for vegetable farming (Fig. 15, 25b); and as fishmeal—an ingredient in fish feed pellets (Fig. 24, 25b).

At the fish 'grow-out farming area', its ring perimeter can be maximized as a farming belt for saltwater crops (25a). Diluted seawater or saltwater is a viable option to freshwater which is used for potable grade applications. Saltwater strains of potato and rice provide good yield for harvest. Presently, saltwater potatoes are grown in the Dutch farmlands of Den Horn, Netherlands [1]. The farms produce 30 tonnes of potatoes per hectare of land with a harvest every four-months. Although potatoes can grow within the temperature range of 10 to 30 degree Celsius, the mean temperature for optimal growth is between 18 to 20 degree Celsius. China's farmlands have been growing 'sea rice' for improved yields to feed its growing population. In the tropics, sea rice can offer yields of five to seven tonnes per hectare of land in four-monthly harvests [23]. This is significantly more than the global average of three tonnes per hectare. Farm crop oil rigs can thus be situated in temperate or tropical waters depending on optimum growth conditions for the crop species.

Table 1 Operational capacities of an offshore fish production and distribution facility

Rig type	3-legged jackup	Surplus fish to feed world population	
No. of workers	120	In 2017	In 2025
No. of workers dormitory	60 [28 m^2 each)	7.4 billion people consume 148 million tonnes	8.1 billion people consume 162 million tonnes
Types of growth chamber [12,900 m^2 total growth area]	4 types: broodstock, hatchery, pre-nursery and nursery	4,400 JU fish farms required to meet demand	4,820 JU fish farms required to meet demand
Seawater bodies [1,298,700 m^2 total water area]	80 net cages in open seawater [1,285,800 m^2 of water area]	***Savings in land area [based on Singapore—5.34 km² of land + water area produces 5,200 tonnes of fishes per year]***	
Fish production yield	33,600 tonnes/year can feed 1,680 people	In 2017	In 2025
		152,000 km^2 of land + water area required	166,000 km^2 of land + water area required
Vegetable production yield	100 tonnes/year [workers consume 12 tonnes/year, 88 tonnes/year for export]	4,400 JUs [5,700 km^2 of water area]	4,820 JUs [6,200 km^2 of water area; takes up 0.002% of earth ocean surface]
Energy source	Wave harnessing generators at 380 million kWh/year [48 million kWh is required to operate the JU fish farm]		
Semi-submersible Fish Wharf			

Rig type	Semi-submersible	**Fish distributed in a year**		33,600 tonnes
Schedule of accommodation	Harvesting • Unloading and sorting bay • Processing facility • Waste to fish feed production plant • Packing, sorting and distribution port • Cold room storage • Office workspace • Wholesale market Sky dining and restaurants	***Area comparison***		
		Senoko Fishery Port	Jurong Fishery Port	SS Fish Wharf
		32,400 m^2	51,000 m^2	33,300 m^2
		Proposed semi-submersible fish wharf is comparable to Senoko Fishery Port, but with the advantage of an integrated processing facility and public recreation spaces, not currently available in existing fishery ports		

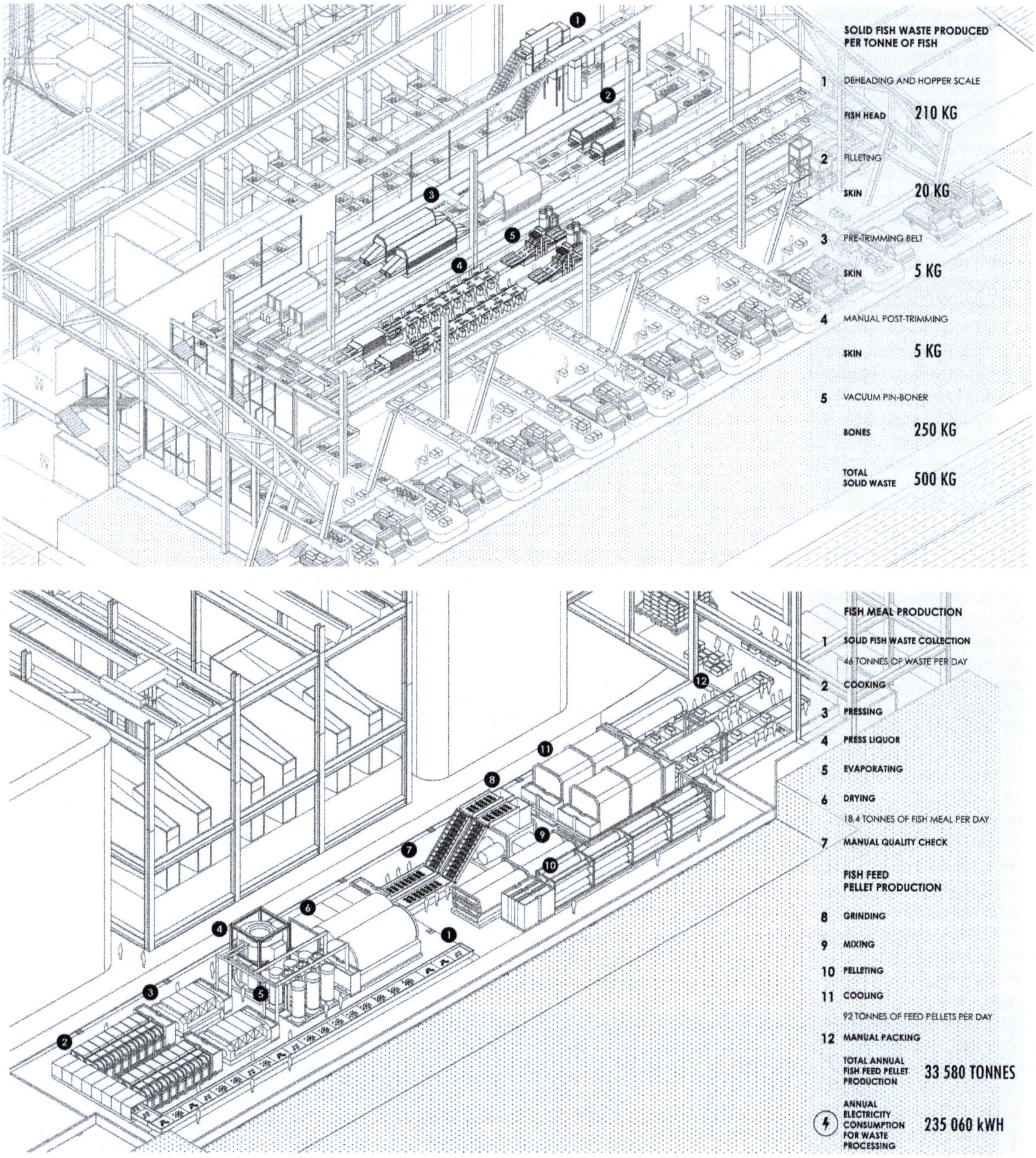

Fig. 24 Waste Loop Strategy Fish waste byproducts can be processed as fishmeal, which is one of the ingredients needed to produce fish feed pellets, used as food supplement for the fishes in the grow-out net cages. The average inclusion rate in feed pellets of fishmeal for salmon, Asian seabass and groupers is 20%. This implies that for every one unit of fishmeal, five units of fish feed pellets can be produced. The annual fish feed pellet production is estimated at 33,580 tonnes. These pellets are packed in boxes and loaded onto the same fishing vessel that transports the net cages for fish harvest. 33,580 tonnes of feed pellets per year is sufficient for the salmon, and the annual feed requirement for Asian seabass and grouper are 50,400 and 60,480 tonnes respectively. This translates to 66% of feed required for the Asian seabass and 55% for the grouper. The fishmeal and fish feed pellet production plants have an annual energy consumption of 235,060 kWh (*calculation is based on an electrical power consumption data for a fishmeal production facility with an evaporation plant. The power consumption is dependent on the amount of raw material in tonnes that enters every 24 h. The kilowatt-hour consumption per ton of raw material is 35* [4])

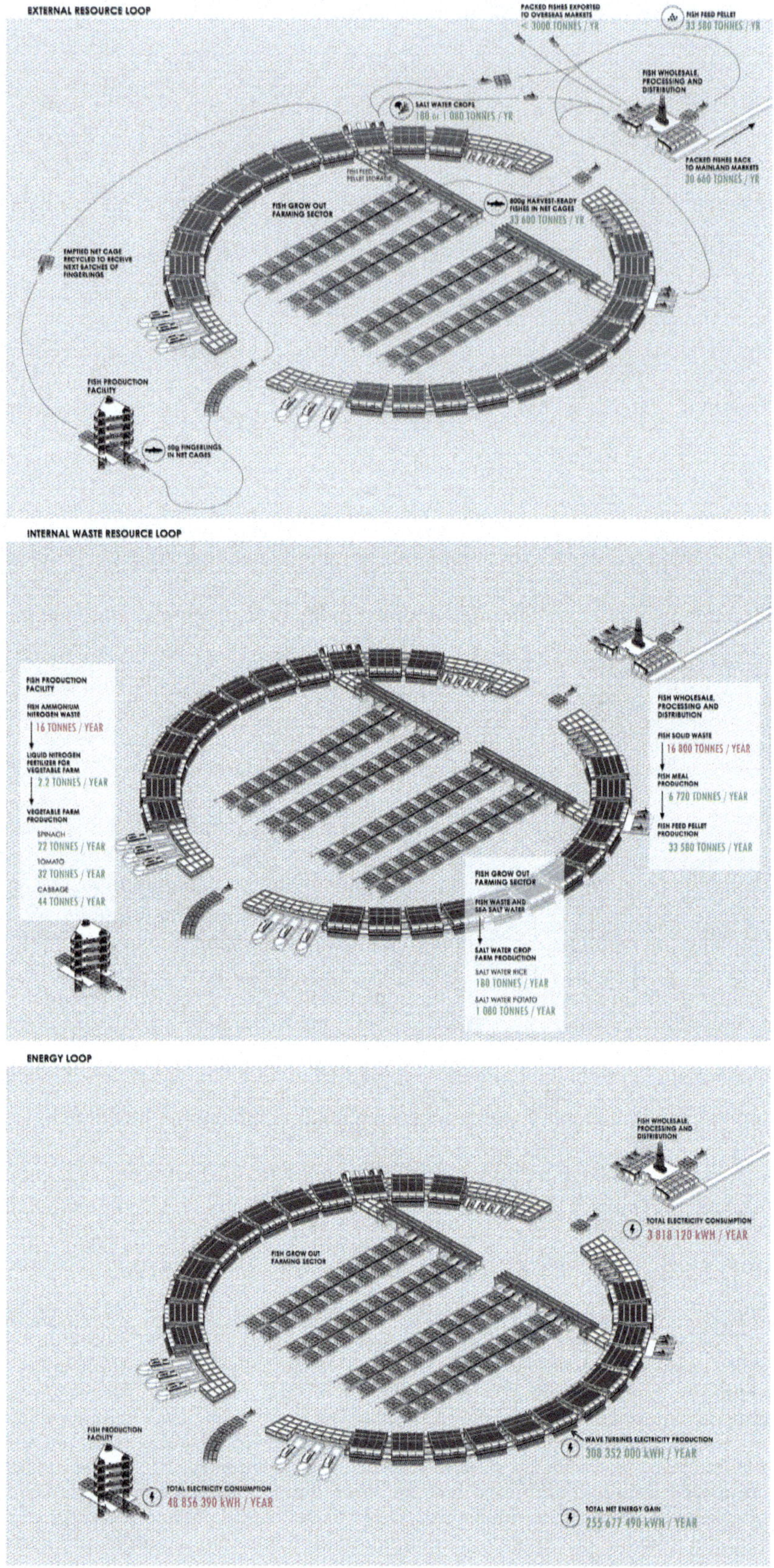

Fig. 25 Resource loop strategies between the three components: jack-up fish farm, fish 'grow-out' area and semi-submersible fish wharf. **a** The external resource loop shows the movement of goods within the offshore fish farm facility. **b** The internal waste resource loop shows the conversion of fish waste into useful byproducts such as fertilizers for the production of crops, fishmeal in fish feed pellets as food supplies for the fishes in the 'grow-out' net cages. **c** The energy loop shows that the energy harvested from the wave turbines circumventing the grow-out farming sector is sufficient to operate an offshore fish production and distribution facility

120 farming terraces can be built above the oil storage containers, with a total farming area of approximately 120,000 m^2. The annual yield is estimated at 180 tonnes of saltwater rice, and 1,080 tonnes of saltwater potatoes. Based on AVA's statistics, the annual consumption per capital for vegetables and rice is 95 kg and 60 kg respectively. Yield from either crop can feed up a minimum of 3,000 people per year.

These resource loop strategies can provide sustenance for the workers manning the fish farm rigs, hence promoting self-sustainability.

4 Climate Change and Sea Level Rise

In 2010, a record 30.6 gigatonnes of CO_2 was released from burning fossil fuel. Increases in fossil fuel use of this magnitude are likely to carry the world far beyond the temperature rise of two degree by 2050 that scientists have estimated is the limit of safety, beyond which climate change is likely to become catastrophic and irreversible [15].

The two major causes of global sea level rise are thermal expansion caused by warming of the oceans and the melting of glaciers and ice sheets. At least partial deglaciation of the Greenland ice sheet, and possibly the West Antartica ice sheet, would occur for a global average temperature increase of 1–4 degrees Celsius (relative to 1990–2000), causing a contribution to sea-level rise of 4–6 m or more (Figs. 26, 27). Worldwide, the threat to coastal cities has led to scenario studies of the extent of inundation with topographic information and mean sea levels.

4.1 Rising Sea Level in Singapore

A 2015 report by Climate Central detailed the implications of different global warming scenarios of 1.5, 2, 3 and 4

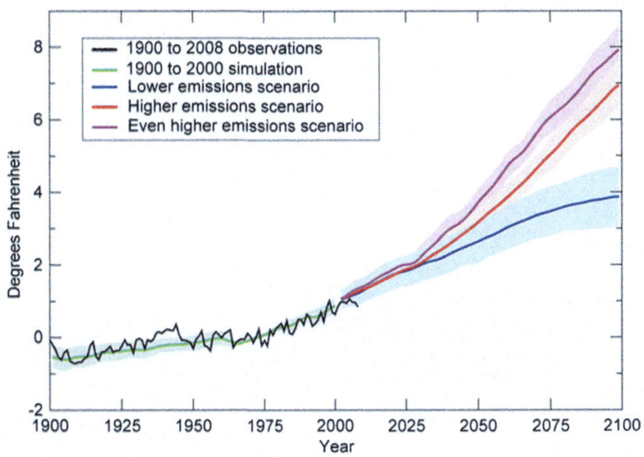

Fig. 26 Temperature rise projection by 2100. (*Source* www.climatecentral.org)

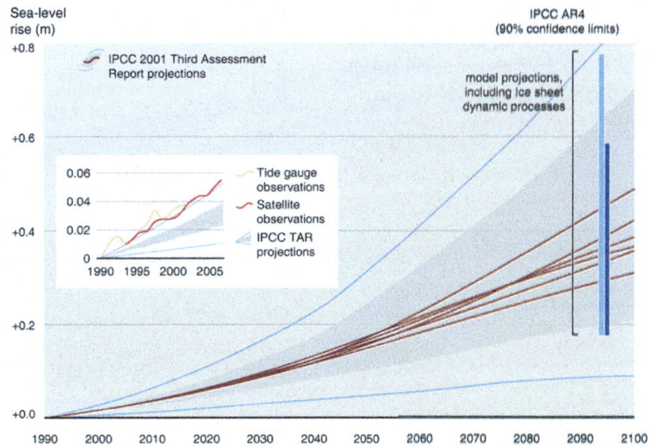

Fig. 27 Sea level rise projection by 2100. (*Source* www.ipcc.ch)

degrees Celsius by 2100. It is found that 4 degrees Celsius of warming from carbon emissions could cause Singapore's median local sea level to rise 9.5 m, submerging the homes of 745,000 Singaporeans. Even if carbon emissions were cut, resulting in the proposed international target of 2 degrees Celsius warming, there would still be 101,000 affected Singaporeans, where median local sea level projected to rise 5.1 m (Fig. 28).

Much of Singapore lies only 15 m above sea level, while 30% of the island is less than 5 m above sea level. If global temperatures continue to rise, many parts of Singapore could eventually be submerged under water. A study of nearshore and offshore proposals for housing on repurposed oil rigs would be a relevant consideration by policy makers working with the nation's Maritime Port Authority, Urban Redevelopment Authority, Police Coastguard and Republic's Navy.

Sea walls Currently, about 70 to 80% of Singapore's coastal areas have monolithic walls or stone embankments [9]. Figure 29 shows a first road-raising project by Singapore Land Transport Authority to stave off rising seas [17]. In anticipation of a 0.25 m and 0.76 m in the Straits of Singapore, New Changi Terminal 5 is being built on a datum level 5.5 m above sea level with canals to divert floodwater from its sea walls [21].

In the early design stages of the Fukushima Daiichi and Fukushima Daini nuclear power plants, the heights of tsunami waves were estimated to be only 6.1 m based on earthquake magnitudes of 7.5 even though 8.6 were recorded previously. Later calculations suggested tsunami wave heights of 8.4 m to above 10 m. But in the 2011 disaster, the tsunami wave height was 13 m and it flooded all the backup generators and power lines of the plant [12].

Hollnagel and Fujita [6] argue that a system cannot be resilient enough to cope with unexpected severe situations, if it is limited to design features alone. The use of barriers along coastlines provides a sense of security based on faith

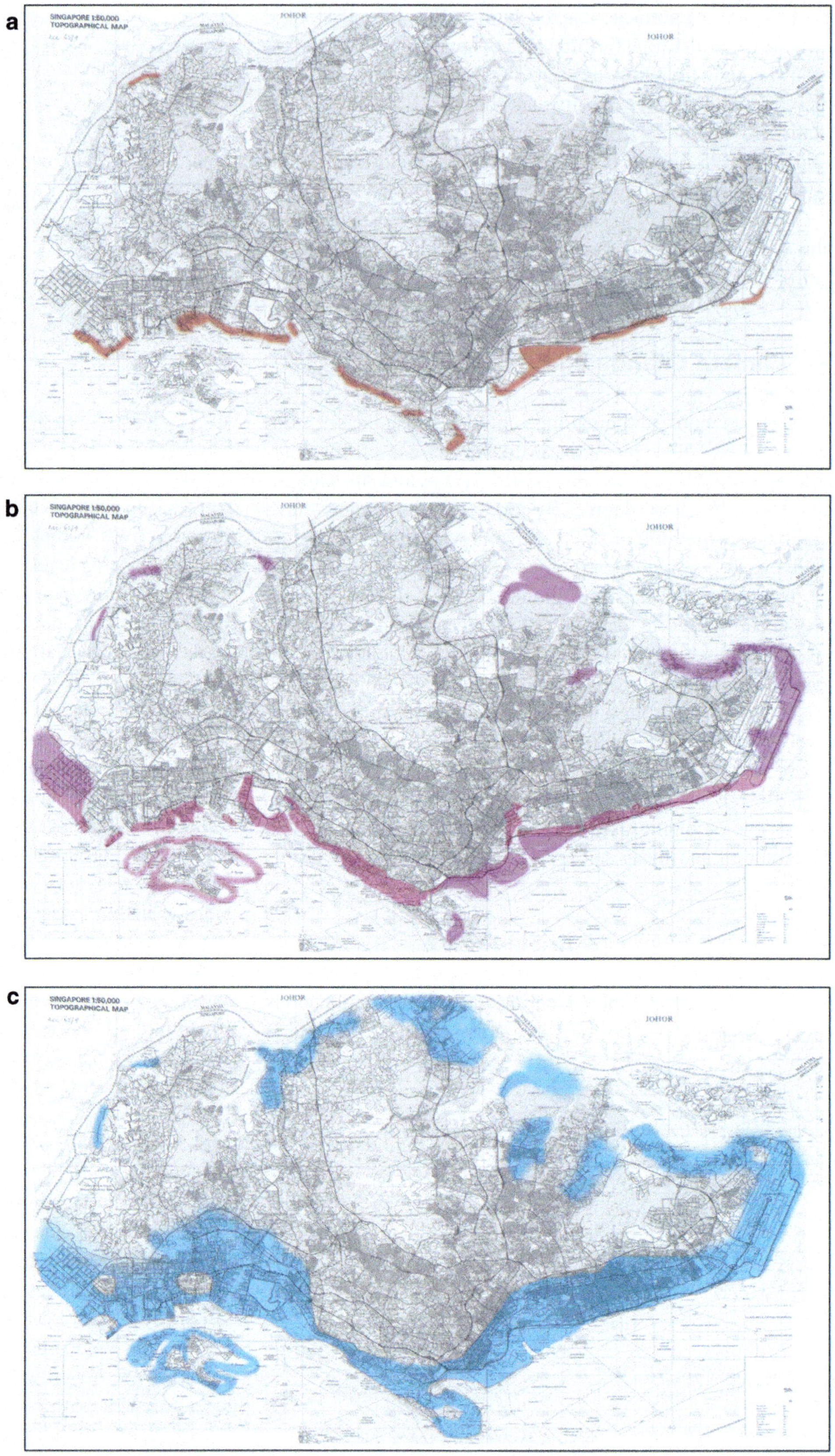

Fig. 28 Flood risk areas in Singapore **a** 1.67 m sea level rise in 2030 **b** 3.95 m sea level rise in 2050 **c** 9.5 m sea level rise in 2100

Fig. 29 One km stretch of two-lane dual carriage Nicoll Drive along the eastern shoreline next to Changi Beach was raised by up to 0.8 m in 2016

in design and may often cut off settlements from the sea, making beaches unusable and affecting wildlife.

While the benefits of floating solutions may seemingly outweigh their costs compared to coastal barriers in terms of effectiveness in protection, much of value assessment depends on the availability of sea space, production and maintenance costs over life cycles. Nevertheless the evaluation of human habitation at sea are not measured only in terms of costs, safety and protection but in respect of their environmental impact on the marine environment and the quality of habitable space at sea.

Land Reclamation Reclaimed land is susceptible to soil liquefaction during earthquakes, which can amplify the damages to buildings and infrastructure. Sedimentation occurs when particles of sand eroded by wave currents and gravity, causes the water to be cloudy, prohibiting sunlight to pass through. Subsidence of compacted filled land occurs also when wetlands are enclosed by levees and drained to create polders. Drained marshes would sink below the surrounding water, increasing the risk of flooding [2]. Figure 30 shows the extent of reclamation in Singapore for housing, entertainment and industrial spaces. Floating settlements therefore explore ways, which avoid the environmental risks of land reclamation in procuring buildable land surface.

5 Oil Rigs as Alternative High-Density Housing Forms

According to the United Nations World Population Prospects [19], the world's population is currently growing by approximately two billion persons in the next 30 years, from the current 7.7 billion to 9.7 billion in 2050 and 11 billion in 2100.

Oil rig structures may be repurposed to offer alternative floating housing typologies in addressing increasing global population. With sea level rise and the long-term threat of inundation of coastal environments, the reuse of oil rigs attempts to explore the idea of a floating settlement in six ways:

(1) What are high-density typologies exceeding those of floating platform systems (e.g. Oceanix City 119,048 persons per km^2, Lilypad 100,000 persons per km^2)?
(2) How would oil rig settlement densities compare with Singapore's public housing neighborhoods (e.g. Pinnacle 367,000 persons per km^2)?

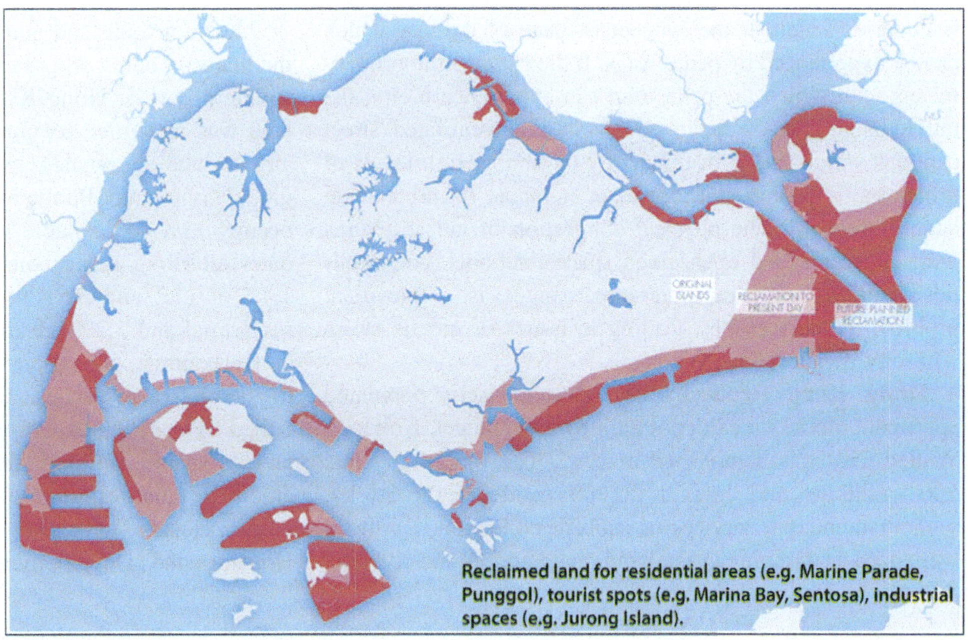

Fig. 30 Reclamation in Singapore. Areas in pink indicate the extent of reclamation in Singapore, while areas in red shows the possible plans for future reclamation. (*Source* https://blogs.ntu.edu.sg/hp331-2014-10/?page_id=7)

(3) What would be the design limits of building densely on structures with floating pontoons that reduce building footprint on water compared to those of floating platforms?
(4) What limits building on tall offshore structures when its building footprint is reduced to isolated supports?
(5) How would these structures be aggregated to form a larger settlement with connected public and vibrant communal spaces above water?
(6) Explore solutions for housing and commercial usage in ways, which result in compact floating typologies without the sprawl of linked platforms floating on water.

5.1 High-Density Housing

In the planning of housing blocks, it is customary to arrange elevator cores to serve a number of apartment units per floor. The most efficient layouts are double-loaded and single-loaded access corridors with elevator cores space at the ends or centers of an apartment block. Tall housing structures face the challenge of living quality when the upper levels are typically zoned for dwelling purposes and privately owned spaced. Where the landlord spaces are typically minimized so as to maximize investment yield with a larger proportion of saleable area per floor plate, there is little or no provision of communal space in these typical floors. When public space is integrated with commercial lettable space for urban vibrancy, command interaction is confined to the lower levels in building complexes [11].

This segregration of the private dwelling from the urban interface is further compounded by a lack of green space at upper levels of the development or when ground level gardens are gated for the exclusive use of residents and these boundaries fragment the connected space of the city which can be experienced by pedestrians. If the street is the primary means with which the pedestrian can experience a city, the tall housing blocks create poorly lit and ventilated street canyons when spaced insufficiently far apart. High levels of urban traffic pollution accumulate in areas of urban heat island effect [8]. The physical congestion of tall structures with underprovided open green spaces, schools, communities facilities and transport interchanges is a prevalent problem in older estates, leading to issues of mental stress with city dwellers [5].

Hong Kong Reputedly the most densely populated apartment blocks on earth not built by an architect, Kowloon Walled City was demolished in 1993. The walled city had 500 buildings and housed 50,000 residents on 2.6 ha, without municipal services or rubbish collection. Rooftops were used as dumping ground and recreation. In the complex of 300 interconnected high-rise blocks were apartments, shops, schools, brothels, gambling and drug dens. A city seemingly without any order or organization but was home to many who lived and who worked in small factories producing plastic goods, toys and food.

To improve the living conditions of modern housing blocks from the 19th century, planning guidelines for usage, safeguarded public health, and city authorities established sanitary conditions. Where industry was co-located with housing, only non-polluting forms of production were permitted and these were planned with businesses and shops to generate places of employment important for a city neighborhood. Open spaces for recreational activity and social interaction were planned in between buildings spaced sufficiently far apart to allow sufficient hours of daylighting into lower level apartment blocks.

Tin Shui Wai is the largest and most problematic of Hong Kong's 'new towns' with a population of 270,000. Majority of its land use is of a residential nature but within a highly dense living environment. Nicknamed 'City of Sorrows', Tin Shui Wai is prone to numerous family and social issues, including domestic violence, mental illness, and suicide as portrayed in the media for years. It is located in a remote location with lack of public facilities in a segregated high-density public housing estate. Its tall and monotonous apartment blocks form a unified barrier to enclose a city in isolation (Fig. 31).

One has to understand the broader context of Hong Kong's challenges. Hong Kong currently has a population density of 22,100 persons per km^2. Its housing stock has lagged behind the pace of population growth of 7.2 million people where 3.3 million persons or 44.7% live in public housing. Waiting time is more than five years with under 270,000 applicants on the waiting list for the public rental housing scheme [22].

74.6% of total land is unsuitable for future development; the majority being woodlands in the New Territories and the southern part of Hong Kong island. 35% of all developed land was reclaimed for port infrastructure, Chap Lap Kong airport, and dominated by infrastructure and services for global commerce. Public amenities and urban transportation occupy 13% of available land and 0.5% of land accommodates all Hong Kong commercial and retail developments. 1.4% of total land use is for public housing, 2.3% for private residential and 3.2% are rural settlements [13].

By contrast, Singapore has one million flats in 24 new towns and three estates. The public housing estates maintained by state funds, are owned by 90% of the population and its spaces are strategically planned to engender social enterprise and community spirit. The People's Association works closely with the residential committees and the Housing and Development Board (HDB) in community

Fig. 31 Tin Shui Wai public housing, Hong Kong

Table 2 Comparing housing between Hong Kong and Singapore

	Hong Kong	Singapore
Population	7.2 million	5.4 million
Land area	1,104 km^2	719 km^2
Residential land area	300km^2	500 km^2
Population density	22,100 persons per km^2	10,600 persons per km^2
Residential building height	Average 40–50 storeys Highest 70 storeys	Average 20–30 storeys Highest 50 storeys
Apartment sizes in public housing	20 m^2 (one-bedroom unit) to 50 (three-bedroom unit)	50 m^2 (one-bedroom unit) to 110 (three-bedroom unit)

building with inter-generational and inter-ethnic groups through recreational, educational and interactive projects. Table 2 compares the housing between Hong Kong and Singapore.

The challenge of vertical typologies for high-density habitation also lies in their spacing between blocks for privacy; to prevent overlooking and to prevent re-entrant airflow which sets up conditions for respiratory diseases in land scarce developing cities. These considerations are relevant to repurposing jack-ups and semi-submersibles, where densities approach under a thousand apartment units in single structures that are equivalent to entire city precincts.

5.2 Singapore Housing

Communal Spaces Successive models in the development of Singapore public housing reveal an increasing emphasis on the provision and design of public and communal space (Fig. 32). These include sky gardens at intermediate and rooftop levels of apartment blocks and ground floor communal spaces integrated with landscaped gardens. This includes the provision of games courts, reading rooms, exercise shelters, exercise corners and children's playgrounds for inter-generation usage. Communal spaces for resident interaction are also provided as horizontal connections in the form of bridges and platforms platform and bridges between vertical blocks.

Housing Model Analysis Ghim Moh housing estates with a medium range density (163,765 to 242,100 persons per km^2) have proportionately modest provisions of commercial and recreational space planned within its estate boundaries as compared with the densest estate at Pinnacle Duxton (367,000 persons per km^2). The green ratio refers to the total area of green divided by the site area, resulting in 52–57% for Ghim Moh estates compared with 80% for Pinnacle Duxton.

Both developments have similar green ratio provisions although the site area of Ghim Moh Valley is double that of Ghim Moh Edge. Ghim Moh Edge accommodates a lesser number of housing units and lesser population than Ghim Moh Valley, but has a higher density of inhabitants per km^2

Ghim Moh Valley

Site area	41,767 m^2
Number of units	1,520
Population	6,840
Density	163,765 people per km^2
Green area	22,000 m^2
Green ratio	52%
Food center	250 m^2
Supermarket	350 m^2
Shops	600 m^2
Education centre	500 m^2
Playground + fitness corner	800 m^2

Fig. 32 Successive models of Singapore's public housing

Ghim Moh Edge

Site area	19,074 m^2
GFA	82,018 m^2
Plot Ratio	4.3
Number of units	1,179
Population	4,600
Density	242,100 people per km^2
Green area	11,000 m^2 (2.4 m^2 per person)
Green ratio	57%
Food center	250 m^2
Supermarket	450 m^2
Shops	200 m^2
Childcare	450 m^2
Playground	450 m^2
7-storey carpark	12,000 m^2
Senior care and activity center	600 m^2

Pinnacle Duxton

Site area	25,172 m^2
GFA	198,859 m^2
Plot ratio	7.9

(continued)

Number of units	1,848
Population	9240 (assuming 5 person per unit)
Density	367,000 people per km^2
Skybridges	Two 500 m long bridges (50th storey and 26th storey)
Green area	20,000 m^2 (2.16 m^2 per person)
Green ratio	80% (high for a premium development)
At 50th storey skybridge	A. Cottage B. Rocky C. Beach D. Lounge E. Hillock F. Fiesta
At 26th storey skybridge	G. Resident committee—180 m^2 H. Sky gym I. Meadow J. Space Net K. Crater L. Padi Field—800 m jogging trek
At 3rd storey environmental deck	M. Playground—800 m^2 N. Basketball court—725 m^2 O. Historical Park/Heritage garden—800 m^2 P. Childcare—600 m^2 Q. Education Center—600 m^2 R. Resident Committee—120 m^2 S. Community Center Annex—450 m^2
At 1st storey (street level)	Food court—300 m^2 Shops and mini mart—350 m^2

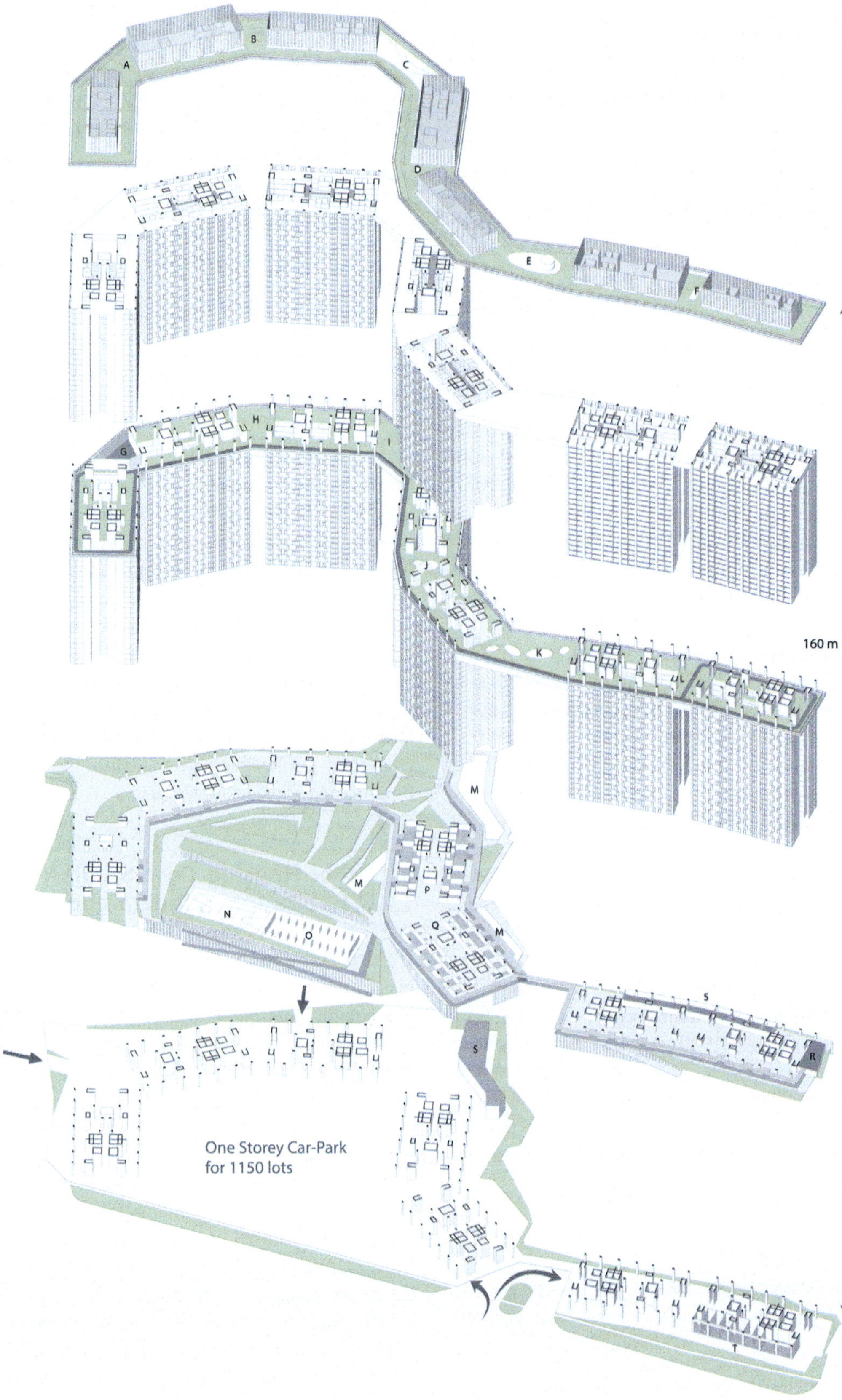

Fig. 33 Pinnacle Duxton

Land Use	Prototype New Town (60,000 Dwelling Units)	
	Land Area (hectares)	Percentage of Land Designated for Given Land Use
Residential*	347	53.4
Major roads	89	13.7
Schools	62	9.5
Industry**	44	6.8
Commercial (town center and neighborhood center)***	30	4.6
Open space	26	4.0
Institutions	15	2.3
Sports complex	7	1.1
Utilities and others	30	4.6
Total	650	100.0
Gross new town density	92 dwelling units per hectare	

** Includes private housing within the town boundary.*

*** Includes nonpollutive industries only.*

**** Includes civic, cultural, and recreational uses and incidental developments in the town and neighborhood centers.*

Source: Housing & Development Board (2000a)

Commercial Facility and Size	Planning Standards
Shops (30 to 400 square meters)	1 per 70 flats
	20% in town center, 50% in neighborhood centers, 30% in precincts
Kiosks (3 to 15 square meters)	1 per 600 flats
	30% in town center, 70% in neighborhood centers
Emporiums (4,500 to 6,500 square meters)	1 to 2 per new town, in town center
Supermarkets (1,200 square meters)	1 to 2 per new town, in town center
Eating houses (450 square meters)	1 per 750 flats
	7% in town center, 23% in neighborhood centers, 70% in precincts
Restaurants (90 to 2,000 square meters)	1 per 1,000 flats
	30% in town center, 70% in neighborhood centers. In addition, 2 or 3 fast food restaurants and 1 or 2 bigger restaurants in town center
Office space	60 square meters per 450 flats
	70% in town center, 30% in neighborhood centers
Cinemas (1,800 square meters)	2 per new town, in town center
Minimarkets (450 square meters)	1 per 6,000 flats
Market produce shops (130 square meters)	1 per 3,000 flats
Market produce lockup shops (40 square meters)	1 per 500 flats

Source: Wong and Yeh (1985)

Fig. 34 Land use distribution, gross density and corresponding planning standards for a new town

Pinnacle Duxton (Fig. 33) is Singapore's first 50-storey public housing project built in 2009 over a five-year period after an international competition was launched in 2004 and picked over 200 entries. It has won several awards for tall buildings of its class and has residential unit types with dissimilar combinations of façade elements such as extended bays, balconies, bay windows and planter areas.

The two 500 m long skybridges provide a total 1500 m^2 of 'inter-block' open space. The 26th storey skybridge is for residents only and is planned for cross-generation usage. It accommodates a residents' committee center, a jogging track and a senior citizen fitness corner, outdoor gym and children playground, community plaza, fitness corner and two viewing decks. It also serves as a refuge floor to evacuate residents in a fire emergency and for firefighters to launch their rescue operations. Residents in all units are privy to programs organized by residents' committees for communal bonding in 2 ha of social leisure and play space in an otherwise constrain city site.

Being a public housing project, the 50th storey skybridge is open to both residents and public for panoramic views of the city. There is a limit of 1,000 people on each skybridge for occupancy load constraints of fire safety and riot control. There is a good provision of private and collective shared spaces in this development. This is achieved with personal gardens/yards in units and shared gardens at block and precinct level.

Table 3 Singapore housing estate organizational structure

	No. of dwelling units	Estimated population
Block	100–200	400–500
Precinct	400–1000	3000–5000
Neighborhood	4000–6000	20,000–30,000
New Town	20,000–60,000	100,000–300,000

Housing Organization and Capacity Studies Plan types, housing standards and land use distribution in Singapore Housing estates is shown in Table 3. The distribution of program types in precincts and neighborhoods (Fig. 34) form the basis of area provisions in the design of flotillas and fleets in the oil rig floating settlements. Oil rig structures may be appreciated as scaffolds for repurposing to support new habitable programs. The challenge is to work with the structural scaffold and find opportunities within constraints of strength and stability under buoyancy conditions when attempting to create a dense housing environment with large areas for public recreation within the constraint of the floating and topside footprints.

6 Other Housing Case Studies: Unit D'Habitation, Silodam and Habitat 67

Unité d'Habitation (Fig. 35) There are 23 different types of residences within the building, varying from small one-person apartment unit to generous family units, and every unit has at least one balcony to overcome the claustrophobia of a small apartment. The invention was the compact arrangement of duplex apartments within the architecture of the housing block, which provided double volume space in a modest floor plan. This provided a sense of spatial height within a small apartment plan. Spaces for the public and private domains would have variegated proportions and heights in passageways and within unit spaces. Most significantly, the access corridors or streets need only be accommodated on every third floor

The idea was to create spacious city homes with the amenities of a city street i.e. a city in a building block with two shopping streets, a hotel and a rooftop terrace. Daily usage is integrated with social, commercial and recreational functions in the housing plan for urban vibrancy within the block. This was important for the well-being of occupants in the post-war period. Mixed-use provisions in one building block of housing and hotel room types would generate a critical mass of occupancy.

Silodam, MVRDV (Fig. 36) The project was initiated to densify the obsolete industrial part of west Amsterdam by transforming a dam structure with a silo above into housing. To be flexible in a changing housing market, a mixed program of 157 dwellings for rent or purchase were planned with commercial and public spaces. The dwellings differ in size, layout, cost and offered panoramic views of the harbor from both sides of the double-height apartments. The architect attempted an improved version of Unit d'Habitation by mixing different apartment types for a range of family sizes and incomes. Different types of housing units can accommodate market changes in the real estate economy.

Social and recreational functions are integrated into the housing to enable vibrancy and social interaction. The openings and communal spaces spread along the building provide opportunities for interaction. Hence, although the form appears to be monumental from the outside, its interior spaces are in marked contrast with domestic scale programs of a public balcony, garden, library, barbeque and fitness areas in a three-dimensional neighborhood. Large and small pockets of spaces with external views to the harbor are dispersed across the building for consistent distribution of communal and interactive spaces throughout the length of the building

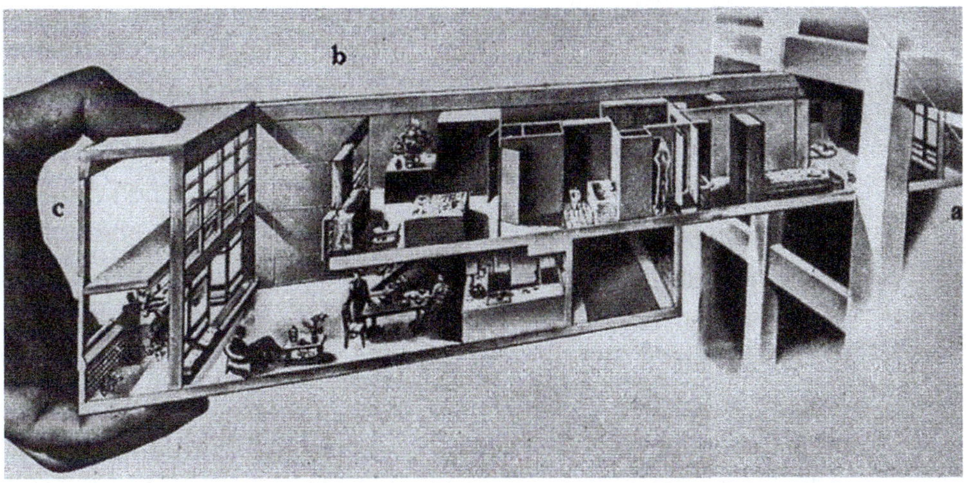

Fig. 35 Unité d'Habitation 3300 m^2 site area 18 storeys 337 apartment units 1600 residents 56 m height, 24 m width, 165 m length m height, 20 m width, 120 m length

Fig. 36 Silodam, MVRDV 2600 m^2 site area 10 storeys 157 apartment units 24 m height, 20 m width, 120 m length 1-, 2-, 3-bedroom apartment units 2-, 3-, 4-bedroom maisonettes Lofts and studios

Fig. 37 Habitat 67 22,160 m^2 gross floor area 12 storeys 158 apartment units 15 dwelling types varying between 1 and 8 cubes 60 m^2 to 465 m^2 apartment units over one to four-storey configuration 21–93 m^2 private terraces

Habitat 67 (Fig. 37) This project was originally intended as an experimental solution for high quality housing in dense urban environments. Its form was a result of rethinking the apartment type by incorporating the spatial quality of a suburban home. The stacked and casually skewed apartments provided a wide range of opportunities to step out onto balcony gardens and terraces. This enabled external views from elevated positions. Where the roof of one unit is the external terrace of the unit above it, apartments are formed by groups of one to four apartment boxes. Each of these 60 m^2 boxes was prefabricated in concrete and stacked in different staggered configurations accessed by streets and bridges from three elevator cores.

The prefabrication process did not reduce the costs of construction due to the reduction in production scale from an original 1,000 units to 158. The stepping back of each box above another provided each apartment with a garden, and with it, effective ventilation and daylight while breaking the traditional form of tower symmetrical blocks.

After the World Expo, numerous habitats were constructed in New York, Puerto Rico, Israel, Rochester and Tehran. Although it has been argued that Habitat 67 was more successful as a research project than a pilot project, it remains a seminal piece of work that influenced later works. The effective provision of daylight and ventilation was not as anticipated and perhaps if aided by simulation tools in contemporary design processes, there would be a scientific method to the stacking composition via a form-finding process based on daylight solar gains, self-shading and ventilation. This process would provide a performative dimension to the aggregation of parts and its multifaceted spatial complexity, and achieve form of sustainability in building.

7 Jack-up and Semi-submersible Housing

7.1 Generating Housing Variants

Housing variants are generated on three types of oil rigs: 14 variants from three-legged jack-up rigs (Fig. 39a), 10 variants from four-legged jack-up rigs and 9 variants from semi-submersible rigs (Fig. 39b). The layout determinants of housing units took into consideration the housing models discussed earlier. In addition, a research into Singapore's housing profile revealed that four-room apartment units of 90 m^2 form the most number in public housing.

As such, comparisons are made between permutations using the area possible for private terraces and public sky gardens, the number of units provided and their densities expressed in terms of the number of inhabitants per km^2. The capacities of each solution to accommodate housing populations provide an indication of the capacity when used in configuring flotillas and settlement fleets for ultimate populations.

Housing permutations are generated from three types of rigs: three- and four-legged jack-ups and four-legged semi-submersibles as shown in Table 4.

Table 4 Types of rigs used to study housing permutation

Model	Type	Dimension (m)	(m^2)	Leg length (m)
Seafox 5	four-legged jack-up	115 × 50	5,750	106
Maersk Intrepid	three-legged jack-up	88.7 × 102.4	4,541	206.8
Diamond Offshore Ocean Great White	semi-submersible	123 × 78	9,594	–
Keppel DSS series	semi-submersible	69.5 × 69.5	4,830	-
		73 × 73	5,329	-
		75 × 75	5,625	-

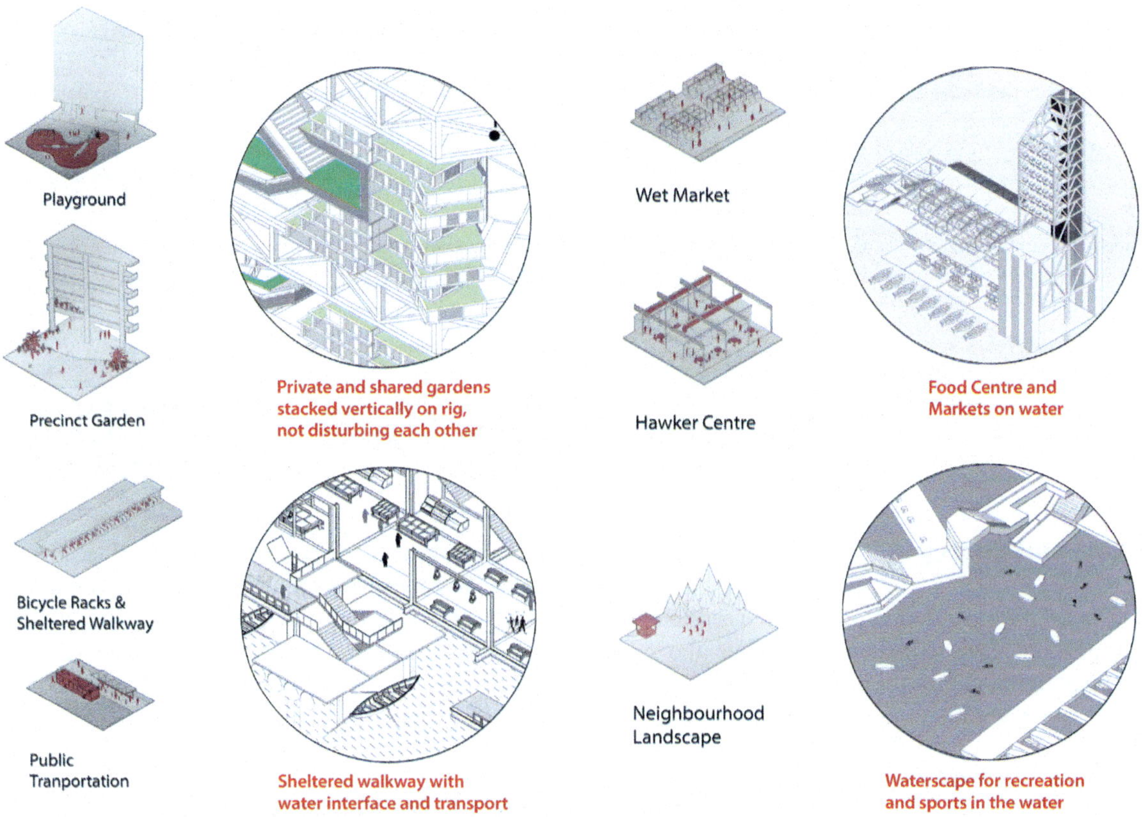

Fig. 38 Inserting urban spaces in oil rig and barge typology

Table 5 Operational capacities of a Jack-up housing

Rig type	3-legged jackup	**Total energy required**	2,907,736 kWh/year [domestic usage] 1,109,600 kWh/year [machinery and service]
No. of apartments	672		
No. of residents	3,360		
Apartment sizes	64, 96, 128 m^2		
Total apartment area	64,512 m^2	**Household foodwaste**	467 kg/day average
No. of private sky terraces	672 [21,504 m^2]		
No. of public sky gardens	12 [12,000 m^2]		
No. of rigs to house population figure in 2050?	2,083,340 rigs to house 7 billion people or 28,000 settlements [38.5 km^2 per 250,000 people]	**Biogas container**	2 × 20 ft containers operates on 380w
Potable water	650 tonnes [domestic usage] 800 tonnes [output]	**Electricity for battery storage (microgrid)**	11,000 kWh/day
Desalination containerized plant	2 × 40 ft containers on the bottom hull [consume 277,250 kWh/year]	**Battery storage container**	4 × 20 ft containers, one on each hull
Greywater	130 tonnes [irrigation] 150 tonnes [output]	**Foodwaste to biogas energy**	563,997 kWh/year
Sewage treatment containerized plant	6 × 40 ft containers on the top hull [consume 831,750 kWh/year]	**PV panels on roof**	6,896 MW/year
		(solar radiation x effective PV area)	[1,667 kWh/m^2 × 4,137 m^2]
		Electricity generated from PV	1,531,000 kWh/year [22.2% of 6,896 MW]
		Total energy generated (PV + biogas energy)	2,094,997 kWh/year [meets 50% of energy requirement]

7.2 Water and Energy

Technologies are incorporated to reduce the need to rely on external water and energy in order to reduce the carbon footprint, creating a more sustainable living community. A wide mix of neighborhood programs is incorporated in a self-sustaining and vibrant community (Fig. 38). Treatment plants of compact configurations with capacities matching load and wastewater volumes and potable consumption quantum are selected for use on the oil rigs (Table 5). Wave energy power systems are integrated into semi-submersibles hulls or anchored in adjacent locations.

7.3 Design in Detail

Two oil rig types, the jack-up for nearshore housing and the semi-submersible for offshore housing are detailed as hybrid typologies for the planning of floating settlements (Figs. 40, 41 and 42). Hybrid typology is a key strategy that combines different programs in ways, which result in new spatial experiences that are not segregated by single use masterplan zoning policies. With this approach, space is saved, travel distances and reticulation systems are shortened liberating habitable space within flotilla and settlement plan. Table 6 summarizes the capacities of a jack-up and semi-submersible housing.

80 to 110 m^2 housing units are stacked in a staggered form, which is cantilevered off jack-up masted towers. The form allows a network of structural plates to be converted into balconies, terraces and gardens that extend from interior living spaces and dissolving interior-exterior spatial relationships. This is a contrast to limited balcony provisions in conventional apartment solutions to form.

The skybridges move the conventionally central located courtyard into the periphery. Floor slabs are cut away at the center to create controlled openings in the form of light wells and breezeways. The skybridges in the jack-up are in marked contrast to the horizontal and linear geometry of the skyparks in Pinnacle Duxton. Whereas Pinnacle skyparks connect two to three floors, jack-up rig skyparks function as ambulatory park connectors. People can walk on parkway belts to leisurely enjoy the panorama, while ascending or descending the vertical structure. Pockets of spaces with vibrant activities serve as nodes for people to gather. These sheltered junctions include a mini cinema, a fitness corner for the young and another for the elderly, while many families can use the gentle slope.

The deck or hull of the oil rigs upholds the structural integrity, transferring the load of the housing units to the three legs of the jack-up rig or to the four legs of the semi-submersible, while the large span of the steel truss is used for co-working and recreational space. This allows the configuration of work, live and play amenities in a superblock with extensive recreational area. The advantage of having minimal supports within the interiors of floor plates allow for flexibility in change of use.

At the water level, the spaces beneath the decks of jack-up and semi-submersible rigs allow for water-courtyards, public sea pools for watersport activities and jetties. These are of generous proportions for a network of neighborhood waterbodies and waterscape at settlement scale.

Comparing Population Densities While jack-up housing can accommodate 2,000 to 4,000 persons equivalent to one HDB precinct, semi-submersibles range from 1,000 to 2,000 persons. A flotilla can accommodate a population in excess of a HDB neighborhood of 30,000 persons, while settlement variants have capacities comparable to a New Town at 300,000 persons. The projected density of the oil rig housing exceeds the HDB types on land. This is achieved in spite of providing larger recreational area, and private terraces to dwelling units of a floating settlement.

Comparing densities in persons per km^2, Ghim Moh has a moderate density of 163,765 to 242,100 persons per km^2 while Pinnacle Duxton has a high density of 367,000 persons per km^2. In comparison, housing densities for jack-up range from 230,000 to 600,000 while semi-submersibles range from 100,000 to 280,000 persons per km^2 respectively.

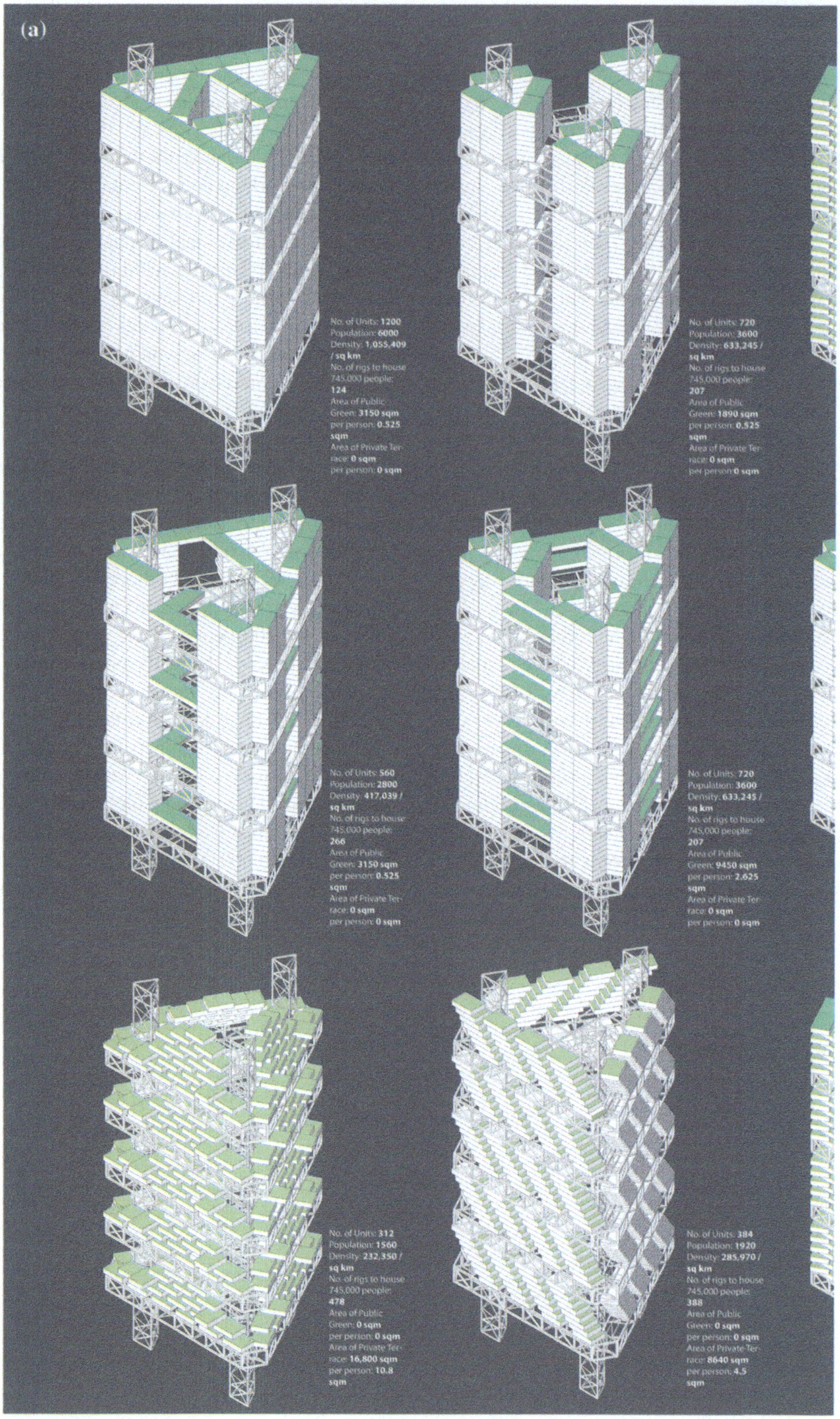

Fig. 39 Housing variants generated on oil rigs **a** 14 variants from three-legged jack-up rigs **b** 10 variants from four-legged jack-up rigs and 9 variants from semi-submersible rigs

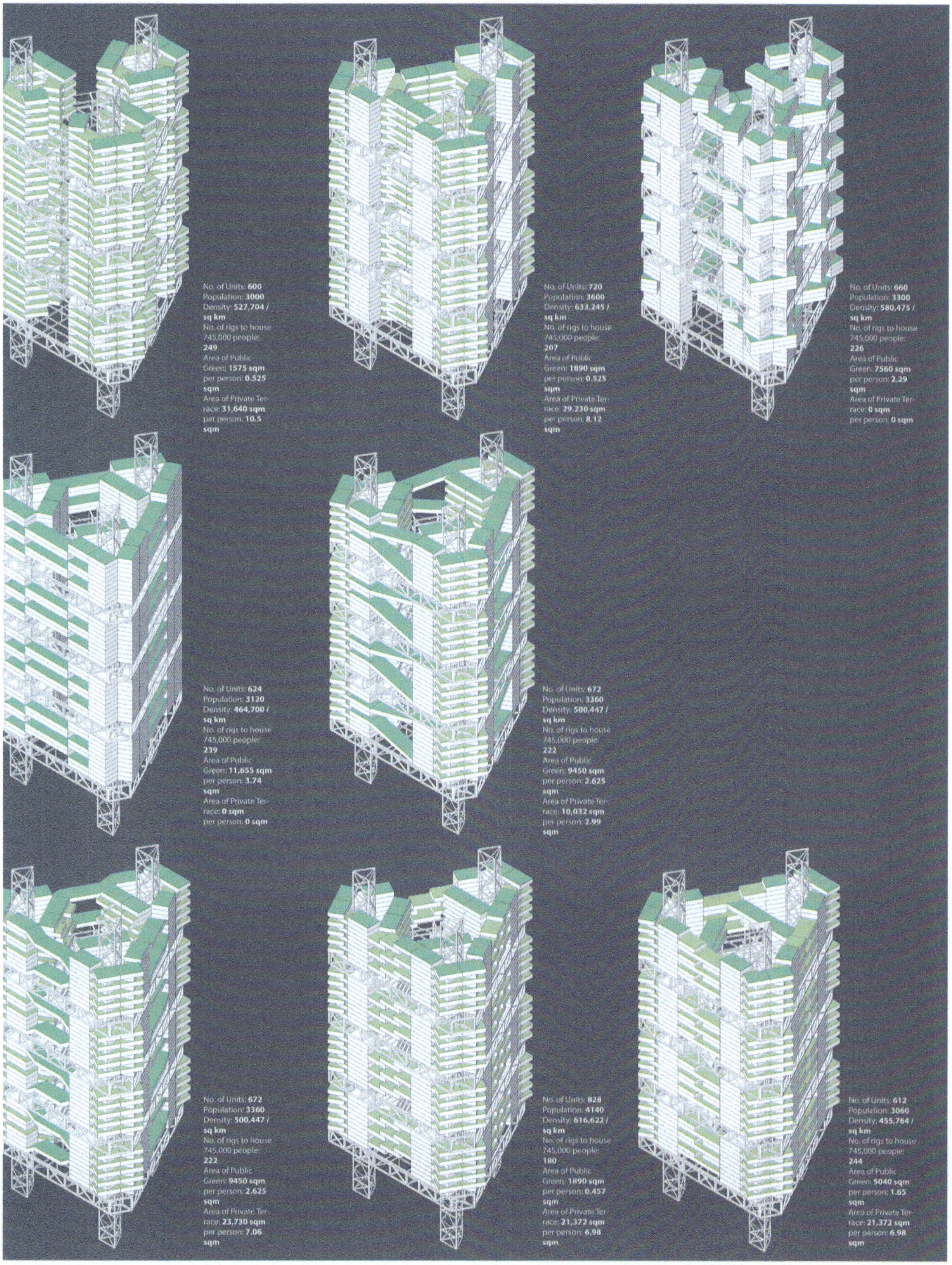

Fig. 39 (continued)

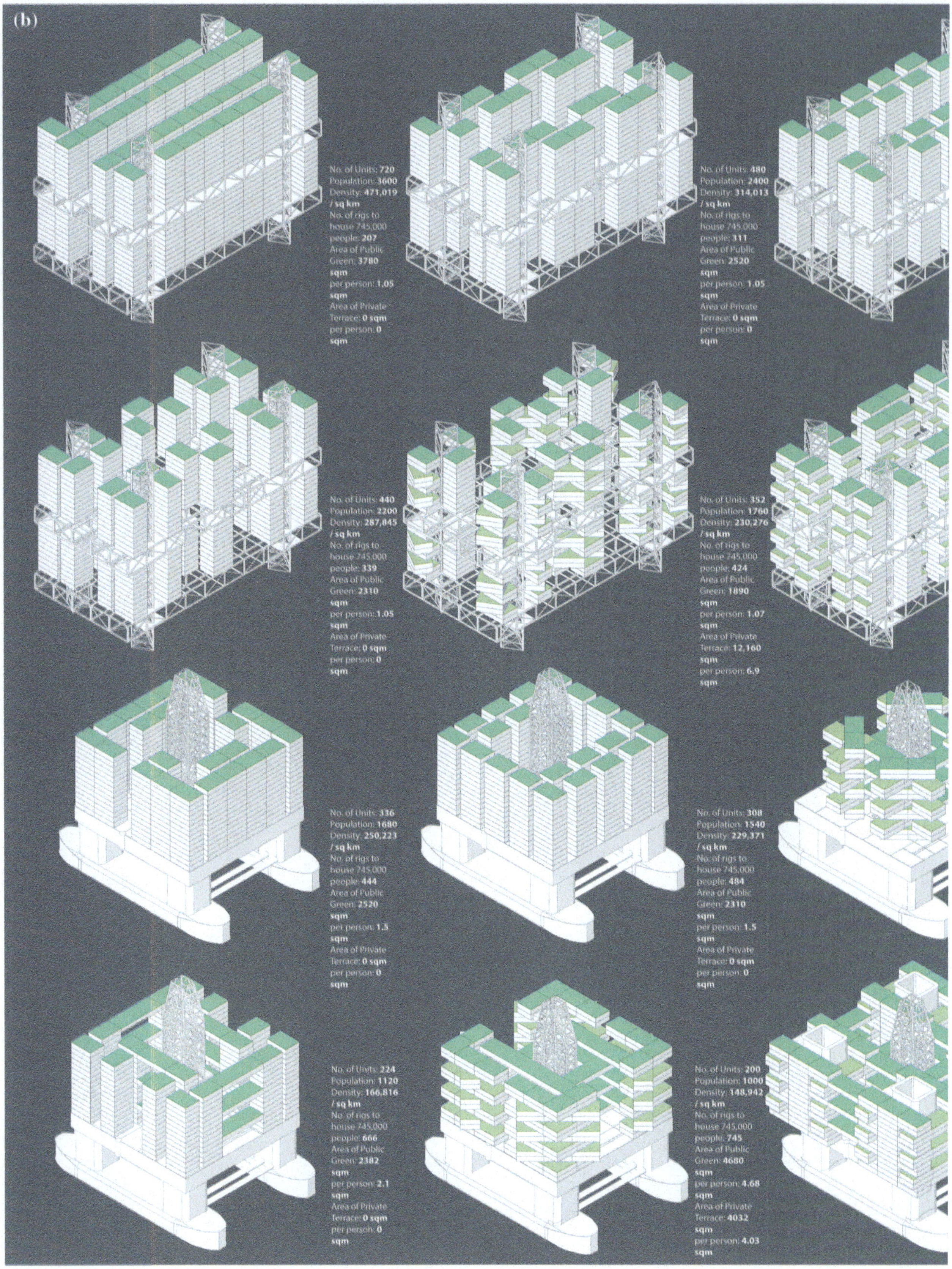

Fig. 39 (continued)

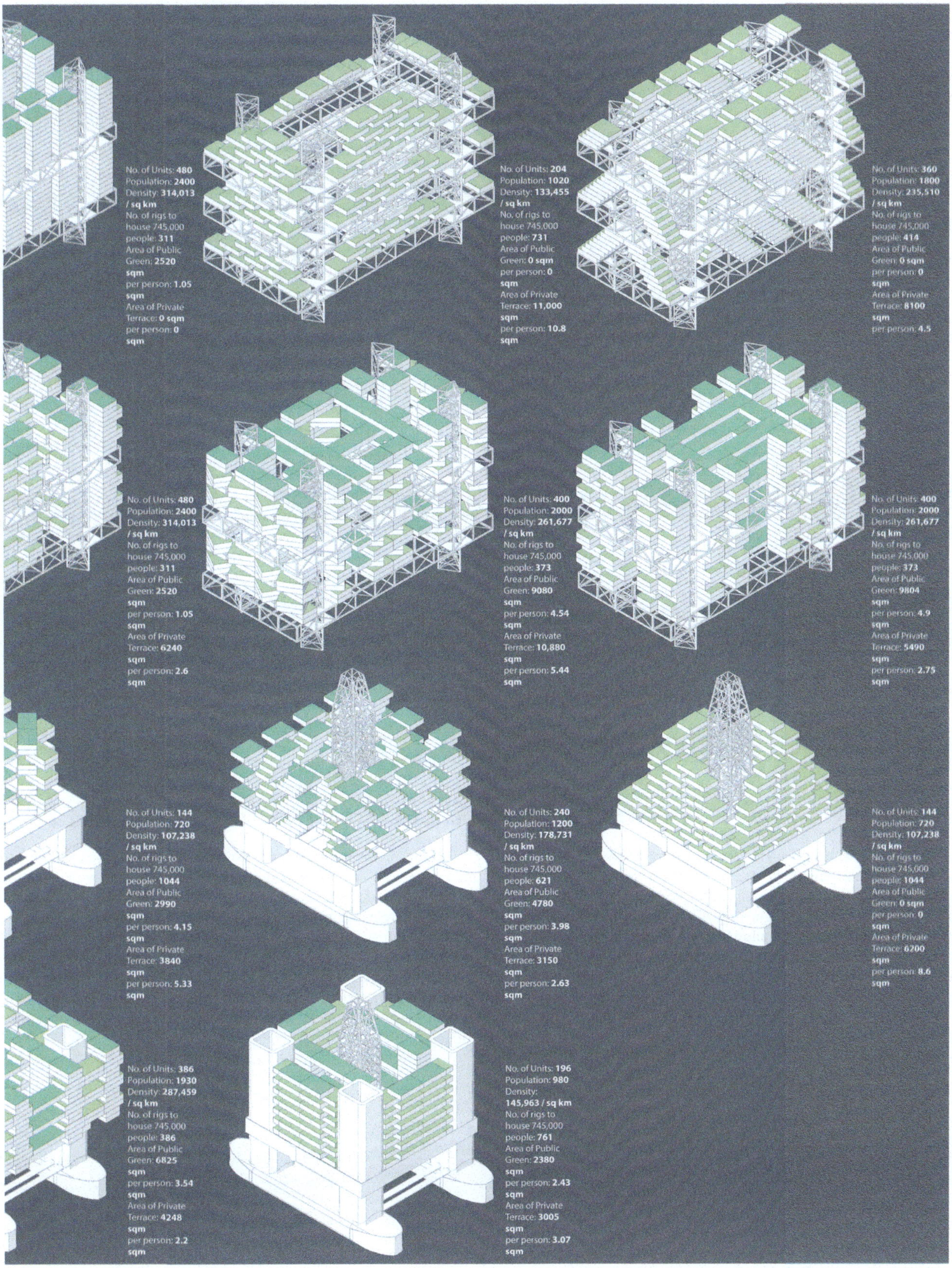

Fig. 39 (continued)

Fig. 40 Jack-up Housing **a** Housing units are arranged around three legs to maximize views of the sea, while reinterpreting central courtyards as skybridges spanning each jack-up leg at the periphery as a transfer structure, opening a vertical breezeway across the built-up section and at the same time allowing light into vertical unit clusters. Four transfer structure decks are repurposed into co-working and recreational spaces other that structural support, allowing residents to live, work and play at their door steps. The derrick tower repurposed into a community center, while the top floors cater to religious uses. The bottom floors accommodate traditional activities such as arena styled gambling with cock-fighting, fortune telling, hall games like mahjong, allowing residents to resettle themselves in a similar socio-cultural context. The derrick, which is cantilevered off the main deck, is a node of gathering spaces with a wide range of recreational programs relevant to the residents. In this illustration, a South Asian setting with worship and prayer spaces, observatories at upper levels with gambling and fortune telling spaces located near the derrick base, linked to market and food centers. Thus planning and design aims at using the industrial structure to create a familiar socio-cultural environment in order to characterize communal space and residential identity. **b** Floor plans

Fig. 41 Sectional perspectives of semi-submersible Housing Housing units are arranged symmetrically on deck (for symmetrical loading) to maintain stability in buoyancy. The skybridges framing the enlarged moon pool allow visual contact with activities on pontoons and on water. The deck with the pontoon is reinterpreted into large urban undercrofts spaces over water courtyards accommodating forms of water sports and recreation, and acting as an atrium/harbor for alighting or temporary berthing

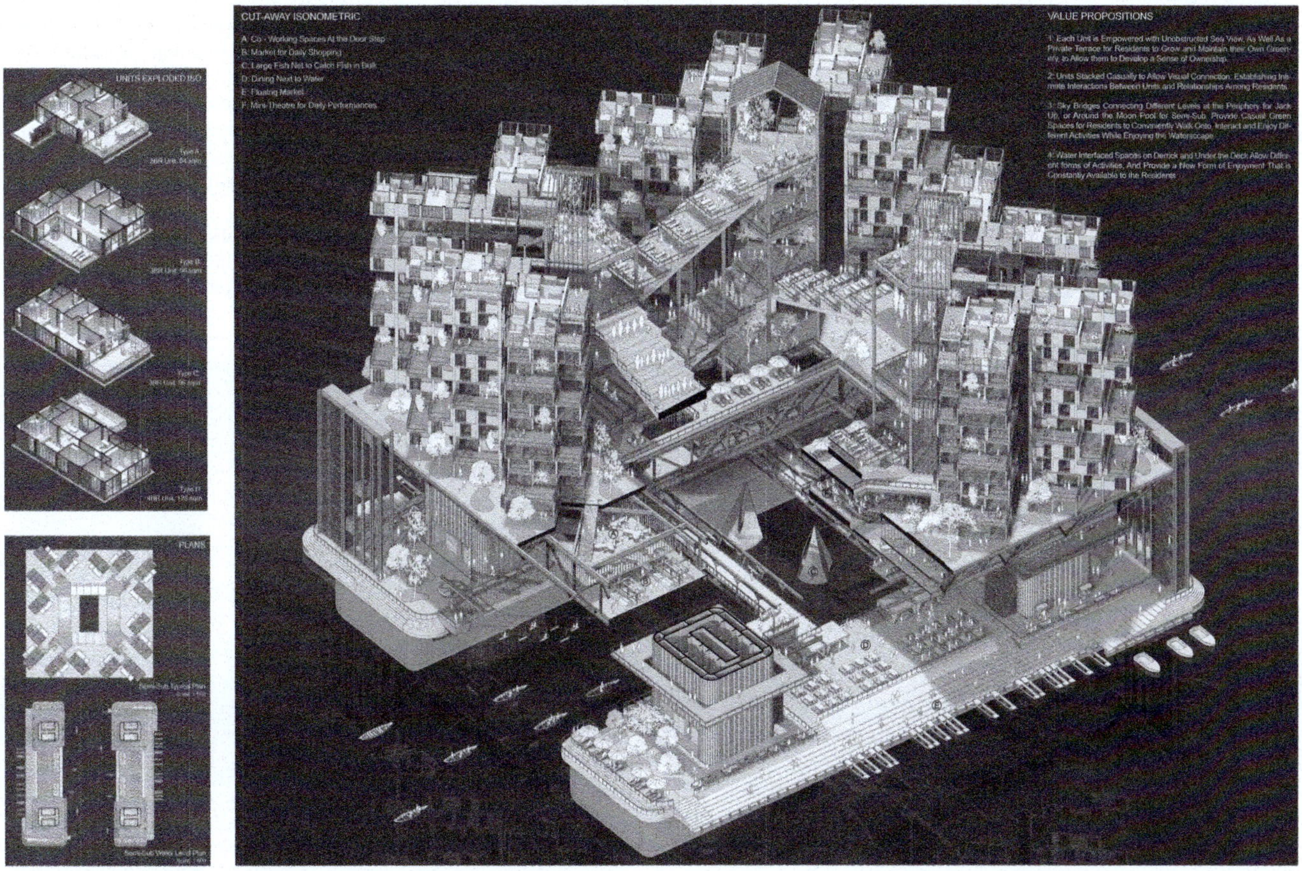

Fig. 42 **a** housing unit types **b** floor plans **c** isometric view

Table 6 Capacities of Jack-up and semi-submersible housing

	4-legged jack-up	3-legged jack-up	Semi-submersible
Footprint (m^2)	7,643	6,720	6,420
No. of housing unit	400	672	246
Housing capacity	2,000	3,360	1,300
Population density (persons per km^2)	261,677	500,447	202,492
Private greenery per person (m^2)	5.44	7.06	7.06
Public greenery per person (m^2) (compared to 2.16 m^2 at Pinnacle Duxton)	4.54	2.625	3.625
No. of oil rigs to house 250,000 people	125	74	192

References

1. Brulard M (2015) Business insider. Dutch saltwater potatoes offer hope for world's hungry
2. Chan CY (2017) Environmental impacts of land reclamation: a case study of the proposed lantau port development. The University of Hong Kong
3. FAO (2016) The state of the world fisheries and aquaculture 2016. Contributing to food and security and nutrition for all. Rome
4. FAO (2017) Consumption of Fuel, Electric Power and Water. http://www.fao.org/3/X6899e/X6899E08.htm
5. Fung F, Wong O (2015) How poor planning leaves Hong Kong's new housing estates with inadequate infrastructure. South China Morning Post
6. Hollnagel E, Fujita Y (2013) The Fukushima Disaster—systemic failures as the lack of Resilience. Nucl Eng Technol 45(1). Science Direct
7. IFFO (2017) The marine ingredients organization. byproduct. https://www.iffo.net/byproduct
8. Loh C, Kilburn M (2011) The costs and benefits of high-density Urban living. LSE cities
9. MEWR. Ministry of the Environment and Waste Resources (2016) A climate-resilient Singapore for a sustainable future
10. Moles R et al (2002) Methodologies for the estimation of sustainable settlement size. Environmental Protection Agency, Ireland
11. Ng WK (2011) Vertical cities Asia. National University of Singapore. School of Design and Environment. NUS Press
12. Perkins R (2015) Fukushima disaster was preventable. USC News Sci Technol
13. Pi J (2019) Public housing in Hong Kong: the challenge of housing the world's Most populated region. The Urban Review
14. Robertson C, Krauss C (2010) The New York times. gulf spill is the largest of its kind, scientists say
15. Rogers S, Harvey F (2012) Global carbon emissions rise is far bigger than previous estimates. The Guardian
16. Tan A (2017) Straits times. science and environment. Singapore beaches hit by oil spill
17. Tan C (2016) Straits times. Nicoll drive being raised to stave off rising seas
18. Teng YM, Gee C (2014) Population outcomes: Singapore 2050. Institute of POLICY Studies. Lee Kuan Yew School of Public Policy
19. United Nations. Department of Economic and Social Affairs, Population Division (2019) World population prospects 2019: highlights
20. United Nations. Department of Economic and Social Affairs, Population Division (2018) Revision of world Urbanization prospects
21. Vasagar J (2016) Singapore airport guards against sea rise. The Financial Times
22. Vetter D (2019) How Hong Kong's public housing system works: costs, waiting times and sales. South China Morning Post
23. Xie C China Daily (2018) 'Seawater rice' rides tide of hope and hype

Optimal Settlement Size and Masterplan Strategies

Abstract

This chapter studies the optimal sizes of flotillas, which attempt a circular economy involving aquaponics production and fish farming to support population sizes adequately sustained by renewable sources of technology. Estimates of capacity are made from specialized vessels designed for accommodation, food production, waste management and medical care. These are then grouped into flotillas for 40,000 persons to be agglomerated into entire settlements. The vessels will each form elements of water-townscapes where city square is interpreted as a water court between vessels docked in specific configurations; water foyers beneath semi-submersible decks used as outdoor living spaces; and wharfs for recreation and commerce. Combinations of flotilla forms generate variants of entire settlements using jack-ups for up to 40 m depth of water and semi-submersibles for deepwater locations. The population densities of these settlements are compared with those of current floating cities, township densities in Singapore and other high-density cities. Using UNESCO guidelines for offshore settlements to estimate the spacing between settlements, these autonomous settlement fleets conceived as sea-cities, occupy 38 km^2 of sea space and are an alternative to land-based mega cities accommodating 100,000 persons per km^2. 20% of the projected population in the year 2025 can be accommodated on 6,510 oil rig settlements spaced 240 km^2 apart over 54.25 M km^2 of sea. Related to high-density settlements is the concern with environmental quality through the provision of public space. One indicator is the rate of recreational area per person. The space standards of each marine inhabitant—at 50 m^2 recreational area per person—exceeds 27 m^2 in London, is equivalent to Amsterdam, with one settlement type reaching 115 m per person equivalent to Vienna. Singapore by comparison, has 65 m^2 of recreational area per person.

1 Concepts and Forms

Current concerns with population growth projected at 9.7 billion in 2050 with the threat of sea level rise has led to a revived interest in floating cities which exist on autonomous systems. However, the population densities of floating proposals are not near the current top four densest countries in the world by inhabitants per km^2, that is, Monaco at 26,150; Macau at 21,420; Singapore at 8,292; and Hong Kong at 7,082 [6]. Designed by Bjarke Ingels, Oceanix City can accommodate 1,650 inhabitants with 12 ha modules of floating platforms [3]. A floating community for 10,000 inhabitants sits on six of these modules with open water in the center, totals 75 ha. Oceanix City has a settlement density of 13,333 inhabitants per km^2, ahead of Singapore.

'Next Tokyo 2045' is a visionary design located in Tokyo Bay by Architect Kohn Pederson Fox. It imagines a resurgent megacity, adapted to climate change through the realization of a high-density eco district that supports transit oriented development for 500,000 people. It offers protection to the low-elevation coastal zones surrounding Tokyo Bay. Distinctive hexagonal infrastructural rings serve as breakwater and clusters of recreational open spaces. The most prominent feature of the design is the Sky Mile Tower, reaching 1,600 m tall and accommodates 55,000 people. The development on water integrates mainland transport infrastructure and manifest as a spine and cluster branch type of urban form.

One observation of the 'Next Tokyo 2045' plan is the collection of mixed-use towers on water. The KPF solution occupies a series of smaller cluster rings of coastal defence on water, while Kenzo Tange's 'A Plan for Tokyo 1960, sought a new urban spatial order based on high-speed urban mobility connecting metropolitan sub-centers. Instead of a civic center in a centripetal plan, Tange formed a civic axis to connect new developments in a linear direction. Its construction plan involved part reclamation and part engineered platforms raised above the reclaimed parts of land. Gas,

J. Lim, *Oil Rig and Superbarge Floating Settlements*, Lecture Notes in Civil Engineering 82,
https://doi.org/10.1007/978-981-15-5297-7_3

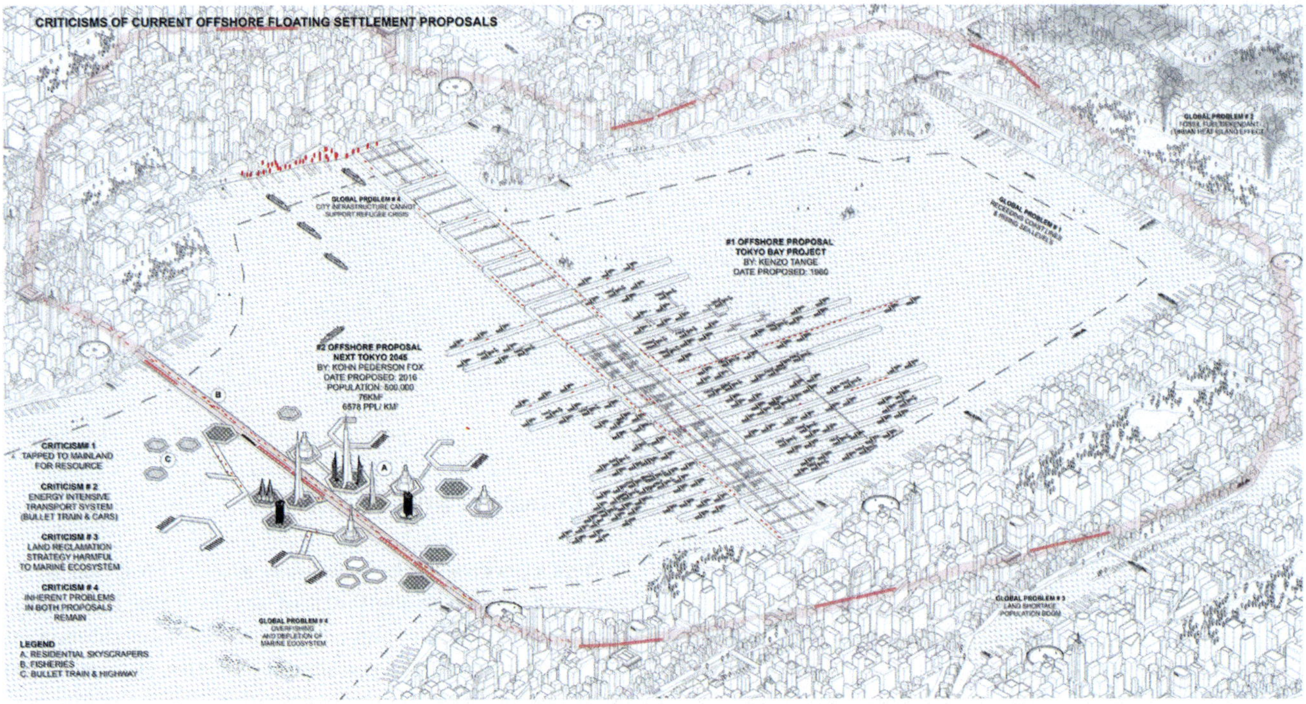

Fig. 1 Comparing 'A Plan for Tokyo 1960' and 'Next Tokyo 2045'

water and electricity lines would be housed within the platforms supporting new residential structures. This dramatic linear form stretched across the bay with bridge blocks enclosing water bodies and green parks in between transport lines and built up areas (Fig. 1).

2 Settlement Size

The idea of an optimal size in a city—in terms of population size—depends on the nature of its economy, and the type of resources engaged in economic production. Early classical view believed that 'the best of all possible worlds would find urban population divided among cities of equal size, that size being optimal [4]. The convergence in size within a range would be applied in public policy for development purposes.

In the case of our oil rigs study, the settlement densities ranged from 9,887 inhabitants per km^2 (with semi-submersibles) to 15,529 inhabitants per km^2 (with jack-ups) in settlement sizes of 28 km^2. Settlement clusters at sea prioritize autonomous systems for food production, wastewater management with renewable forms of energy in closed loops to minimize environmental impact on marine ecology. The output of this essential infrastructure is estimated from population quantums to minimize wastage [2], for example, the capacity of specialized vessel to treat wastewater in the same facility for hydroponic farming and a fish farming facility to mitigate overfishing.

Decentralized wastewater treatment plants for 40,000 people at precincts or centralized wastewater treatment plants for entire settlements of 250,000 people are incorporated with farms and orchards designed on semi-submersibles (see chapter "Post-disaster Applications for Displaced Populations", Fig. 19). Wastewater treatment technology today enables compact systems to be up-scaled to serve 1.2 million population equivalents while the smallest ones can be in the hundreds of population equivalents. Membrane batch reactors (MBRs) combine a suspended growth biological reactor with solids removal through filtration. The compact membranes are designed for operation in small spaces with high removal efficiency of contaminants. In this respect, the membrane filtration system can replace the secondary clarifier and sand filters in an activated sludge system. MBRs produce good effluent quality with automation within smaller space requirements as they operate at high volume loading rates with lower hydraulic retention times. However, there are higher capital and operating costs compared to activated sludge systems of the same through-put. Energy is needed for pressurized systems to push water through the membranes in aeration processes to control bacteria growth on the membranes. Nevertheless, MBR systems provide operational flexibility by adding or subtracting units in adjusting to capacity.

250,000 population is a minimum threshold of high-order tertiary functions that can be exploited by the urban size [1]. A higher percentage of these activities in a city guarantee a

Fig. 2 **a** Town center comprising a town hall, hospital, crematorium orchard; connected by market bridges with bustling water activities. These programs are co-located based on their resource loops to maximize energy efficiency. **b** Sectional perspective. Legend: A. Town Hall, A1. Cupola—Command Centre/Observatory, A2. Parliamentary Chamber, A3. Stadium, A4. Seed Vault/Ammunition Depot. B. Crematorium Orchard. C. Town Hospital. D. Shopping Belt. E. Market Bridge

lower level of social issues resulting from lack of employment opportunities. Beyond a 361,000 population, the expected congestion and diseconomies of scale prevail in large settlements. Large settlements pollute more and generate more environmental damage than medium-sized settlements due to higher levels of production. The latter appear to have a greater internal capacity to keep social, economic and environmental costs under control.

3 Flotillas and Settlements

Specialized vessels illustrated in Fig. 3 are arranged into flotillas to accommodate the number of inhabitants required to optimize the system in an efficient scale of operation. In this respect, they form building blocks of a settlement (Fig. 2). Flotillas are also planned in configurations, which exploit promenade and water courtyard relationships with water-edge pathways enclosing waterside activities. The population size of a settlement fleet would be capped at an efficient maximum scale (Table 1) for wastewater treatment with a compact footprint of one semi-submersible and the provision of one hospital for 250,000 people. Tertiary education is sized at one college per 100,000 population.

35,000 people can be accommodated on a flotilla of nine jack-up housing rigs (Fig. 4), each of varying typology and which are connected to three parklands repurposed from barges, converging to a central neighborhood water courtyard. Large waterbodies flanked by circulatory pathways facilitate recreational water activities, while the parkland on the decks of barges provide greenspace to attract migratory airborne species.

Connected to the parkland barges are the semi-submersible oil rigs where vertical farm production spaces are integrated into the dwelling zones. Recreational spaces are planned as outcomes of communal activity in all forms of production and consumption. To reduce settlement footprint, key systems such as wastewater treatment plants are integrated with orchards/crop production facilities and conceived as urban landscapes. This neighborhood ring can be used to hold events for large gatherings involving resident and visiting spectators.

While growth facilities in jack-ups are located in the bodies of water dedicated for farmways, the fishing port is located at junctions with promenades linking to settlement centers and perimeter parkways. At this strategic location, the fishing port is also ideal for a wholesale market and food center forming a vibrant node for settlement inhabitants at all

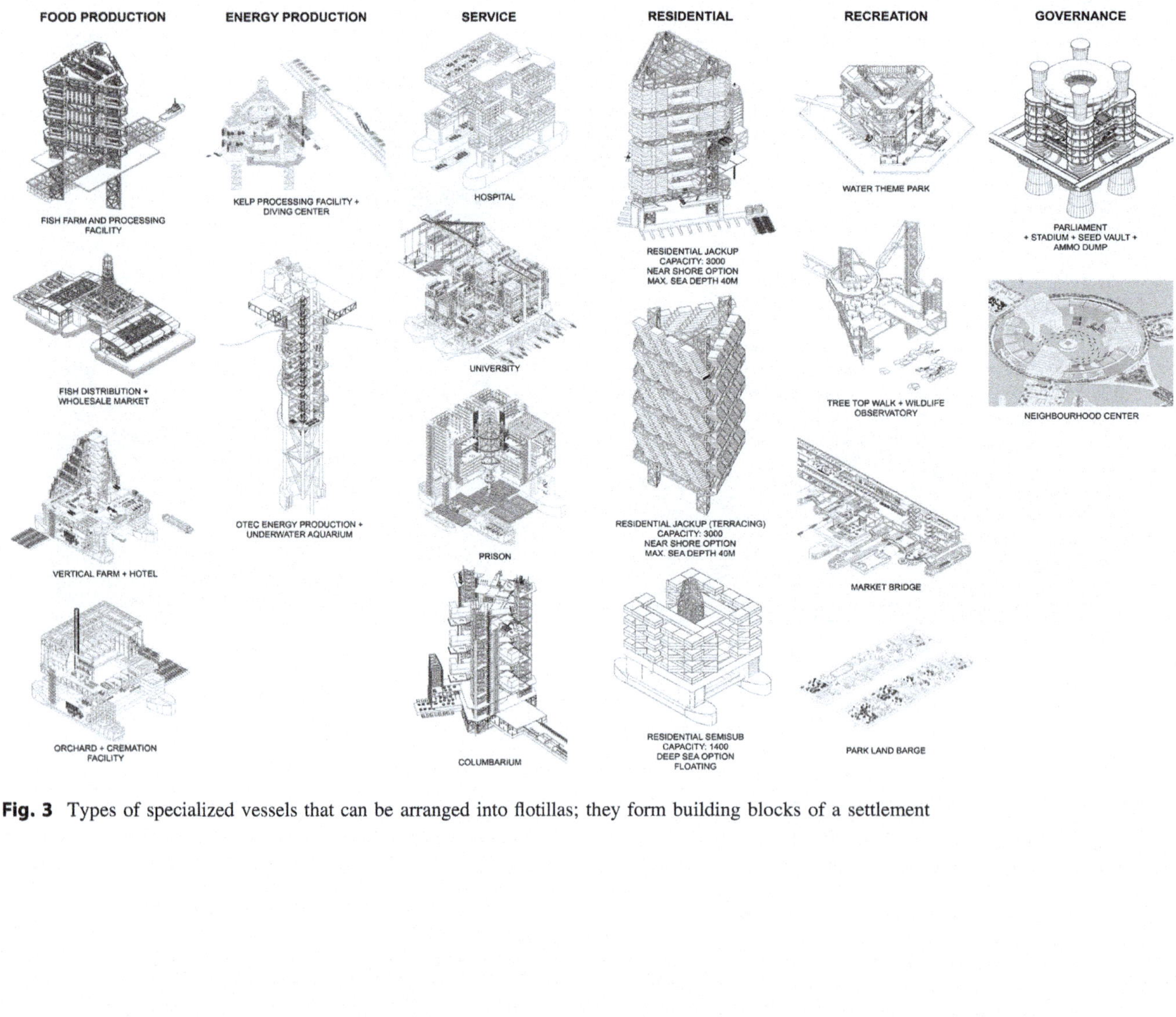

Fig. 3 Types of specialized vessels that can be arranged into flotillas; they form building blocks of a settlement

Table 1 Population equivalents of infrastructure types

Wastewater treatment plant	Decentralized WWTP	40,200 ppl
	Centralized WWTP	120,000–250,000 ppl
Recycling of general waste	Plant size for efficiency	>100,000 ppl
Hospital	Smallest size and fully equipped 500 beds (two beds per 1,000 ppl)	250,000 ppl
Education	Primary school	1000–2000 ppl
	Secondary school	15,000–20,000 ppl
	Higher education institutes	50,000–100,000 ppl

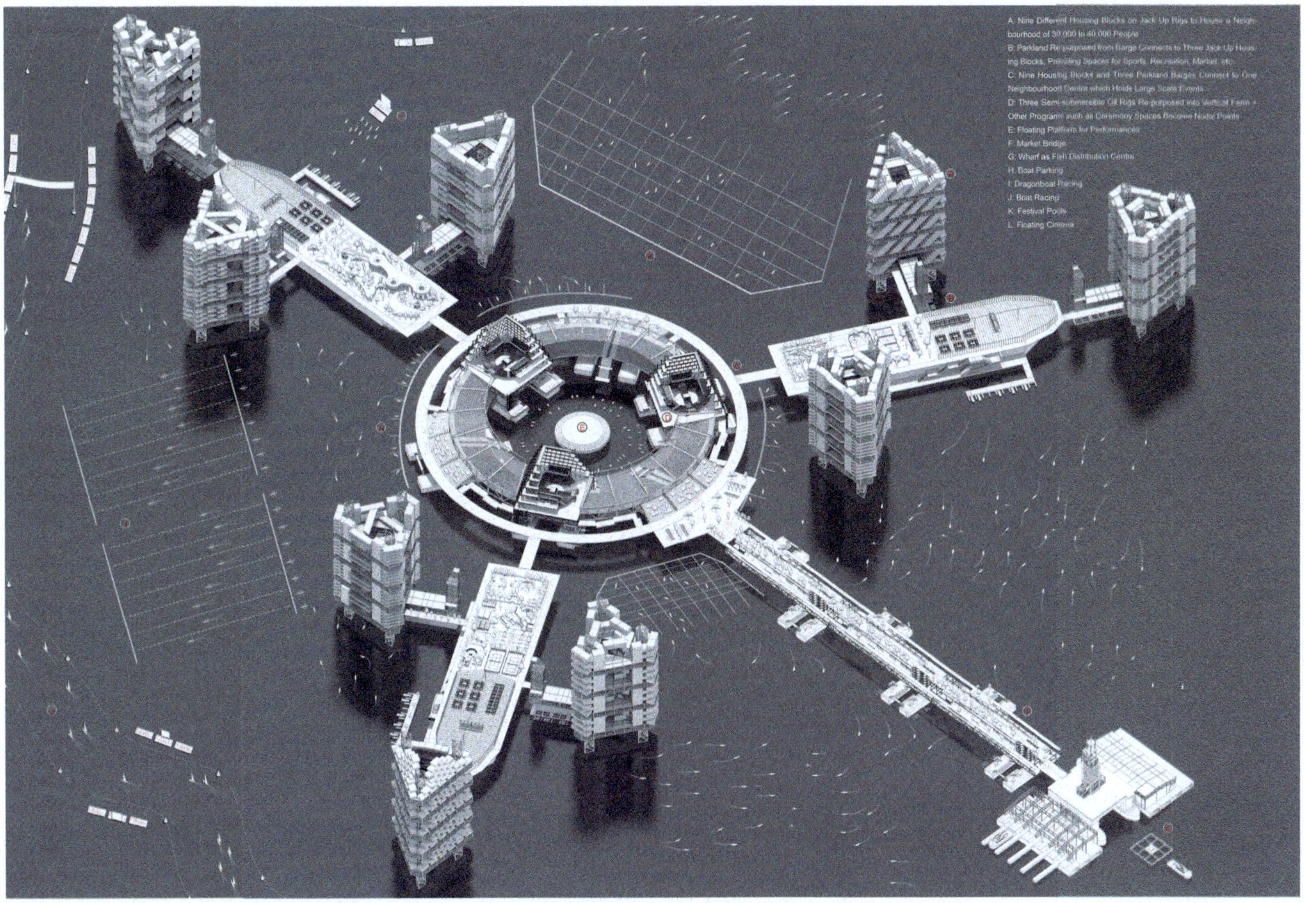

Fig. 4 A flotilla of 30,000–40,000 population

times. Special fishing ports for export occupy the outermost ring of the settlement for fishing transfer and barge traffic. Here, the oil rig structures are of a scale ideal for large gatherings of settlement inhabitants forming a landmark for recreation and commerce (Figs. 5, 6 and 7).

In the settlement variant 'Snowflake' (Fig. 8), its total number of 226,800 inhabitants can be accommodated by seven flotillas each carrying 32,400. Snowflake has an area occupying 1,519 ha of sea with a population density of 14,923 persons per km^2.

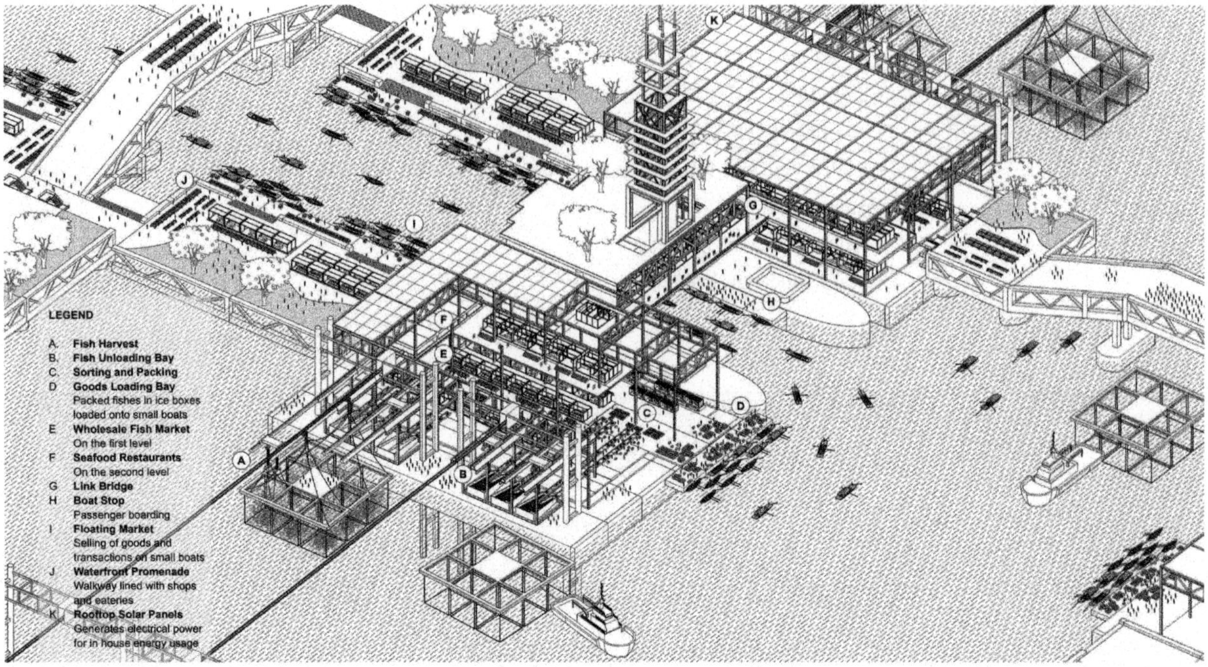

Fig. 5 Fish wharf and a floating market at deck level, by the waterfront promenade with shops and eateries

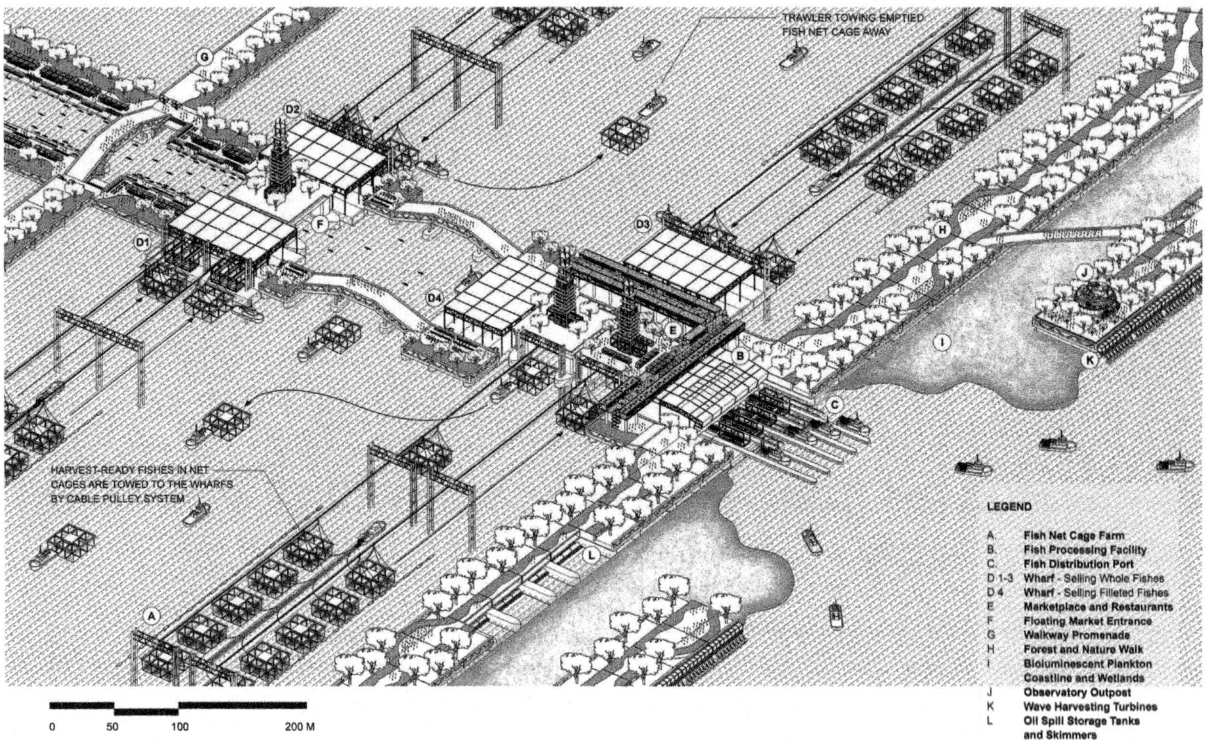

Fig. 6 Concentration of fish wharfs and fish farms within a settlement, linked by waterways to the town's green periphery of forest and wetlands

Fig. 7 Market bridge connecting various nodes in the floating settlement and maximizing recreational activities

Fig. 8 Settlement variant 'Snowflake' comprising seven flotillas accommodating 226,800 persons

4 Settlement Patterns

A floating settlement has the challenge of not diminishing biodiversity and open space, with a concentration of population enabling the advantages of specialization in forms of economic production, pedestrian accessibility and efficient energy and waste treatment systems. Future settlements have to look beyond infrastructural solutions to sustain their existence but also to sustain natural assets beyond their physical boundaries. The challenge is to put back more than what we take from nature and; one attempt with marine settlements is replenishing food supply to mitigate fish wars. Physical configurations and settlement programs which generate a surplus of water and energy resources need to be tested in efforts to diminish water wars and the environmental impact of new technology on ecological footprint.

Figures 9, 10 and 11 is a study on radial, grid and linear configurations of floating settlements, with contemporary (Centrum in Amsterdam, Eixample in Barcelona) and theoretical (A Plan for Tokyo 1960) cities of similar city patterns. Table 2 compares the population density. The radial configuration has a density of 22,235 persons per km^2 in an area of 10.2 km^2 compared to Centrum Amsterdam of 10,625 persons per km^2 in an area of 8 km^2. The grid configuration has a density of 27,000 persons per km^2 in an area of 9.6 km^2 compared to Eixample Barcelona 30,330 persons per km^2 in area of 7.48 km^2. While the linear configuration has a density of 25,920 persons per km^2 compared to Tange's Tokyo Bay Plan 15,000 persons per km^2.

The limitations of the settlements are inherent in their geometric patterns in accommodating networks of transportation and living space for future expansion. Radial and grid patterns have effective configurations. A radial plan provides ease of orientation and is easily defensible while a grid plan offers accessibility and opportunities for diversification. A linear plan, on the other hand, is ideal for nearshore settlements such as a disaster relief application. It provides quick access from the mainland and easy navigation allows for faster evacuation. Quick installation of township components can be carried out simultaneously as it extends linearly.

For offshore settlements, a uniform grid layout enables ease of setting out, assembly, installation and expansion. Standardized modular units allow easy replacement and reconfigurations. In remote and harsh environments, the redundancy of a grid plan is useful in withstanding localized damage to infrastructure.

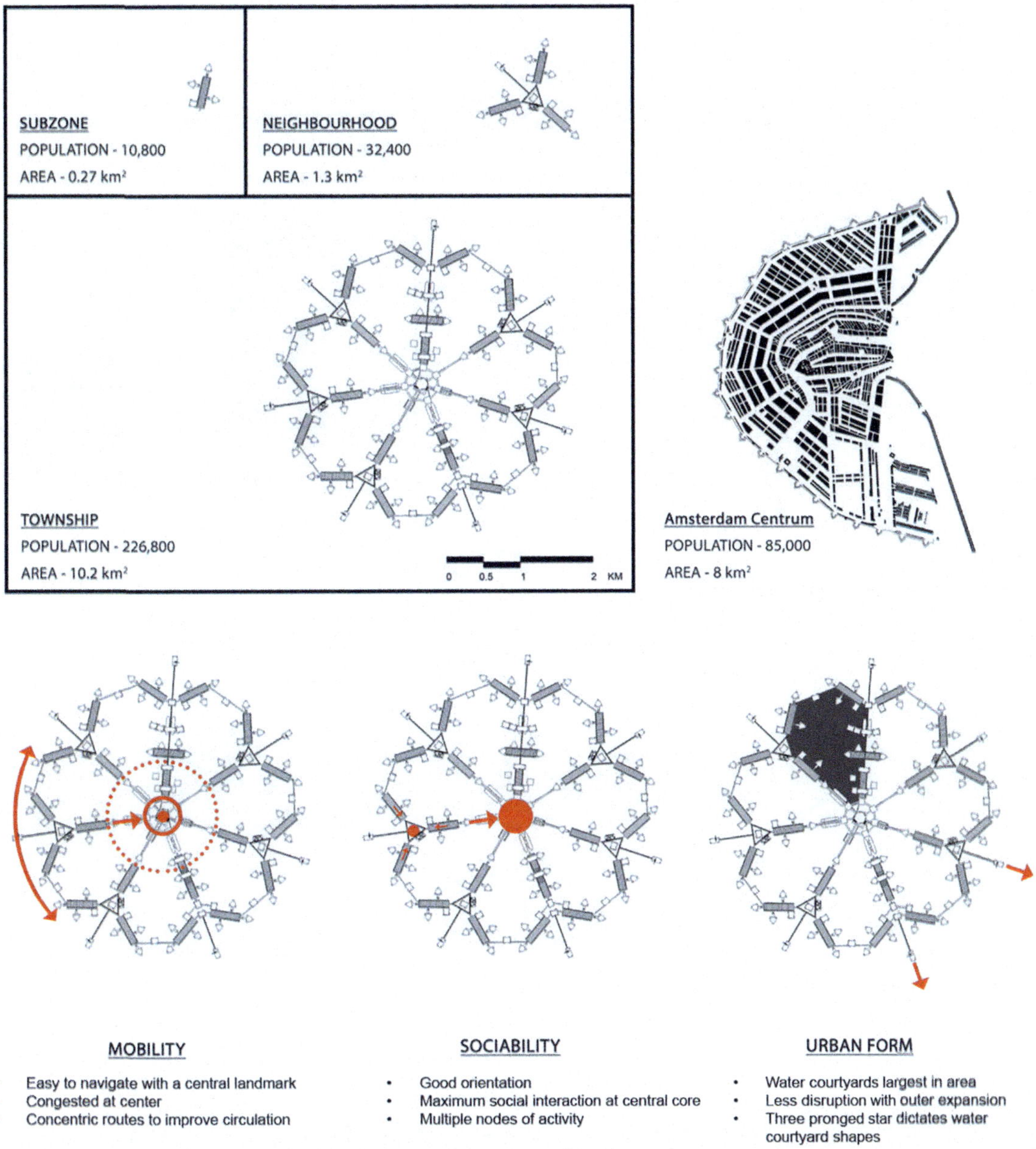

Fig. 9 A radial settlement pattern comprising a central hub surrounded by seven flotilla neighborhoods. Using Singapore HDB terminology planning: one HDB precinct = 600–1000 dwelling units, one jack-up housing = 672 units equivalent to one HDB precinct, one flotilla = one neighborhood, one floating settlement = one HDB new town

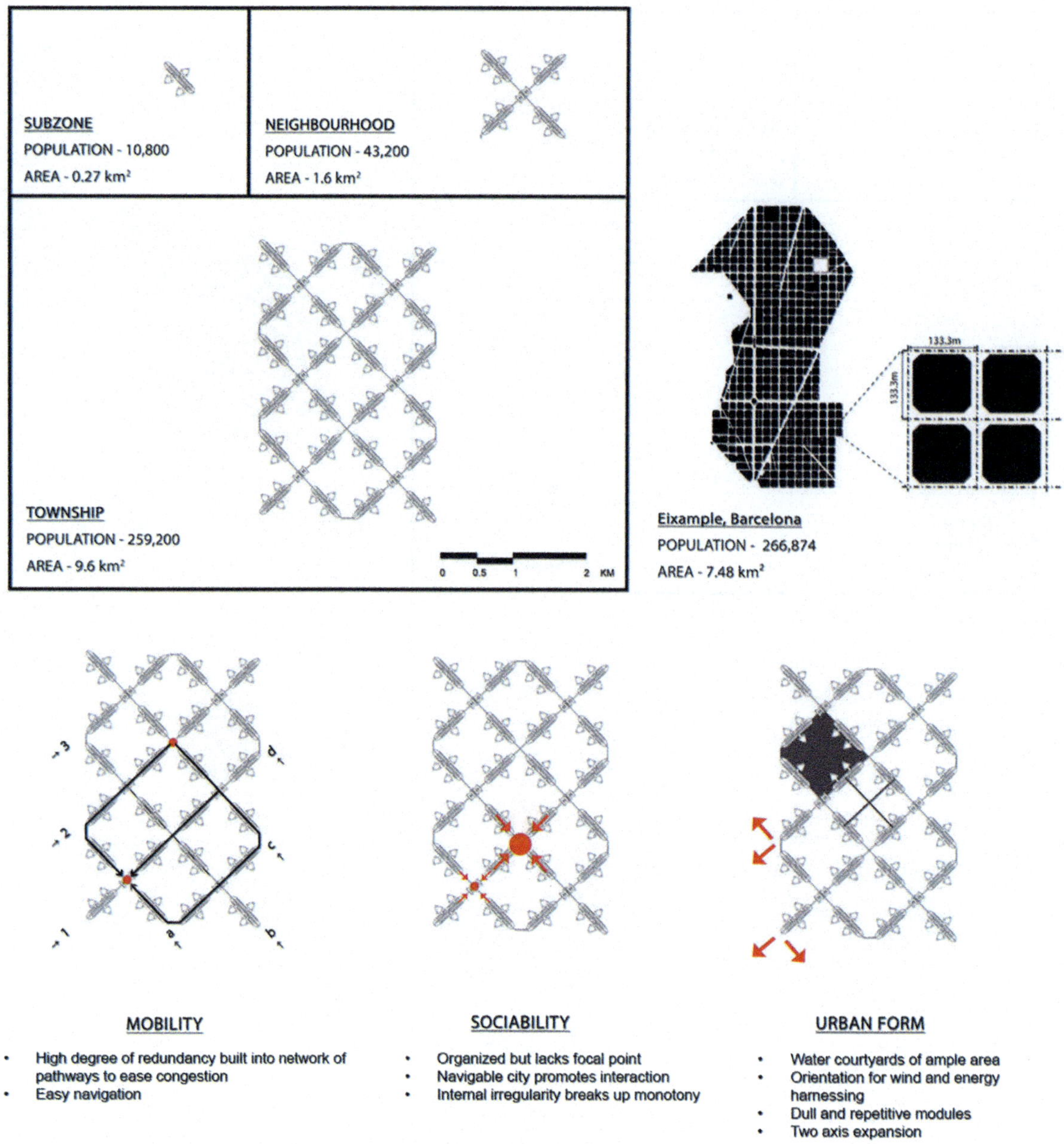

Fig. 10 A grid settlement pattern where streets run at right angles to each other, forming a diagrid of flotilla neighborhoods

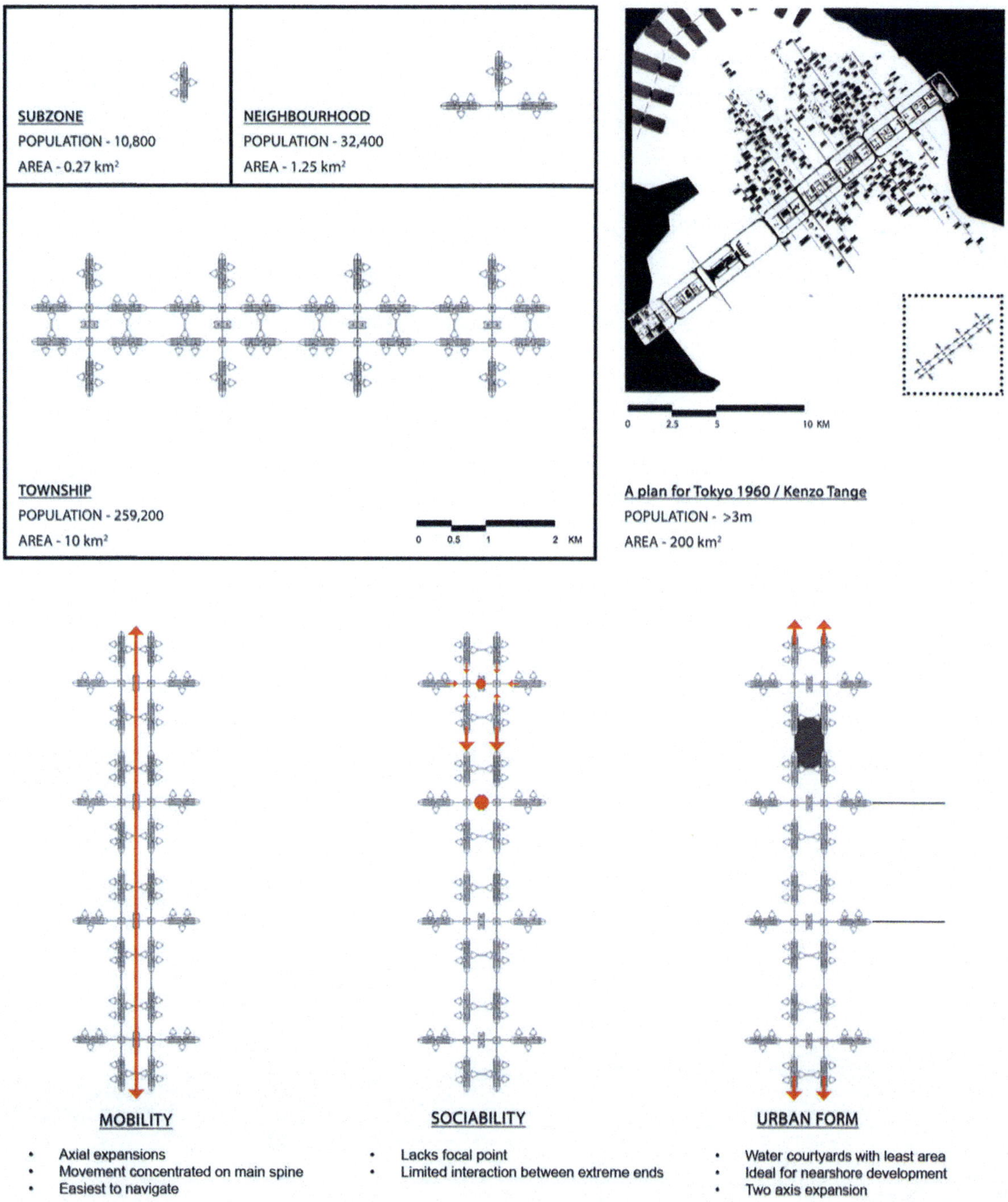

Fig. 11 Linear settlement pattern: an urban plan for an elongated formation comprising a series of functionally specialized parallel flotilla neighborhoods

Table 2 Comparing population density

	Population	Area (km^2)	Density (persons per km^2)
Centrum Amsterdam	85,000	8	10,625
Radial pattern	226,800	10.2	22,235
Eixample Barcelona	266,874	7.48	30,330
Grid pattern	259,200	9.6	27,000
A Plan for Tokyo 1960	3,000,000	200	15,000
Linear pattern	259,200	10	25,920

5 Masterplan Strategies

5.1 Settlement Permutations

The following key considerations guided permutations of alternative settlements on water. They are broadly categorized into three branches: Landscape Integration, Zoning and Water Separation, and Transport Plan.

Landscape within settlement emphasizes on proportion of open recreational spaces such as greenery and water courtyards to form urban vibrant nodes. Whereas landscape as defense mechanism uses floating wetlands along the periphery of the masterplan as a buffer from tidal waves and tsunamis which also acts as the outermost protective ring to the settlement. Luminescent plankton forms a landscape strategy for intrusion detection while creating a unique ambience for the settlement during the night. A significant proportion of landscape in the settlement serves the greater ecology of marine wildlife and migratory species. This also creates a unique experience for the settlement in the form of opportunities for wildlife watching of migratory birds and marine wildlife (Figs. 12, 13 and 14).

Zoning and water separation accommodates daily recreational activities within the inner settlement of less turbulent water. The landscape buffer also acts as a primary filter to

Fig. 12 Unique experience for wildlife watching of migratory birds and marine wildlife

Fig. 13 Luminescent plankton forms a landscape strategy for intrusion detection while creating a unique ambience for the settlement during the night

Fig. 14 Open recreational spaces such as water courtyards form urban vibrant nodes

maintain water quality around and within the settlement. No-wake zones strategy partitions the waterscape within the settlement so that the quality of water for farming is kept in good quality and is separated from those for recreational and circulation purposes. Outside no-wake zones, wave energy is harnessed at the outer boundaries of the settlement where the wave conditions would be optimal for harvesters. The wave energy harnessers also serve as wave attenuators. The extent to which the perimeter of the settlement may be deployed for wave energy harvesters with efficient transmission networks is a key consideration for urban occupancy and enjoyment.

In the transport plan, integrating promenades and waterways with nodes in settlement infrastructure and landscape elements intensifies the network of urban accessible space. Ambulatory networks and water taxis are coordinated with scenic vistas to arrive at urban destinations ranging from nodes of intense activity to wildlife locations of solitude. Service and waste disposal networks are to adopt efficient layouts to maximize space. Programs and hybrid typologies of related waste loops could be in proximity to reduce service distance.

Settlement fleets may be distinguished by the quality and quantum of water integrated spaces provided for recreational activity, living space per person, and the accessibility to all parts of its boundaries with a promenade network. Settlement fleets may specialize in knowledge and other forms of production, local climate and migratory species sighted from offshore locations thereby distinguishing themselves from one another. Living and recreational spaces have unique qualities when integrated with production activity. Specialized vessels do not just house mechanical plant and equipment but they also provide the spatial backdrop for settlement vibrancy as a counterpoint to the city. While settlement fleets are kept to less than 4 km diameter in size, they have different population capacities, production outputs and proportion of recreation areas per person. The 4 km diameter factors in a pedestrian range in the planning of transportation forms, which operate on small-scale renewables and do not pose significant energy demand.

5.2 Siting Offshore Settlements

Settlement fleets are also planned on a spacing based on UNESCO minimum biosphere reserve area of 9.7–21.9 km radius and a 22 km radius based on territorial waters for full sovereign rights. Biosphere Reserve areas of 150,000 hA are believed to enable a balanced relationship between man and nature. In proliferating settlements, the UNESCO guideline for Biosphere Reserve Area of 44 km is observed [5]. As shown in Fig. 15, instead of repeating 200 settlements at 44 km apart within 500 by 500 km of water space, 30 settlements spaced at 240 km apart were used in the proposal. This is an attempt to lessen the impact of settlements on the marine ecology and differentiate settlements by distance beyond the UNESCO guideline.

Where each of the oil rig settlement can accommodate 245,000 population equivalent, 30 such settlements can be organized on 250,000 km^2 of sea, thus 6,510 settlements spaced 240 km^2 apart over 54.25 M km^2 of sea may be occupied by 1.6 billion persons. This is an estimated 20% of the projected population of 8.1 billion in 2025. Figure 16 illustrates three settlements spaced apart at sea.

The location of these settlements depends on 'safe zones' as shown in Fig. 17 with the following characteristics: (a) shows areas with a sea air temperature range of 5–30 °C for human thermal comfort and food farming, (b) away from marine wildlife and migratory pathways, both airborne and water, (c) away from hurricane pathways and waves higher than 6 m, and (d) away from pirate and commercial fishing zones, rubbish gyres and shipping routes.

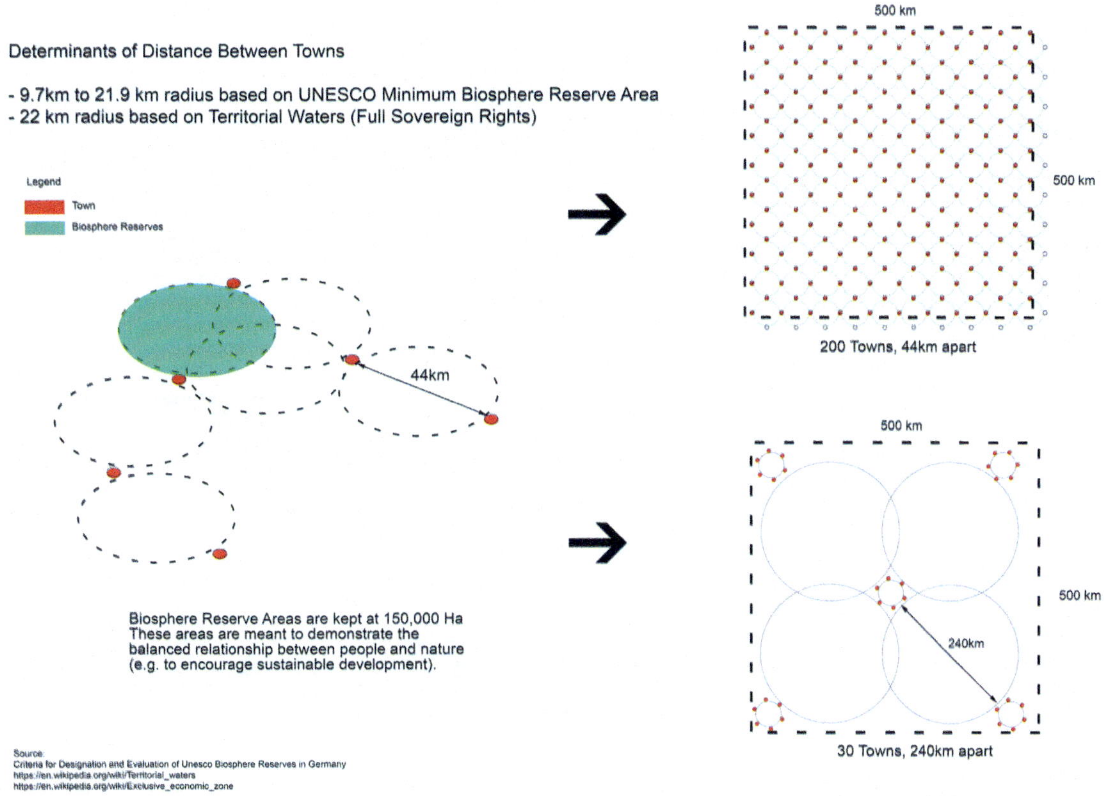

Fig. 15 Siting an offshore settlement

Fig. 16 Settlement variant 'Horseshoe' with population of 228,800 and two 'Snowflake' settlements in the background

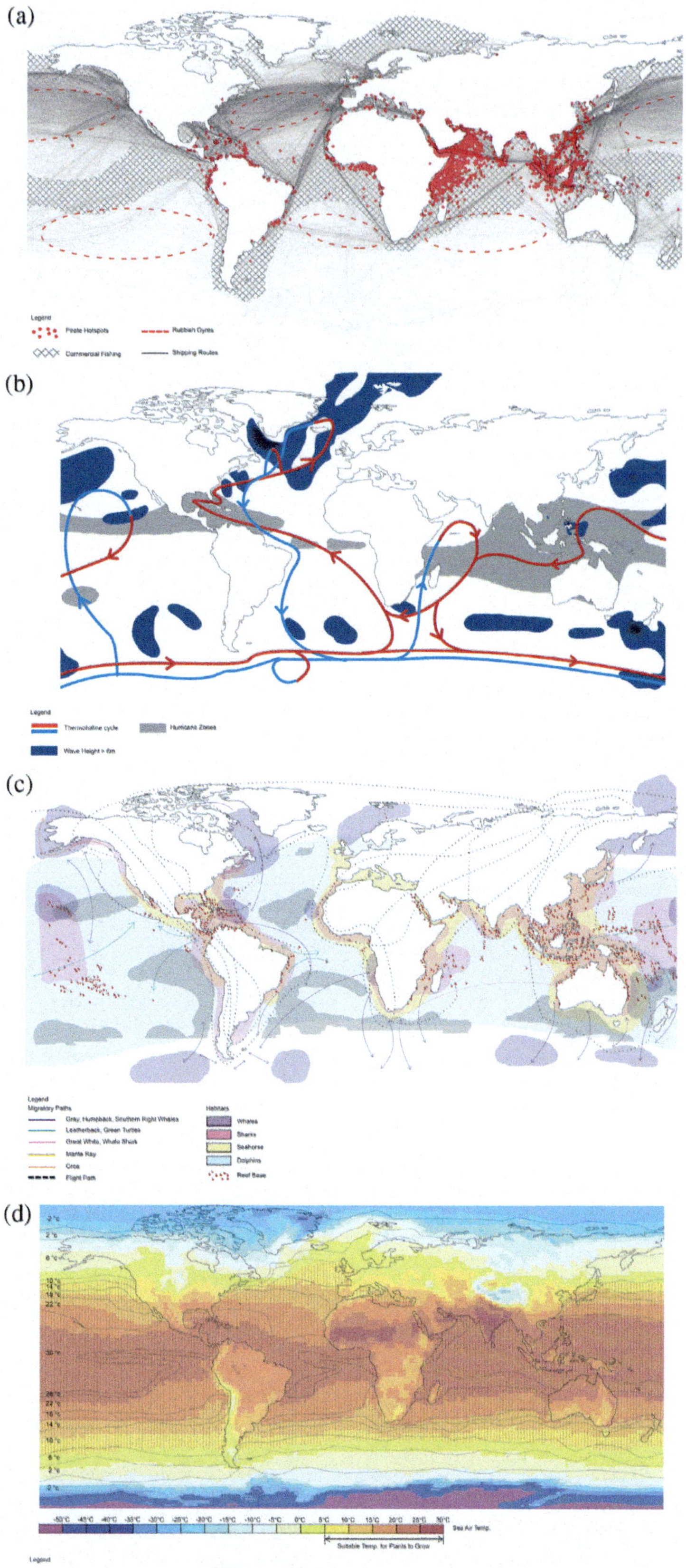

Fig. 17 'Safe zone' considerations for floating settlements. **a** Avoiding zones of pirate operations, commercial fishing, rubbish gyres and ocean shipping routes. **b** Avoiding pathways of hurricanes. **c** Avoiding marine migratory routes; and 'pit-stop for bird migratory routes. **d** Zones of suitable air temperature 5–30 °C; and sea surface temperature suitable for assisting fish and marine food growth

6 Capacities and Advantages

Figure 18 summarizes the results of various permutations, comparing settlements outputs in respect of sustainability in fish farming, harnessing energy and population density. Studies on settlements circulation routes, town greenery, production and water spaces are included in the Appendix.

The rectilinear geometry of the semi-submersible allows for straight-forward connections that naturally extend into a grid pattern. Although the semi-submersible town looks more cluttered and takes up more water footprint, its water courtyards may be linked sequentially as a continuum and in a linear axis. By contrast, the jack-up town has configurations of water courtyards, which are radial and branch-like. However area of water courtyards are comparable when enclosed by both rig types.

Across all the permutations (Graph 1), there are surpluses in fish production with the highest coming from the 'Seastar' and 'Quadrant' settlements at 130 kilotons per year. 1,247 numbers of each 'Seastar' and 'Quadrant' settlement could produce enough fish to meet the world's population in 2025. These settlements with the highest amount of fish production had their fish production facilities within its periphery. This strategy does not confine the growth of the fish production facility and allows for easy expansion outwards. The planning is also suited for exporting the surplus fish at wharf/market nodes along its coastal border.

In contrast, settlements that place their fish production facilities in the middle are confined by the water area and thus limiting the capacity of their fish produce. However the strategic benefit for such placement comes from the food security and protection from external or environmental threats. Therefore there exists a tradeoff between production capacity and food security when it comes to the planning of fish production facilities within a settlement.

The bulk of energy supply for all settlements comes from wave energy harnessing devices that are placed along its security perimeters. These security perimeters also double-up as green spaces for the enjoyment of residents as well as a habitat for migratory species. Recreational area and wave energy generation is created simultaneously.

A longer perimeter will allow for more wave energy to be harnessed. However an interesting observation arose from the permutations (Graph 2). The 'Seastar' permutation, which does not have a single large boundary perimeter, is still able to achieve a high wave energy yield of 819,450 MWh from its multiple short circumferential borders enclosing ten different farming clusters. In deep-sea locations, 'L-corner', 'Axial', 'Island', 'Lagoon' and 'Concentric' settlements have energy, which make ocean thermal energy conversion plants viable. This is due to the extent of length of settlement boundary that can be used for harnessing wave energy. All other settlements of 208,000–268,000 population equivalent have an energy yield ranging from 500,000 to 900,000 MWh.

In closing, the physical advantage of the oil rig structures to planning settlements lies in the following aspects: (1) their minimal water footprints with semi-submersible pontoons and jack-up legs, (2) the ability to enclose large communal spaces under elevated platform decks and in between their legs, (3) the ability to define open water courtyards for production and recreational programs when planned in clusters viz. they form nodes of urban space and communal activity and for connection to urban networks, (4) a proportion of usable space within the platform deck structure and the semi-submersible hulls. They are in essence elevated tower and podium types in the urban form of cities; and (5) the substantial extent of airspace above platform decks to accommodate varying housing densities, hybrid farming and recreational spaces with infrastructure, the integration of transportation nodes with promenade networks, mixed usage including sky parks and workspaces in vertical configuration to increase green surface area per person.

These allow the floating settlement the advantage of being configured in different ways to integrate both polycentric and monocentric urban forms to flexibly change urban scale and hierarchy. The use of cluster flotillas specializing in forms of production or recreation increases open water space. These facilitate the experiments with re-reefing and recharging marine ecologies. It also overcomes the limitations of landed cities with a hierarchy using only the vessel and the flotilla as units of a settlement plan to accommodate global population increase.

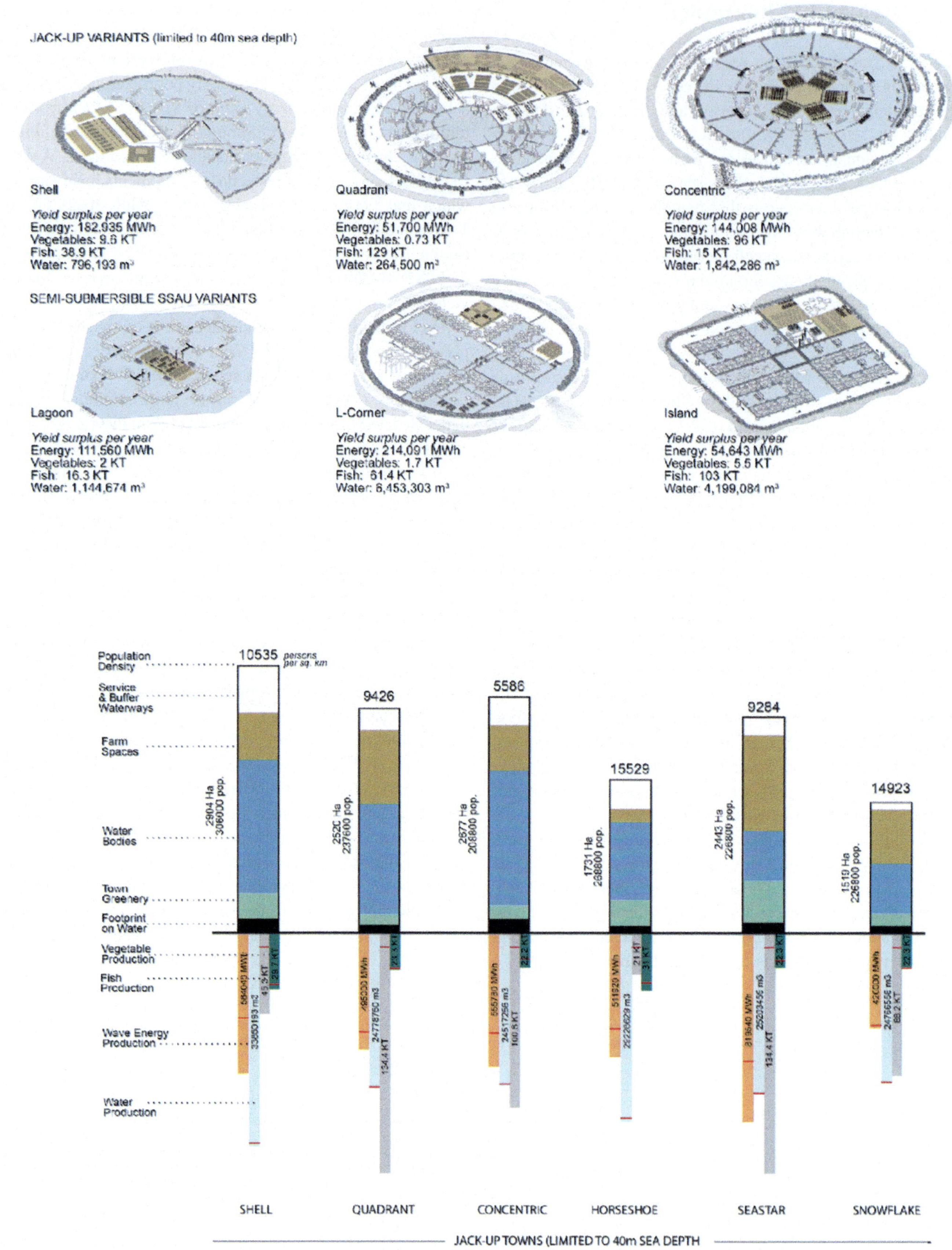

Fig. 18 Overall comparison of settlement variants

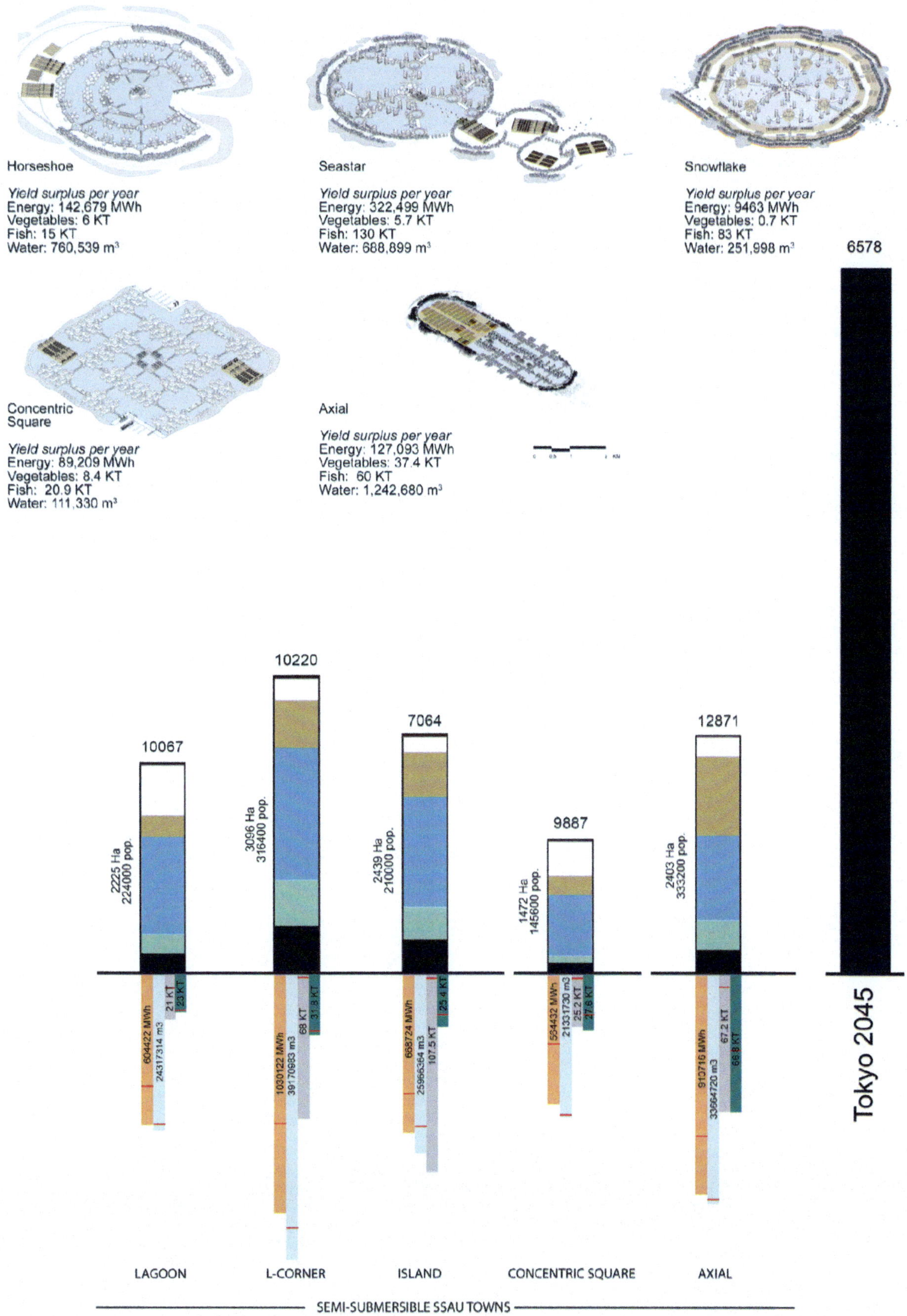

Fig. 18 (continued)

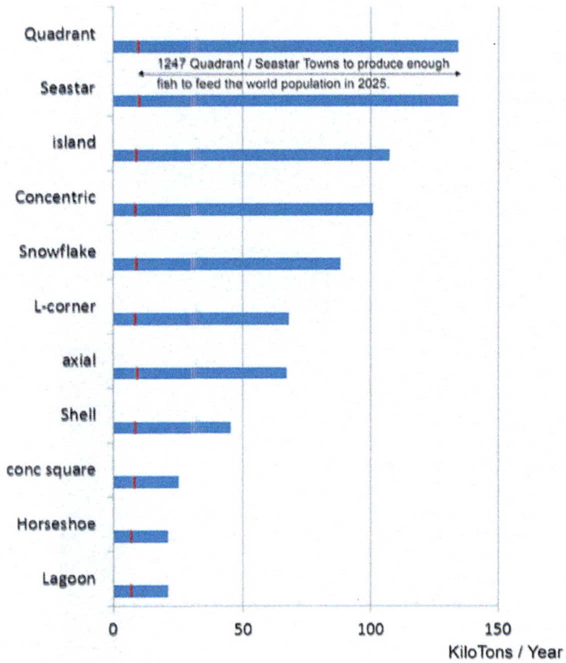

Graph 1 Comparing settlement variants in fish production

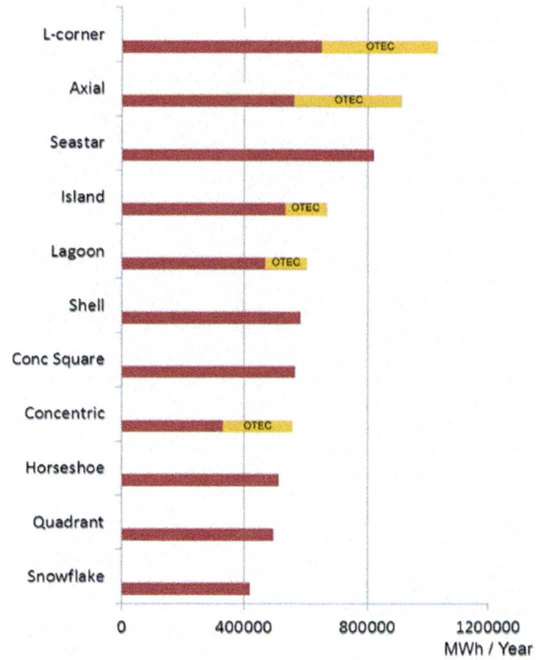

Graph 2 Comparing settlement variants in wave energy

Appendix

Shell Permutation - Overview

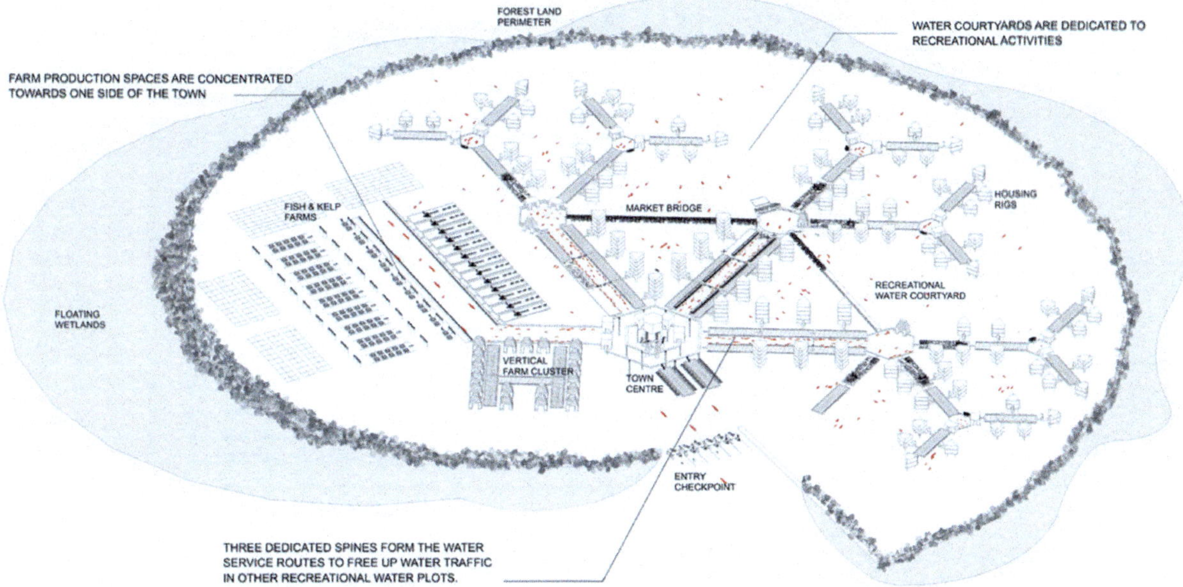

Population: 306,000
Enclosed Area: 29045000 sqm
% Foot Print: 5.35%
% Water Body: 49.1%
% Production Area: 14.3%

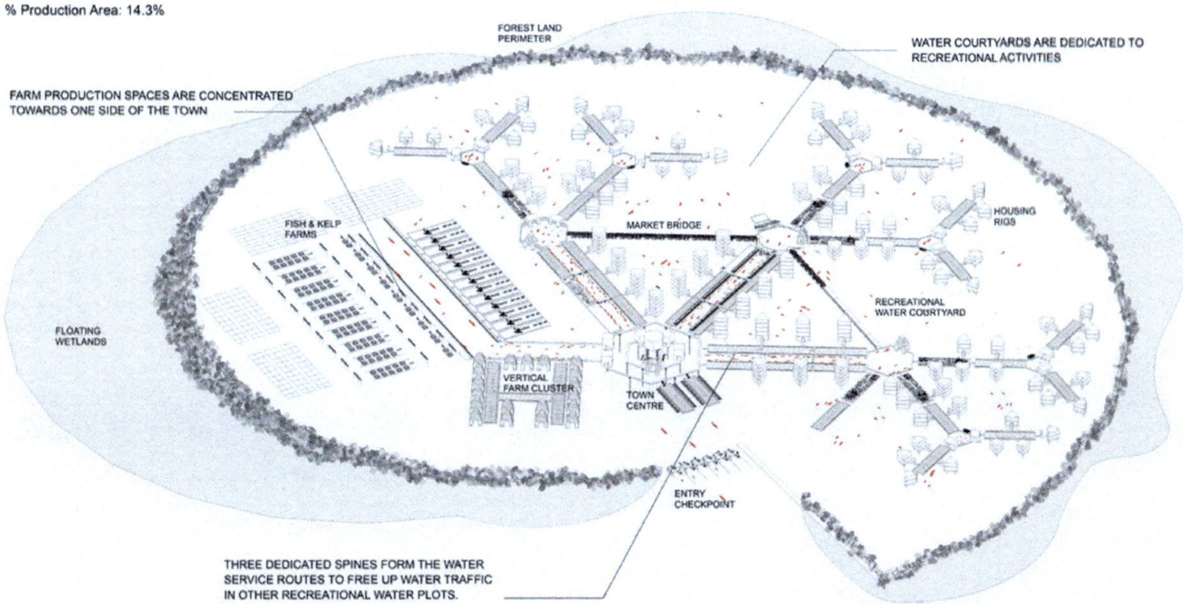

Production Spaces

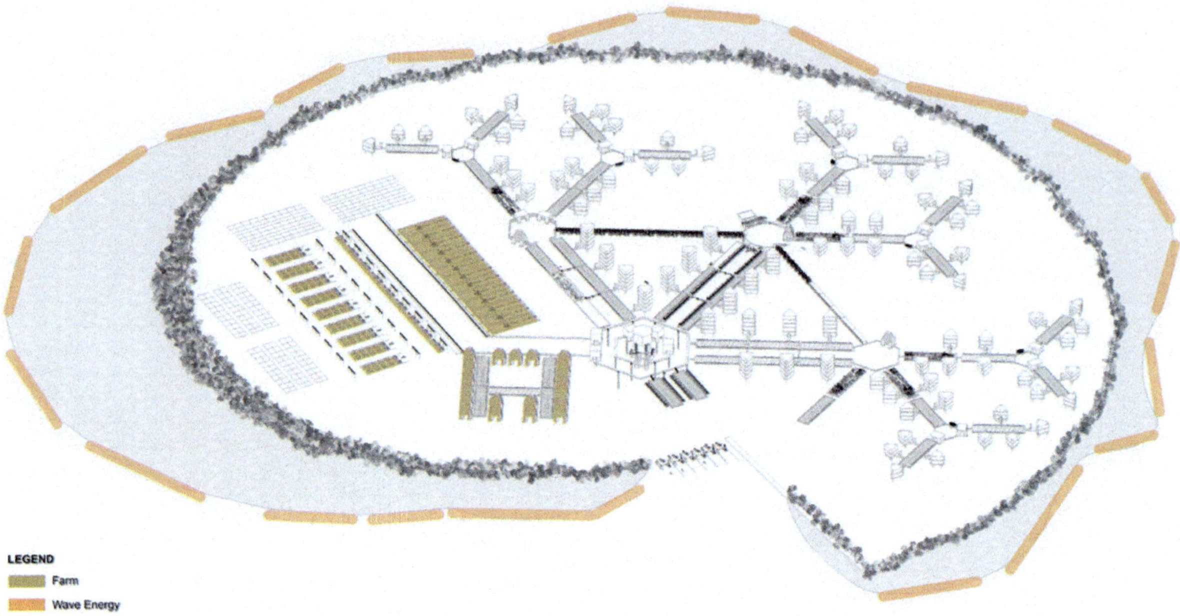

Town Greenery

Total Town Greenery Area: 2972888 sqm
% Town Greenery Area: 10.2%

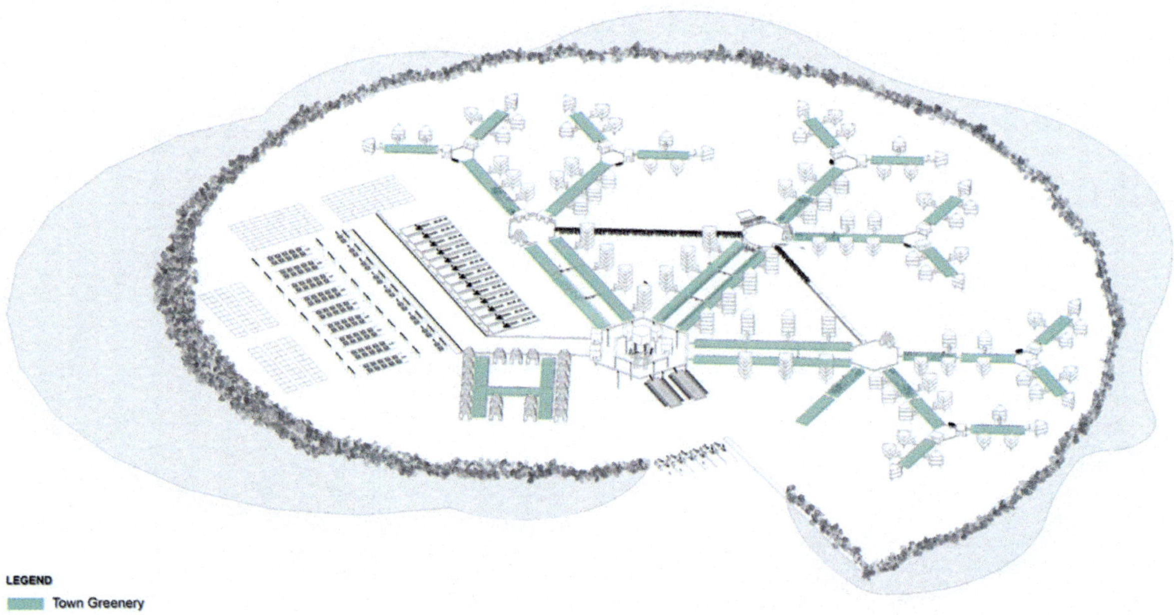

Quadrant Permutation

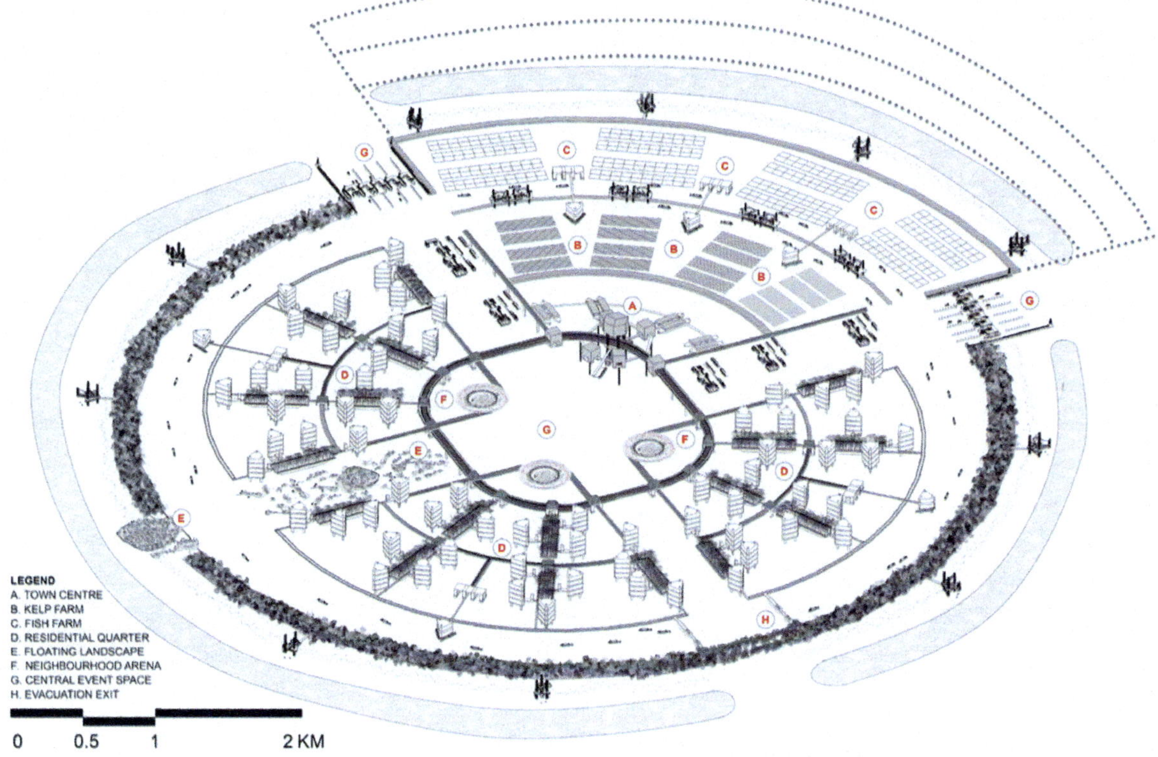

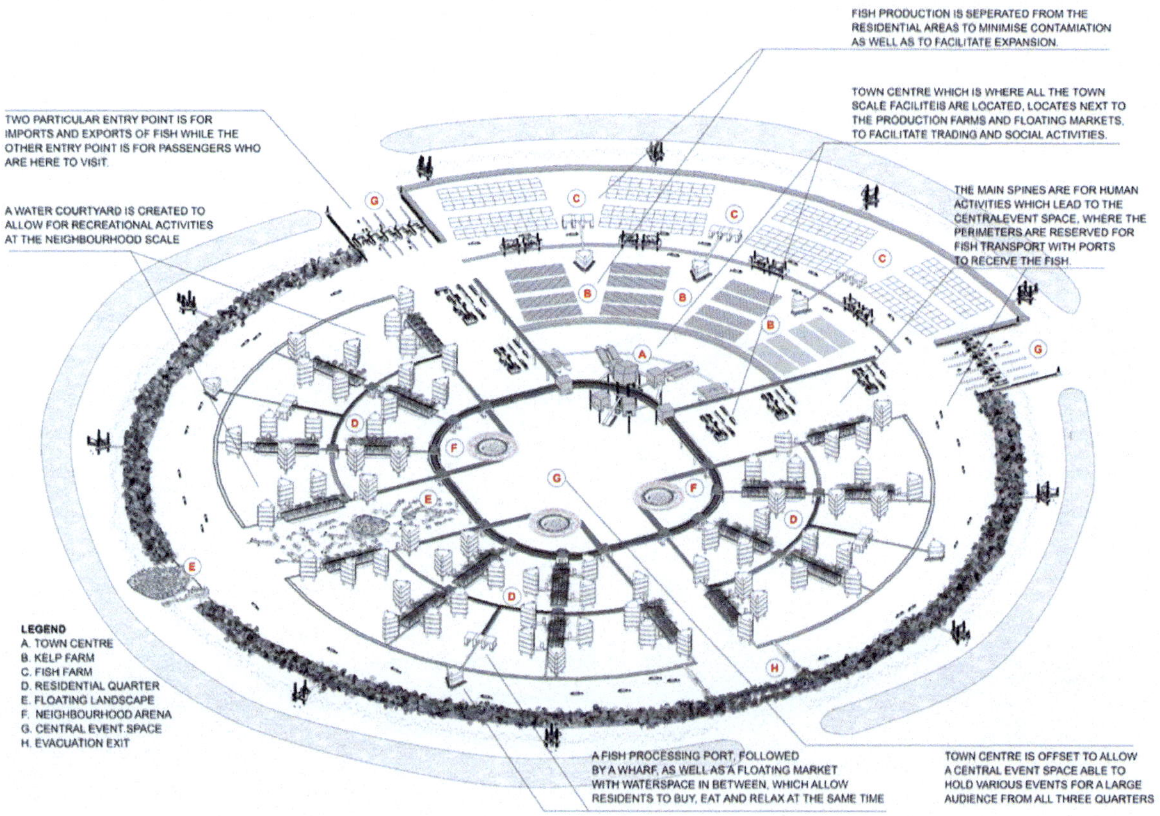

Production Spaces

Fish & Vegetable Farm
Wave Energy Generator

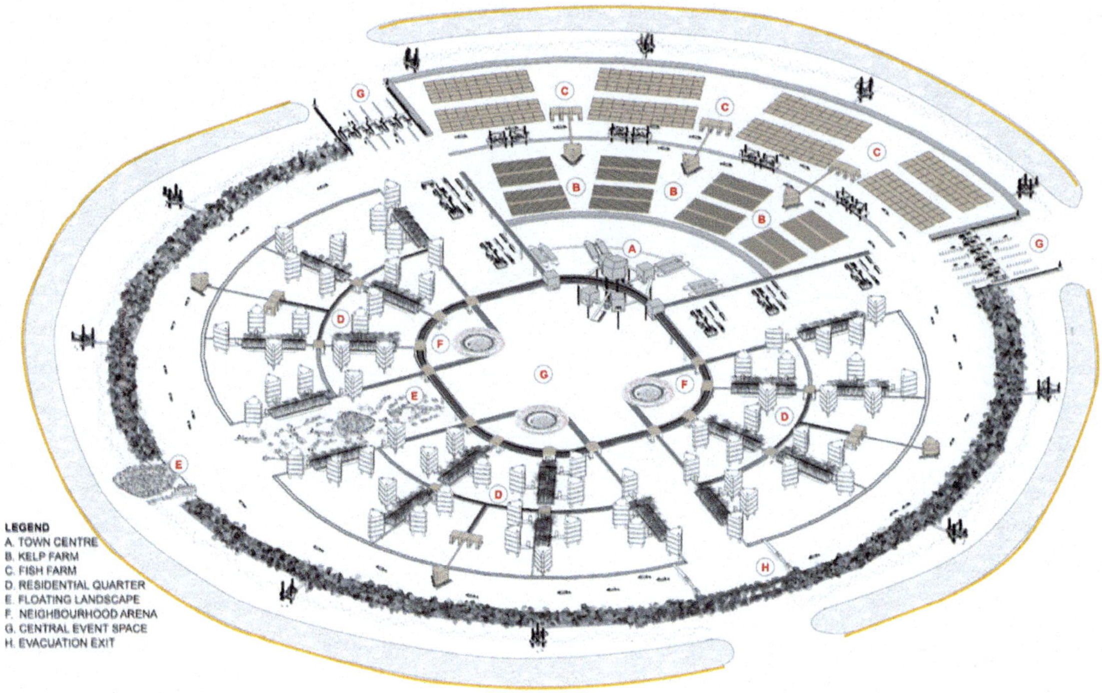

Waterspaces

Total Waterscape Area: 9,234,789 sqm
Percentage of Waterspace Area: 25%

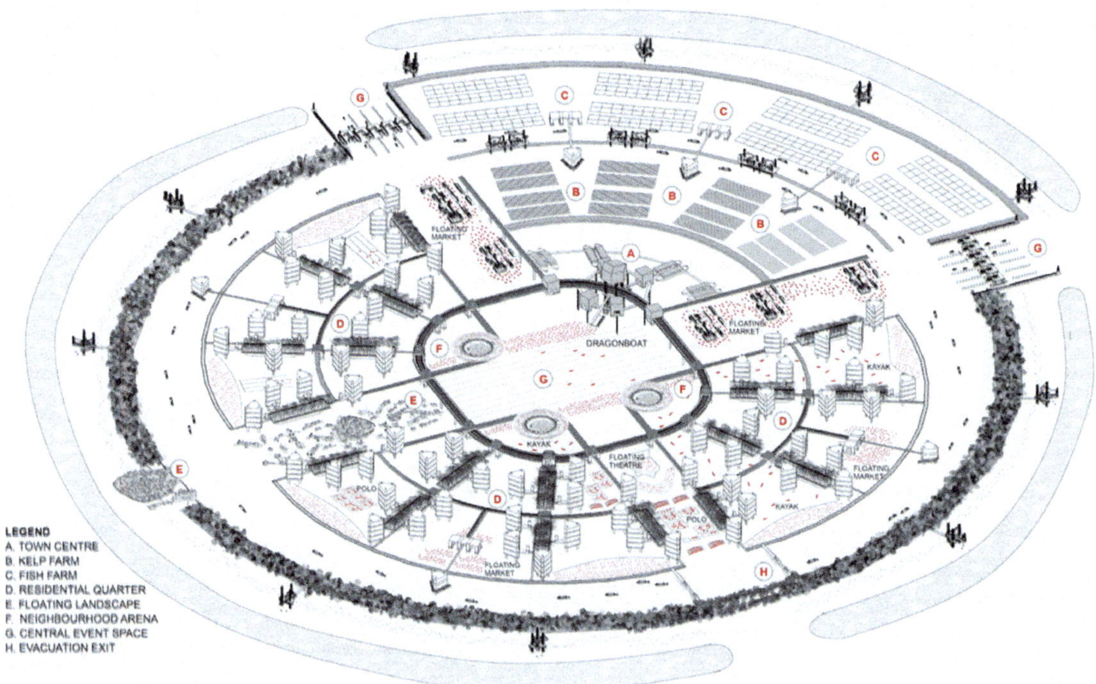

Concentric - Circulation Routes

Water Taxi Ways: 8401233 sqm
Water Taxi Stops: 55
Market Bridge and Walkways: 76800 m

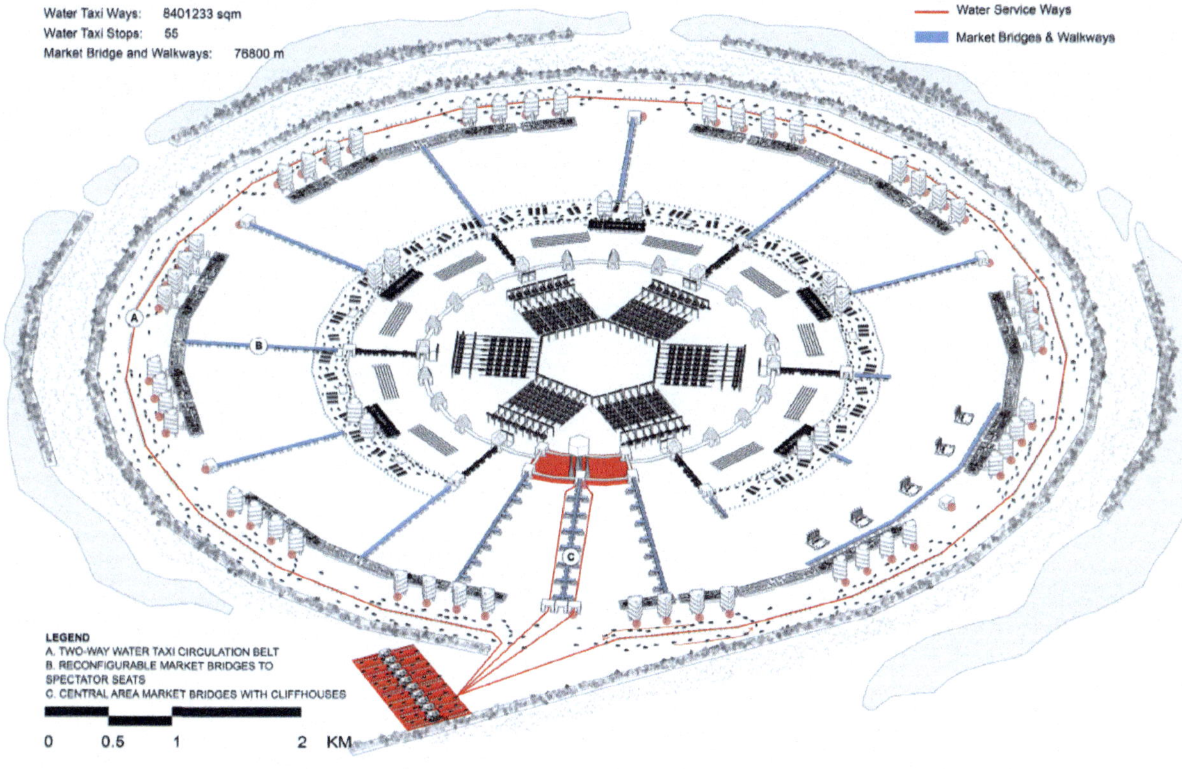

Town Greenery

Total Town Greenery Area: 2,590,716 sqm
% Town Greenery Area: 9.68%

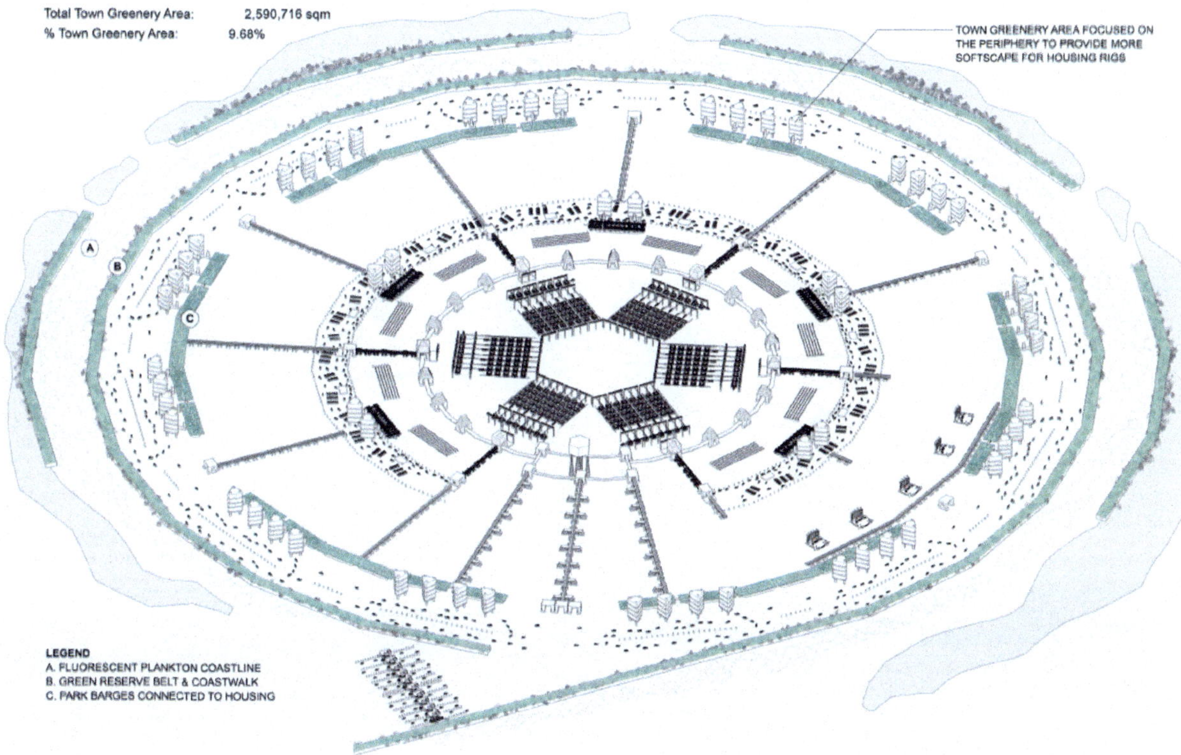

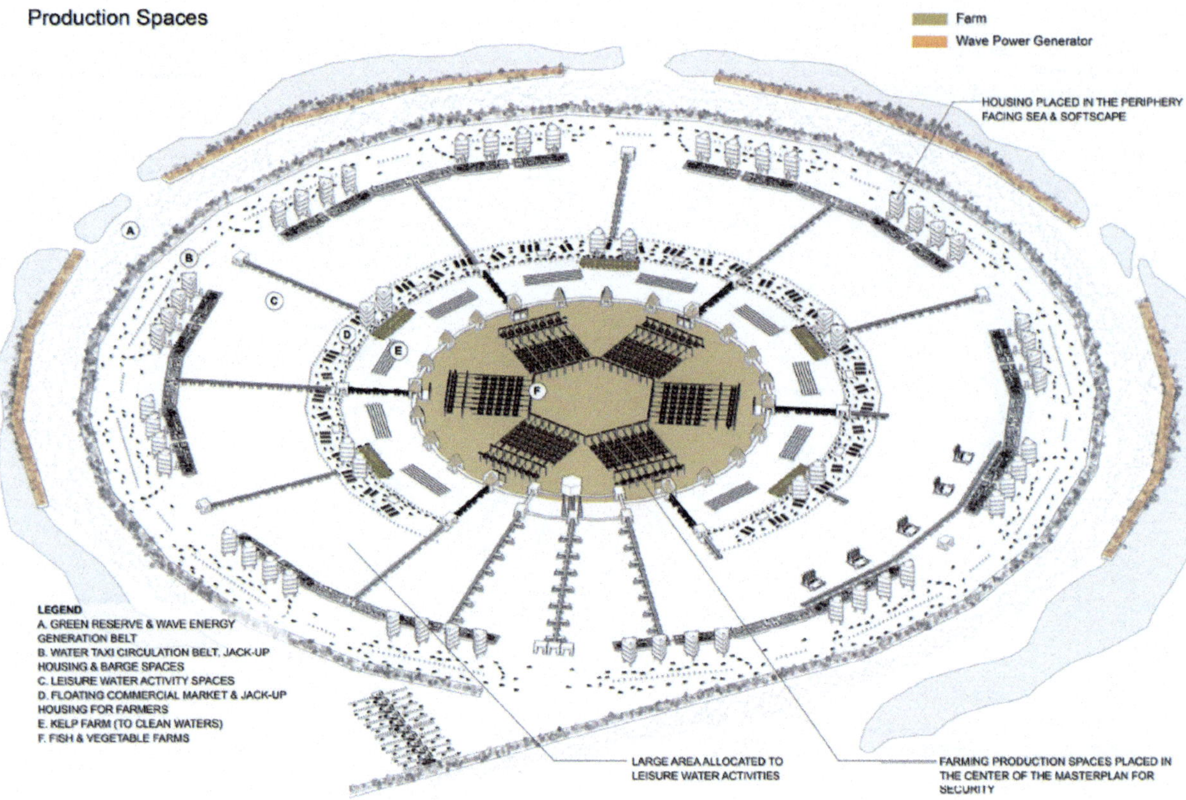
Production Spaces
Farm
Wave Power Generator
HOUSING PLACED IN THE PERIPHERY FACING SEA & SOFTSCAPE
A
B
C
D
E
F
LEGEND
A. GREEN RESERVE & WAVE ENERGY GENERATION BELT
B. WATER TAXI CIRCULATION BELT, JACK-UP HOUSING & BARGE SPACES
C. LEISURE WATER ACTIVITY SPACES
D. FLOATING COMMERCIAL MARKET & JACK-UP HOUSING FOR FARMERS
E. KELP FARM (TO CLEAN WATERS)
F. FISH & VEGETABLE FARMS
LARGE AREA ALLOCATED TO LEISURE WATER ACTIVITIES
FARMING PRODUCTION SPACES PLACED IN THE CENTER OF THE MASTERPLAN FOR SECURITY

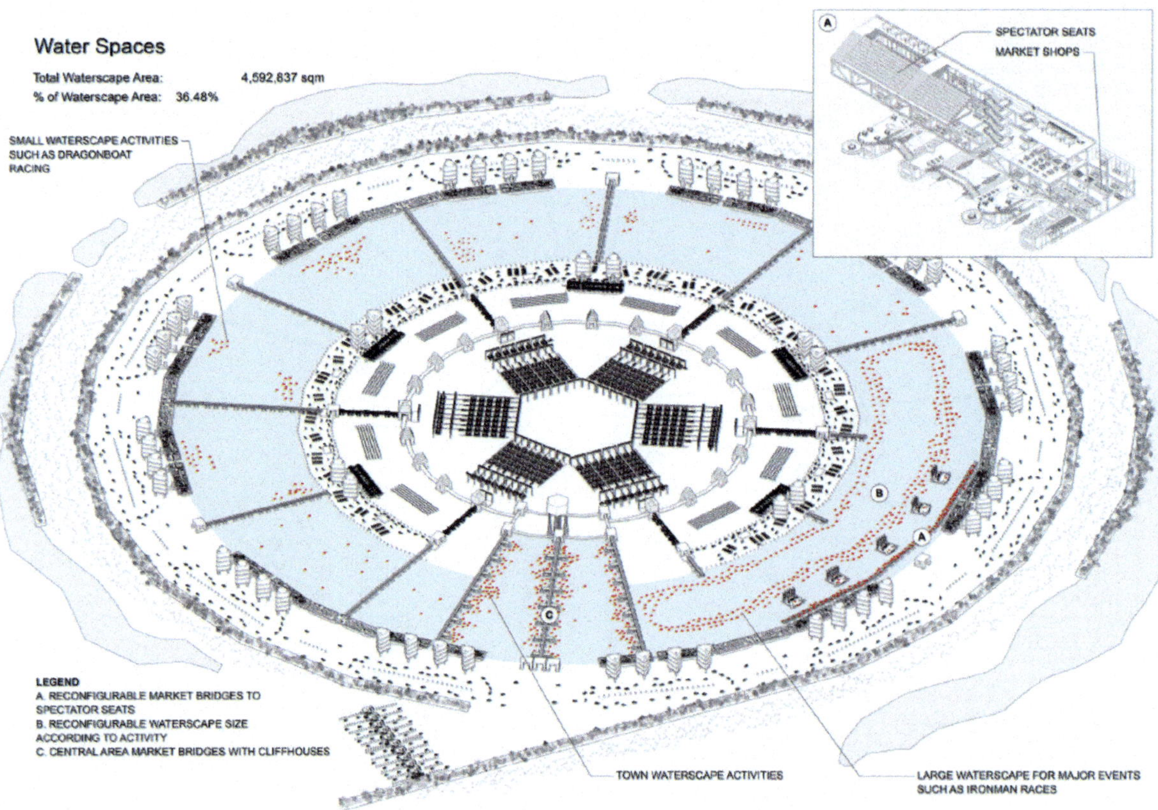
Water Spaces
Total Waterscape Area: 4,592,837 sqm
% of Waterscape Area: 36.48%
SMALL WATERSCAPE ACTIVITIES SUCH AS DRAGONBOAT RACING
A
SPECTATOR SEATS
MARKET SHOPS
B
A
C
LEGEND
A. RECONFIGURABLE MARKET BRIDGES TO SPECTATOR SEATS
B. RECONFIGURABLE WATERSCAPE SIZE ACCORDING TO ACTIVITY
C. CENTRAL AREA MARKET BRIDGES WITH CLIFFHOUSES
TOWN WATERSCAPE ACTIVITIES
LARGE WATERSCAPE FOR MAJOR EVENTS SUCH AS IRONMAN RACES

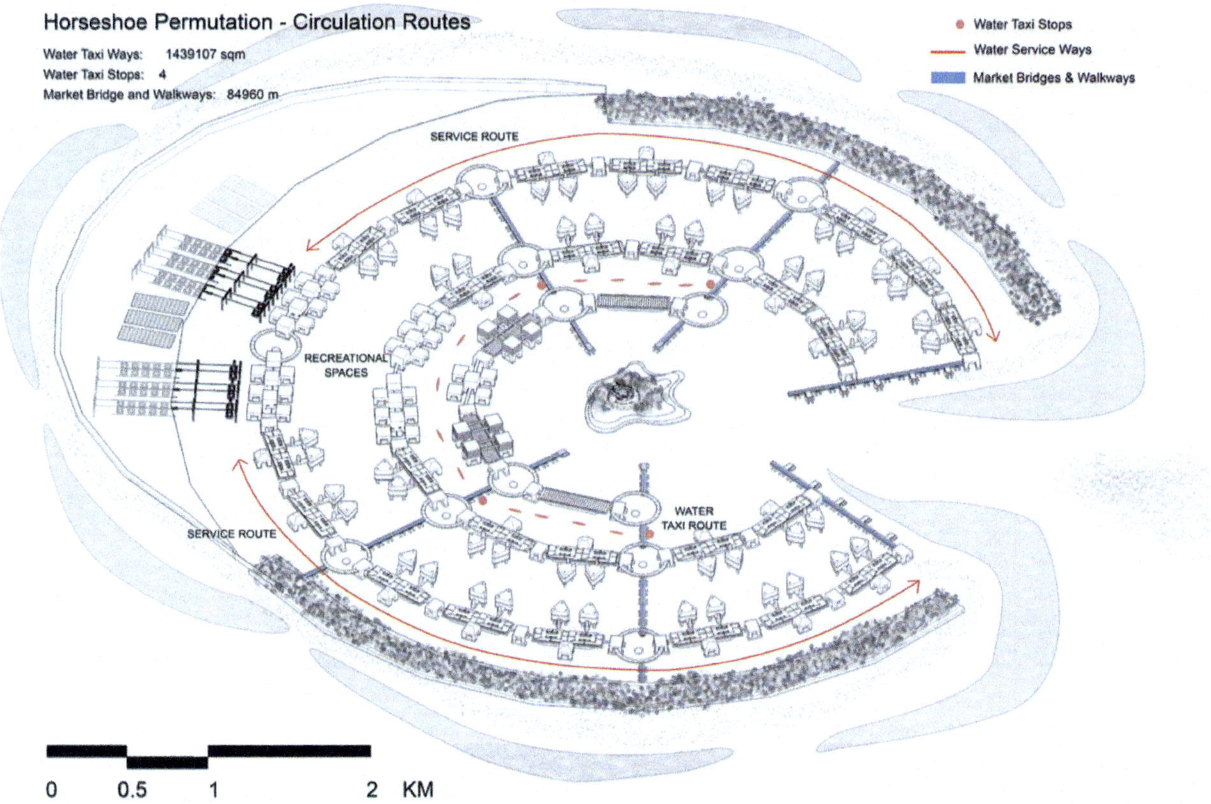
Horseshoe Permutation - Circulation Routes
Water Taxi Ways: 1439107 sqm
Water Taxi Stops: 4
Market Bridge and Walkways: 84960 m
Water Taxi Stops
Water Service Ways
Market Bridges & Walkways
SERVICE ROUTE
RECREATIONAL SPACES
WATER TAXI ROUTE
SERVICE ROUTE
0 0.5 1 2 KM

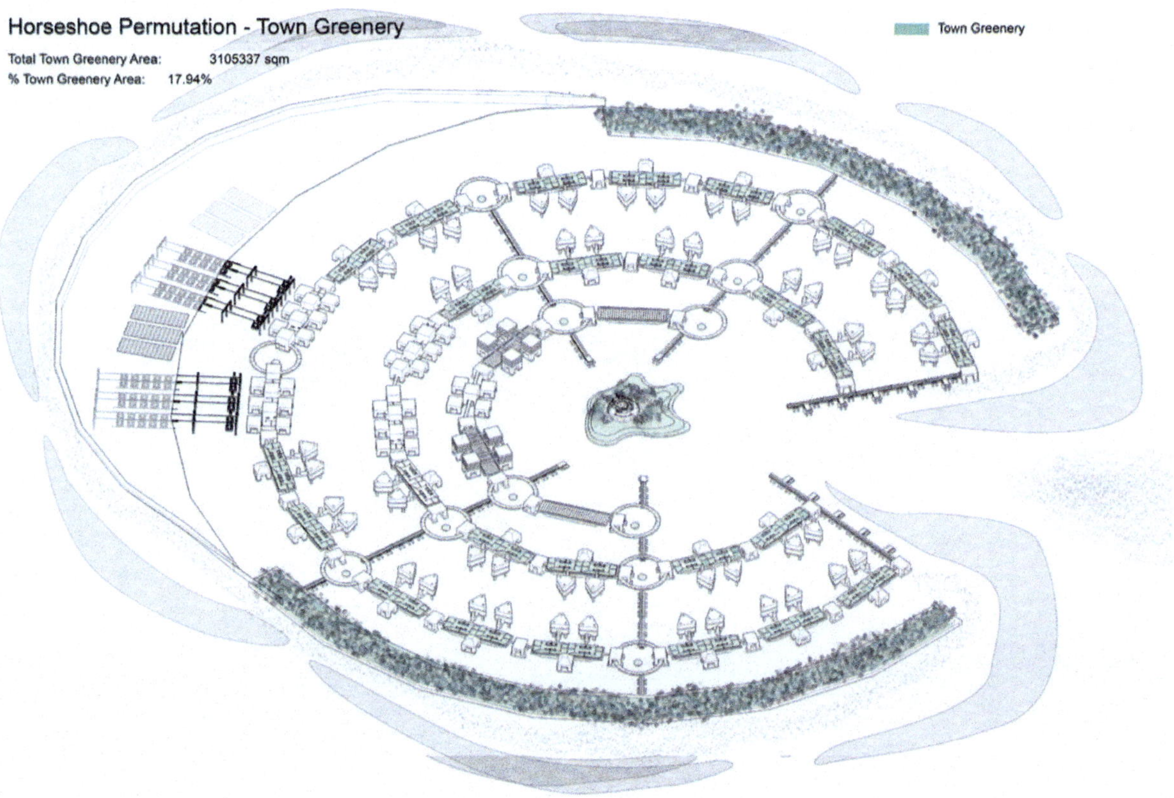
Horseshoe Permutation - Town Greenery
Total Town Greenery Area: 3105337 sqm
% Town Greenery Area: 17.94%
Town Greenery

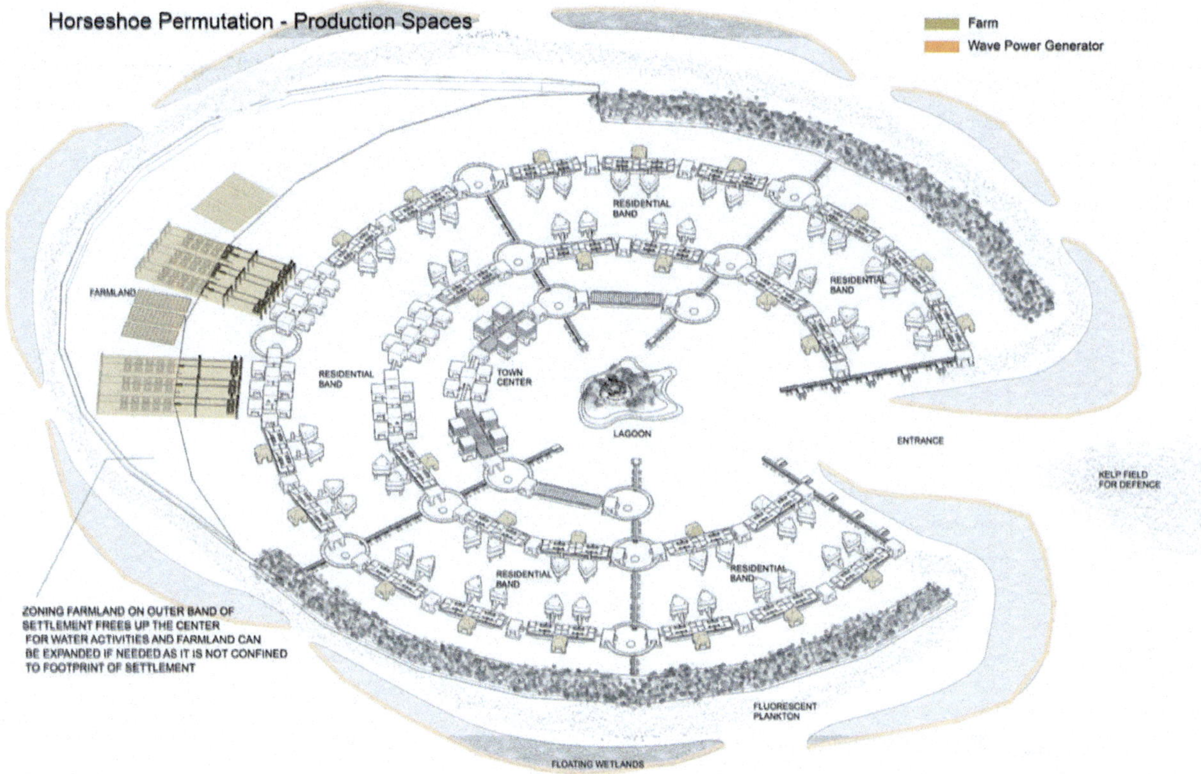
Horseshoe Permutation - Production Spaces
Farm
Wave Power Generator
RESIDENTIAL BAND
RESIDENTIAL BAND
FARMLAND
RESIDENTIAL BAND
TOWN CENTER
LAGOON
ENTRANCE
KELP FIELD FOR DEFENCE
RESIDENTIAL BAND
RESIDENTIAL BAND
ZONING FARMLAND ON OUTER BAND OF SETTLEMENT FREES UP THE CENTER FOR WATER ACTIVITIES AND FARMLAND CAN BE EXPANDED IF NEEDED AS IT IS NOT CONFINED TO FOOTPRINT OF SETTLEMENT
FLUORESCENT PLANKTON
FLOATING WETLANDS

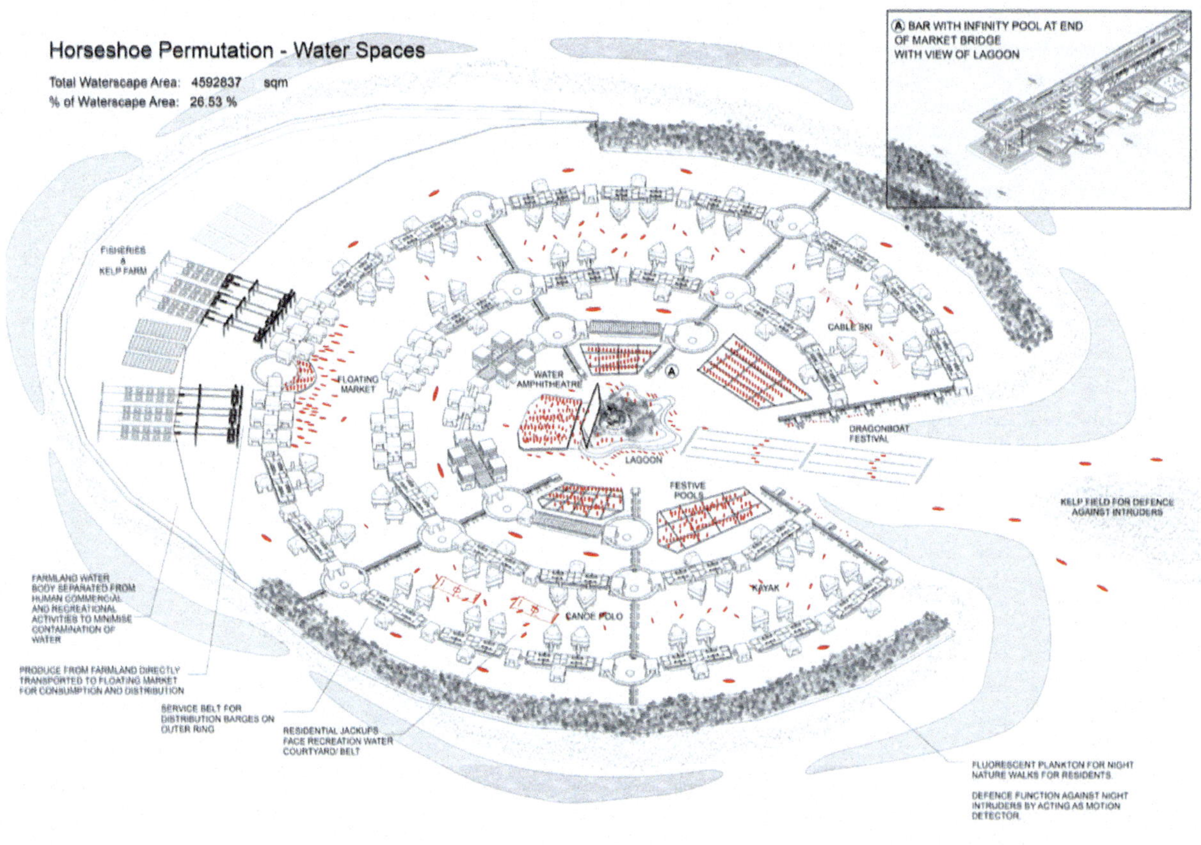
Horseshoe Permutation - Water Spaces
Total Waterscape Area: 4592837 sqm
% of Waterscape Area: 26.53 %
Ⓐ BAR WITH INFINITY POOL AT END OF MARKET BRIDGE WITH VIEW OF LAGOON
FISHERIES & KELP FARM
CABLE SKI
FLOATING MARKET
WATER AMPHITHEATRE
Ⓐ
DRAGONBOAT FESTIVAL
LAGOON
FESTIVE POOLS
KELP FIELD FOR DEFENCE AGAINST INTRUDERS
KAYAK
CANOE POLO
FARMLAND WATER BODY SEPARATED FROM HUMAN COMMERCIAL AND RECREATIONAL ACTIVITIES TO MINIMISE CONTAMINATION OF WATER
PRODUCE FROM FARMLAND DIRECTLY TRANSPORTED TO FLOATING MARKET FOR CONSUMPTION AND DISTRIBUTION
SERVICE BELT FOR DISTRIBUTION BARGES ON OUTER RING
RESIDENTIAL JACKUPS FACE RECREATION WATER COURTYARD/ BELT
FLUORESCENT PLANKTON FOR NIGHT NATURE WALKS FOR RESIDENTS.
DEFENCE FUNCTION AGAINST NIGHT INTRUDERS BY ACTING AS MOTION DETECTOR.

Sea Star Permutation - Circulation Routes

Water Taxi Ways:	**1 115 440 sqm**
Water Taxi Stops:	**11**
Market Bridge and Walkways:	**598 720 m**

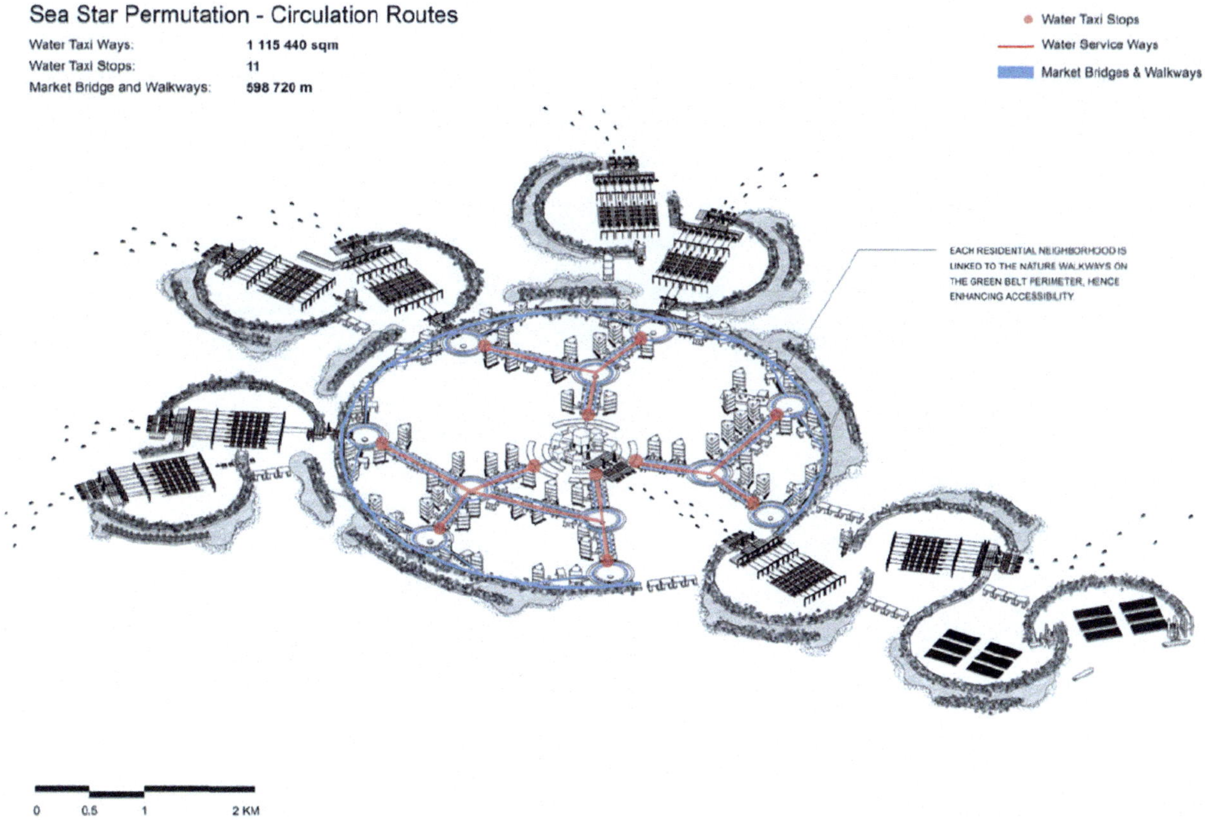

Town Greenery

Total Town Greenery Area:	**4 077 760 sqm**
% of Town Greenery Area:	**16.69 %**

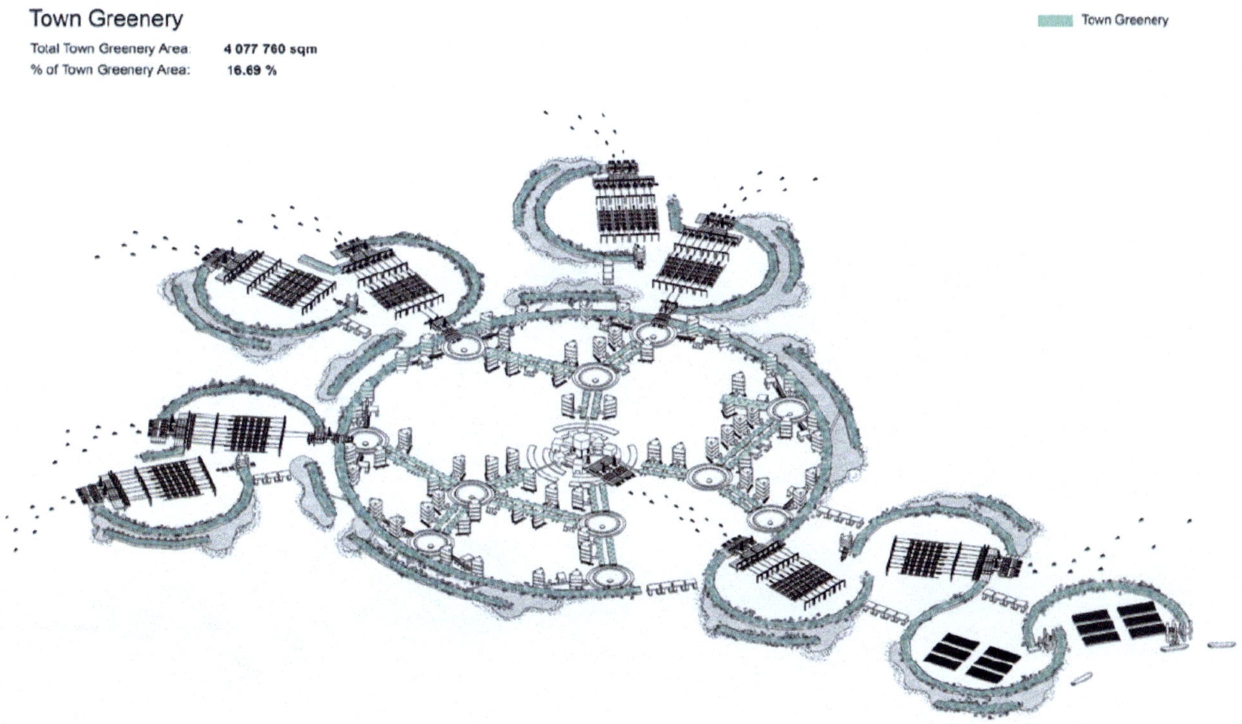

Production Spaces

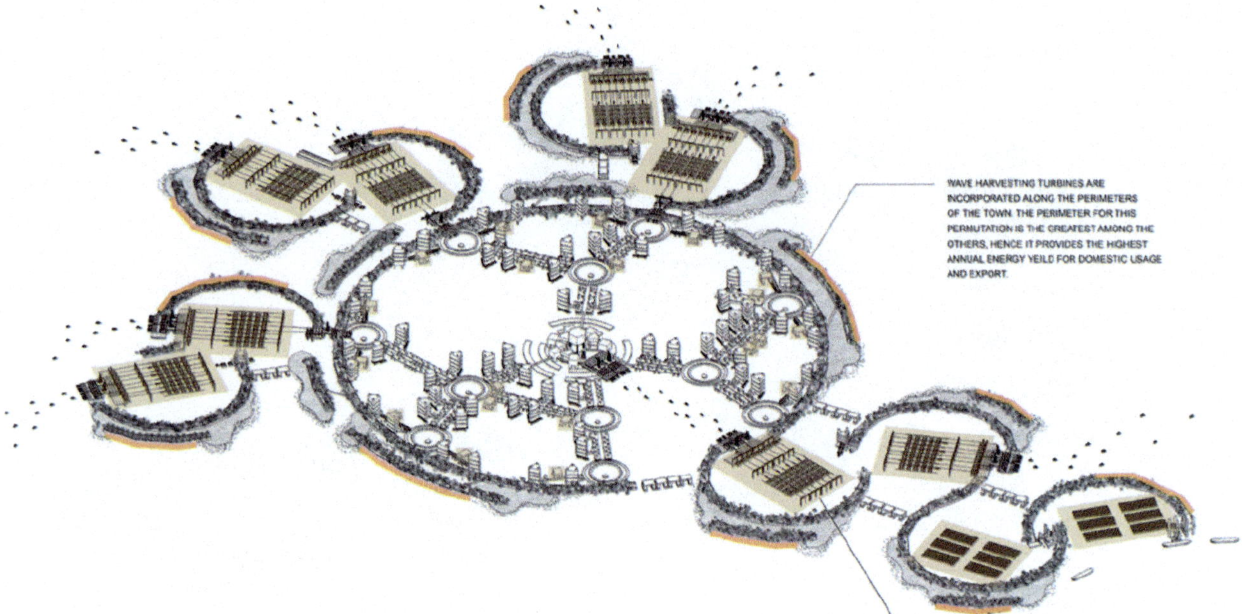

Water Spaces

Total Waterscape Area: **6 247 550 sqm**

% of Waterscape Area: **33.38 %**

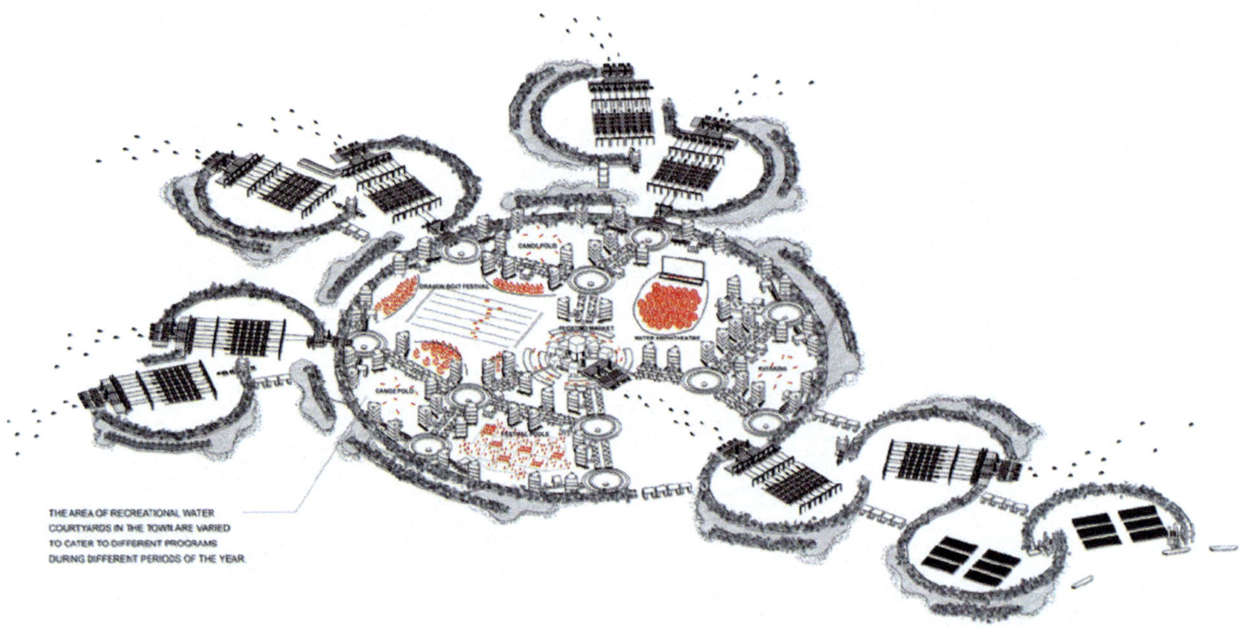

Snowflake Permutation

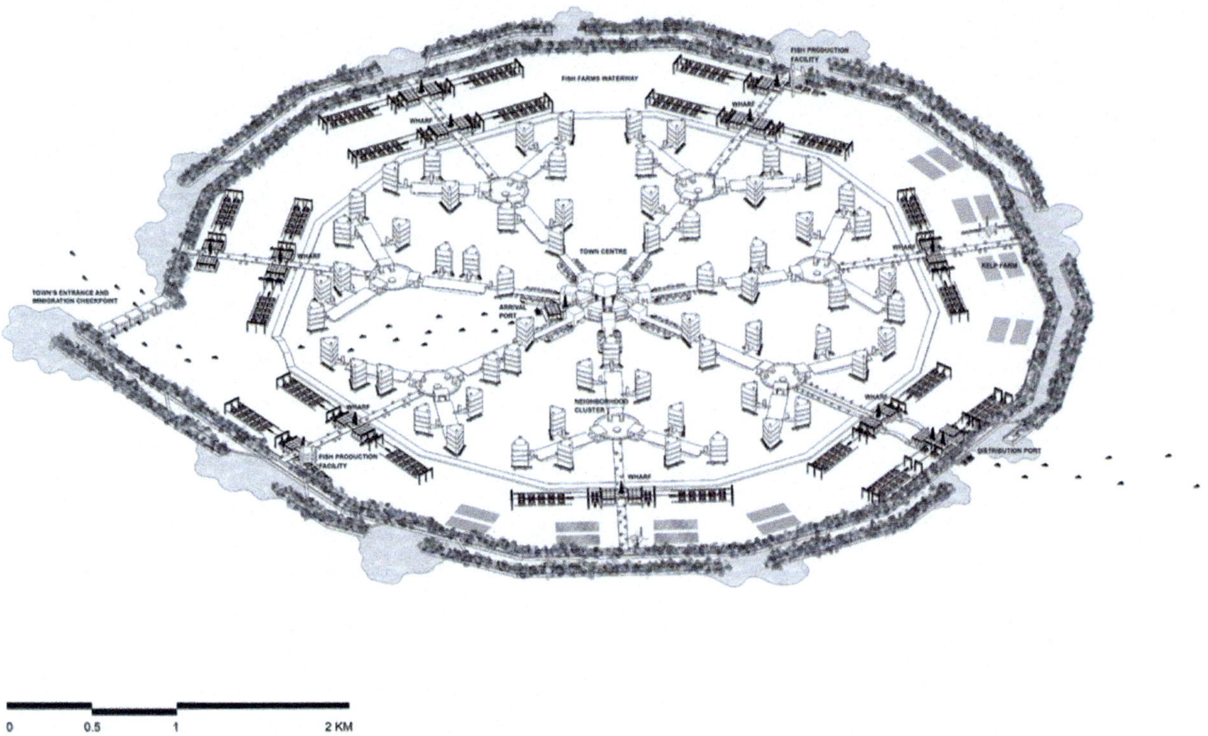

Circulation Routes

Water Taxi Ways:	3 315 540 sqm
Water Taxi Stops:	13
Market Bridge and Walkways:	750 720 m

Water Taxi Stops
Water Service Ways
Market Bridges & Walkways

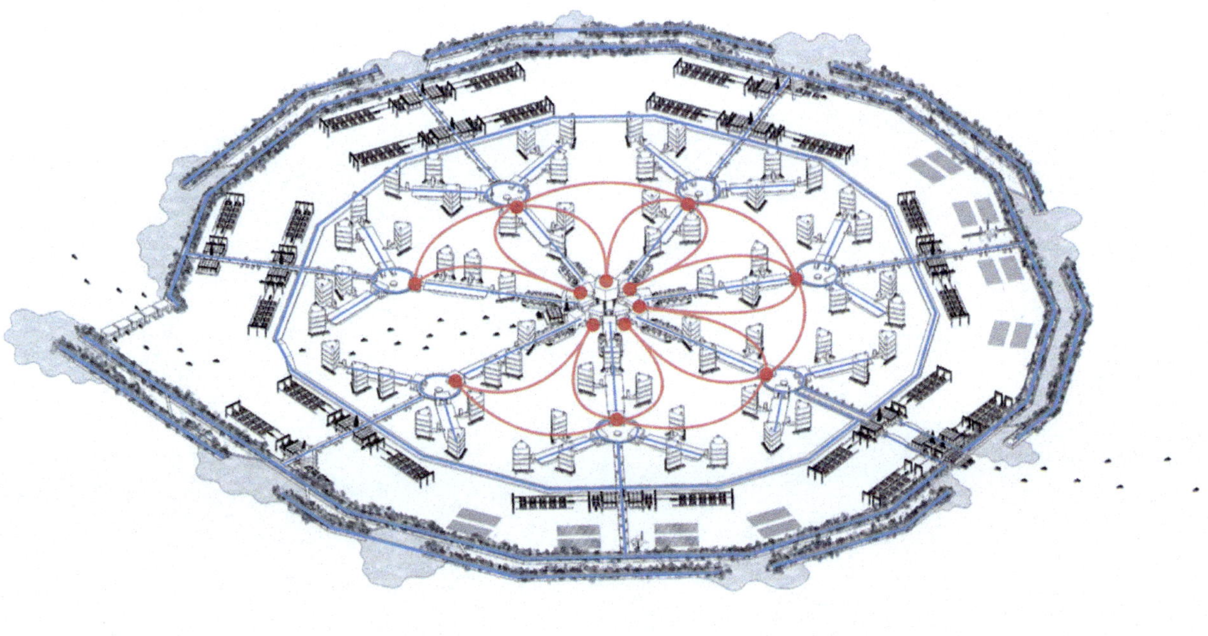

Production Spaces

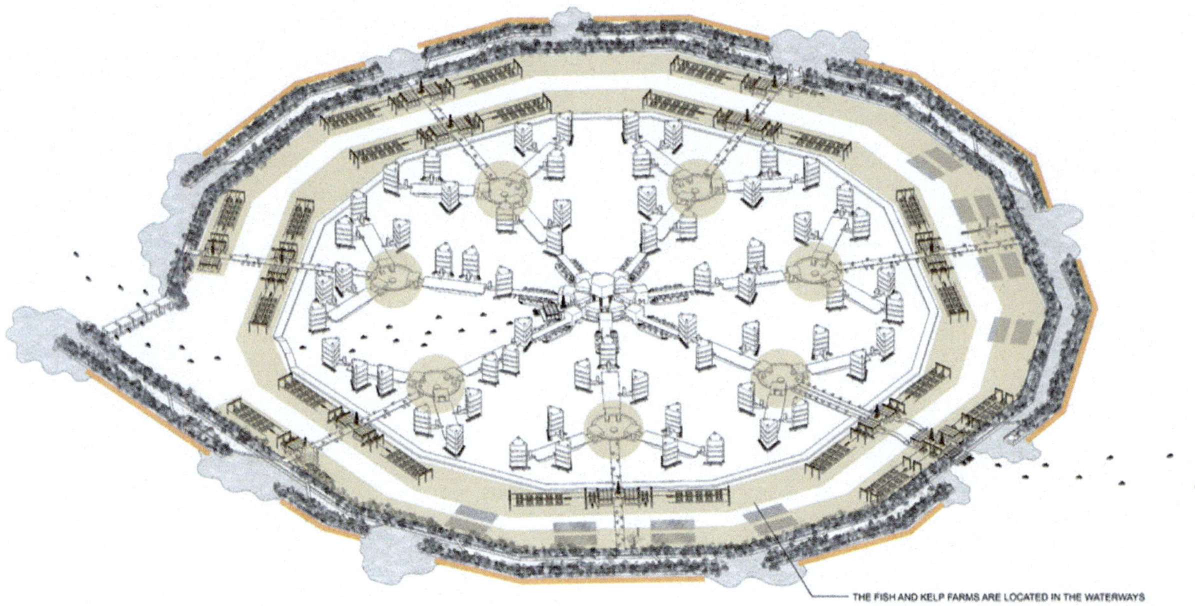

THE FISH AND KELP FARMS ARE LOCATED IN THE WATERWAYS SANDWICHED BETWEEN THE RECREATIONAL WATER COURTYARDS OF THE NEIGHBORHOOD CLUSTERS AND THE EXTERNAL GREEN BELT PERIMETER.

THE FARM WATERWAYS PROTECTED BY THE PERIMETERS AGAINST THE CHOPPY WATERS AND POTENTIAL POLLUTION FROM THE SEA.

Town Greenery

Total Town Greenery Area: **1 668 408 sqm**
% of Town Greenery Area: **12.05 %**

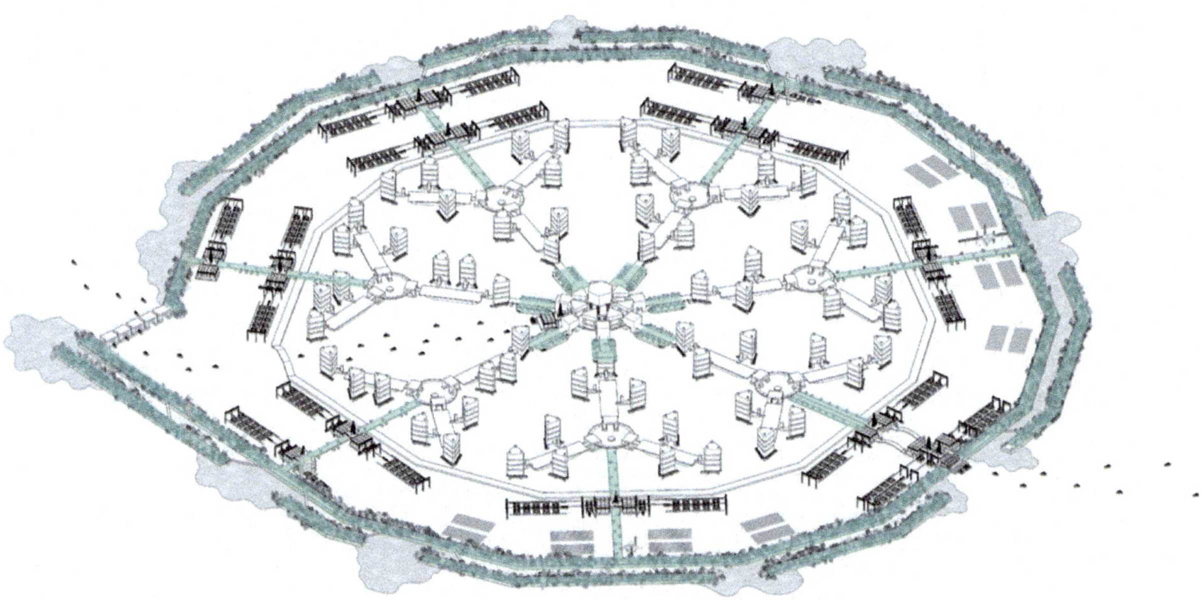

Lagoon Grid Permutation - Overview

Population: 224,000
Enclosed Area: 22783680 sqm
% Foot Print: 15.55%
% Water Body: 28%

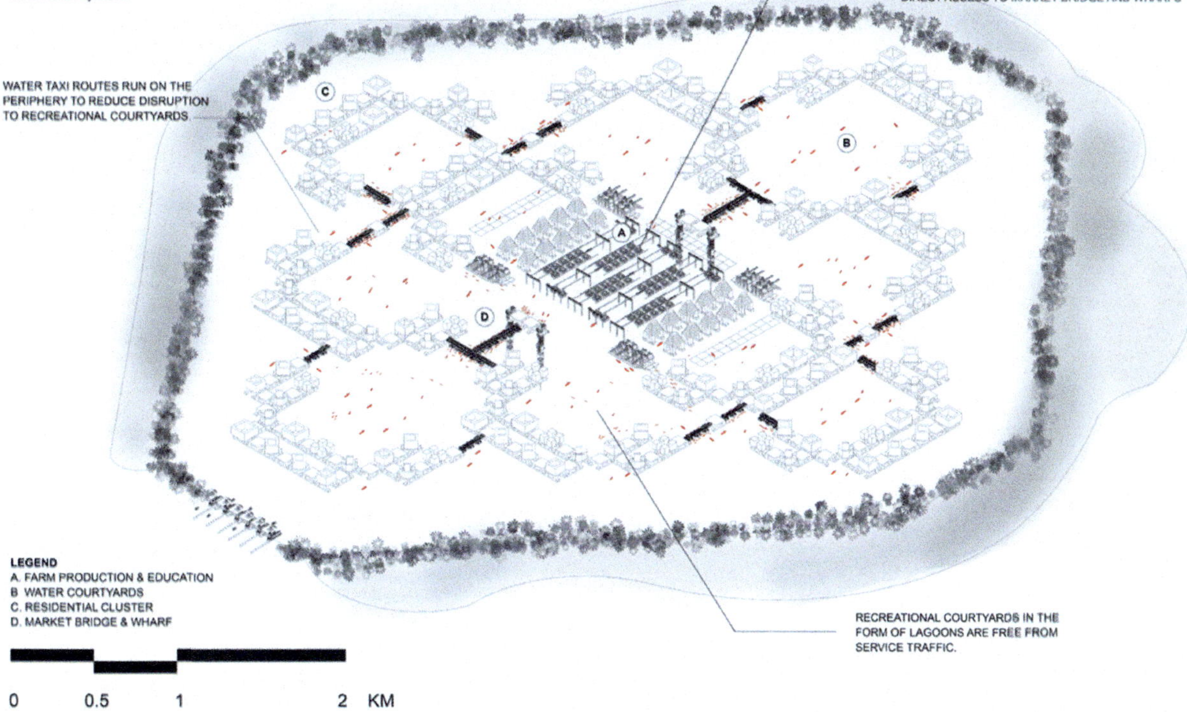

Circulation Routes

Water Taxi Ways: 128952 sqm
Water Taxi Stops: 30
Market Bridge and Walkways: 214168 m

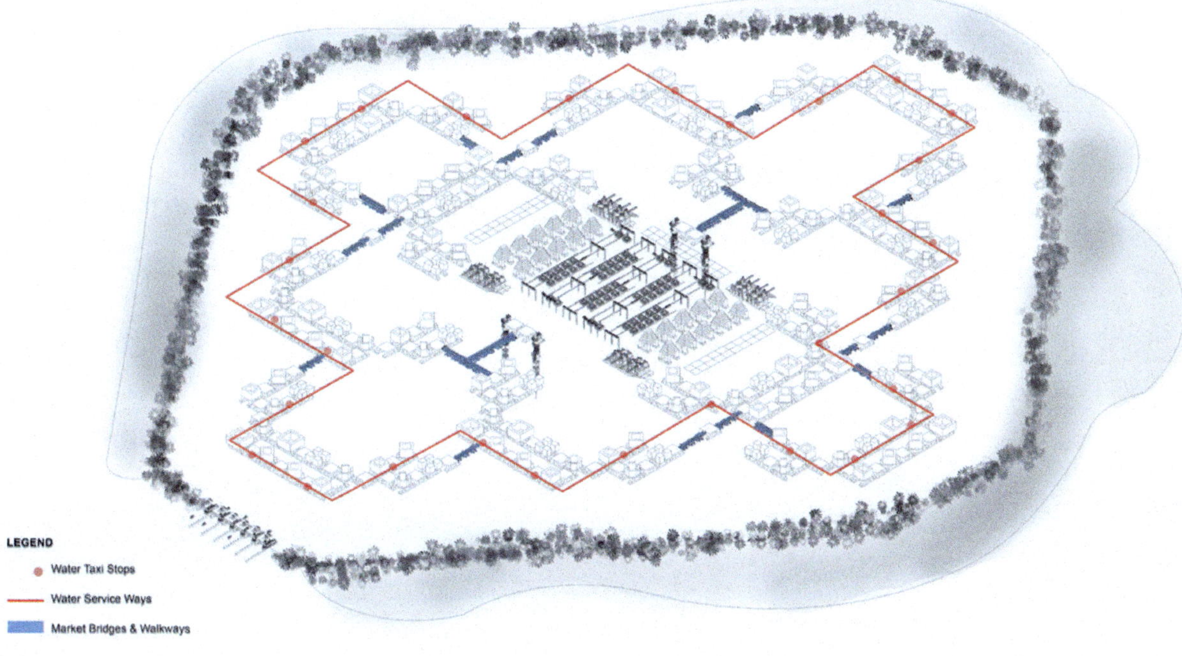

Production Spaces

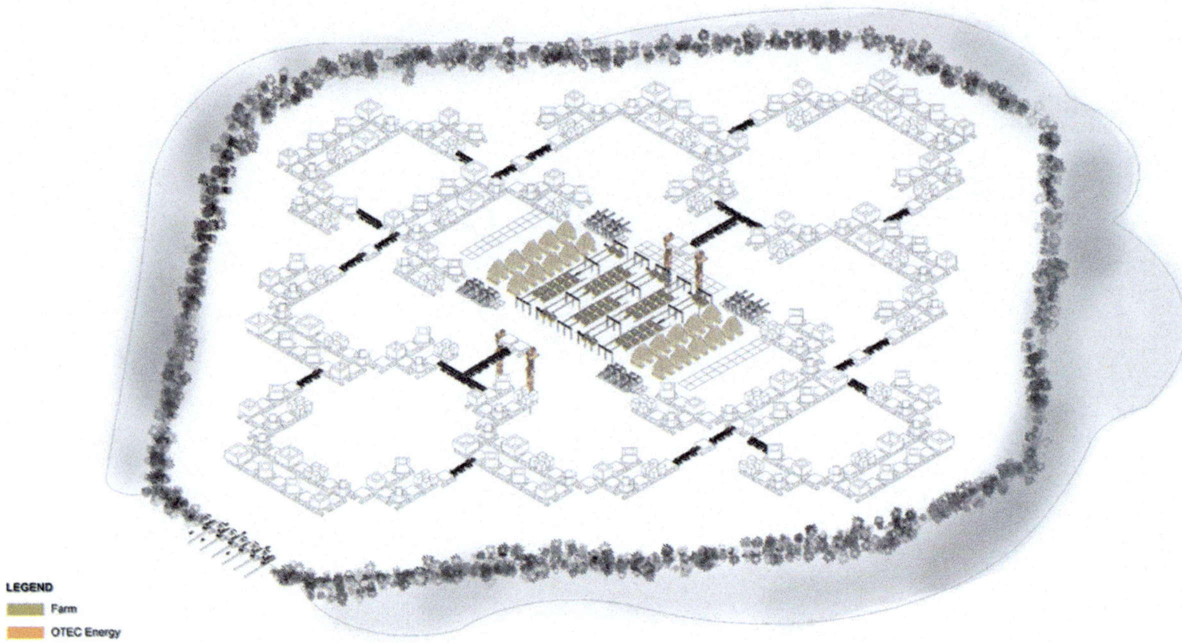

Greenery

Total Town Greenery Area: 7447854 sqm

% Town Greenery Area: 56.42%

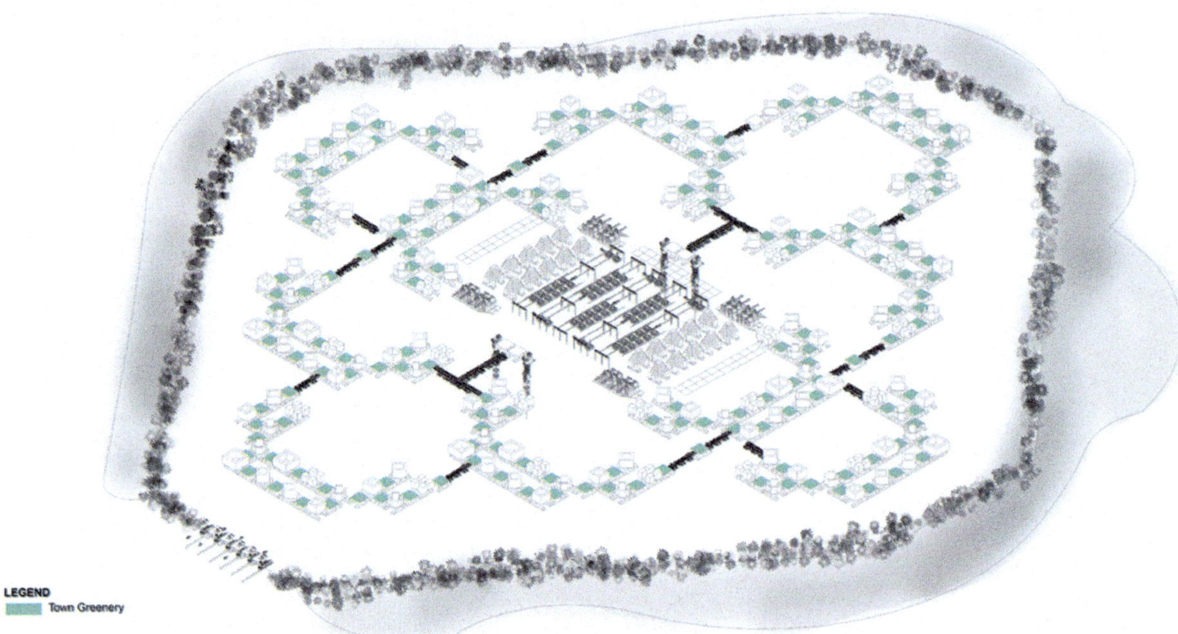

L-Shapes Permutation - Circulation Routes

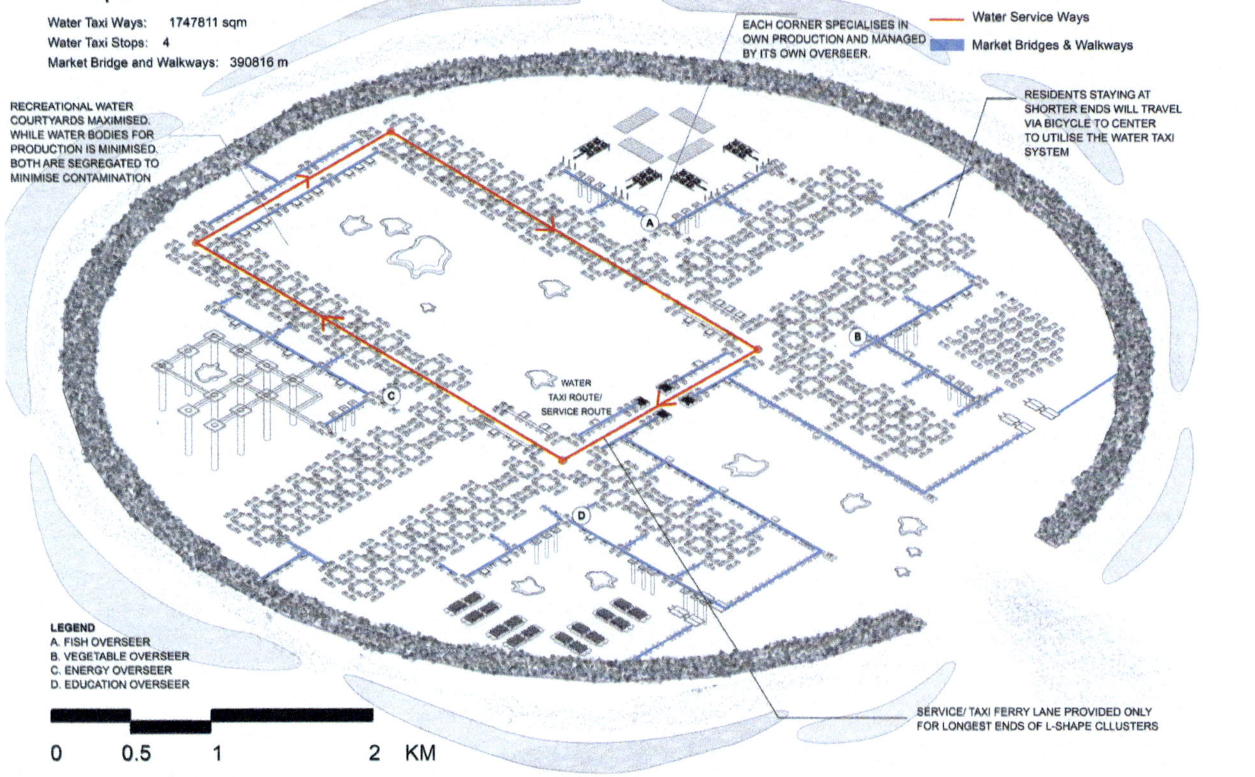

L-Shaped corners Permutation - Town Greenery

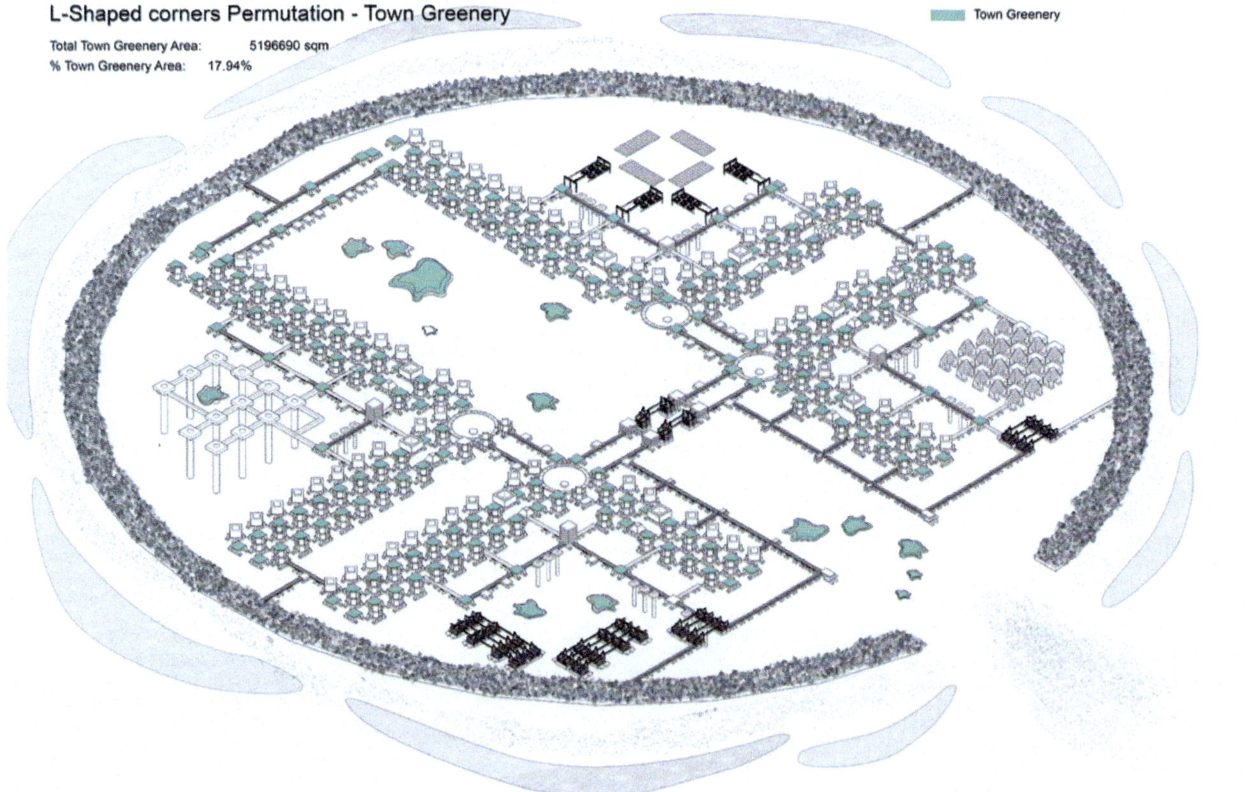

L-Shaped Permutation - Production Spaces

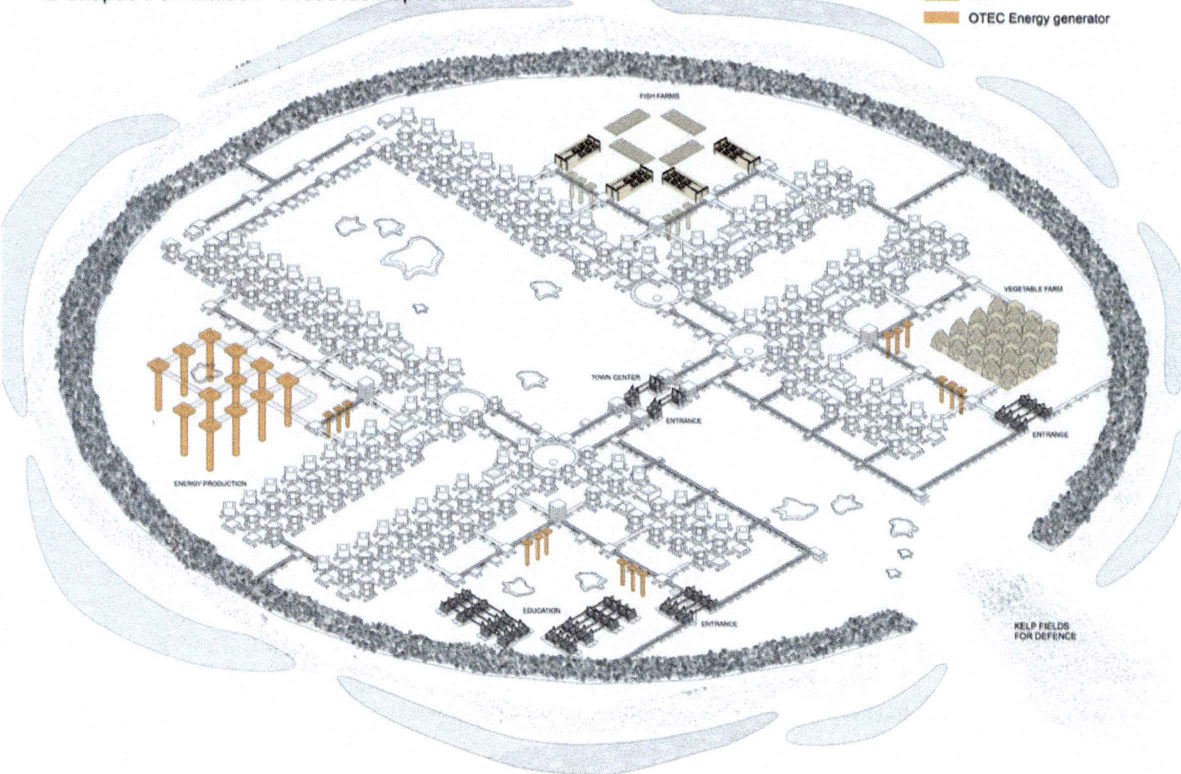

L-Shape Permutation - Water Spaces

Total Waterscape Area: 14986196 sqm

% of Waterscape Area: 48.40 %

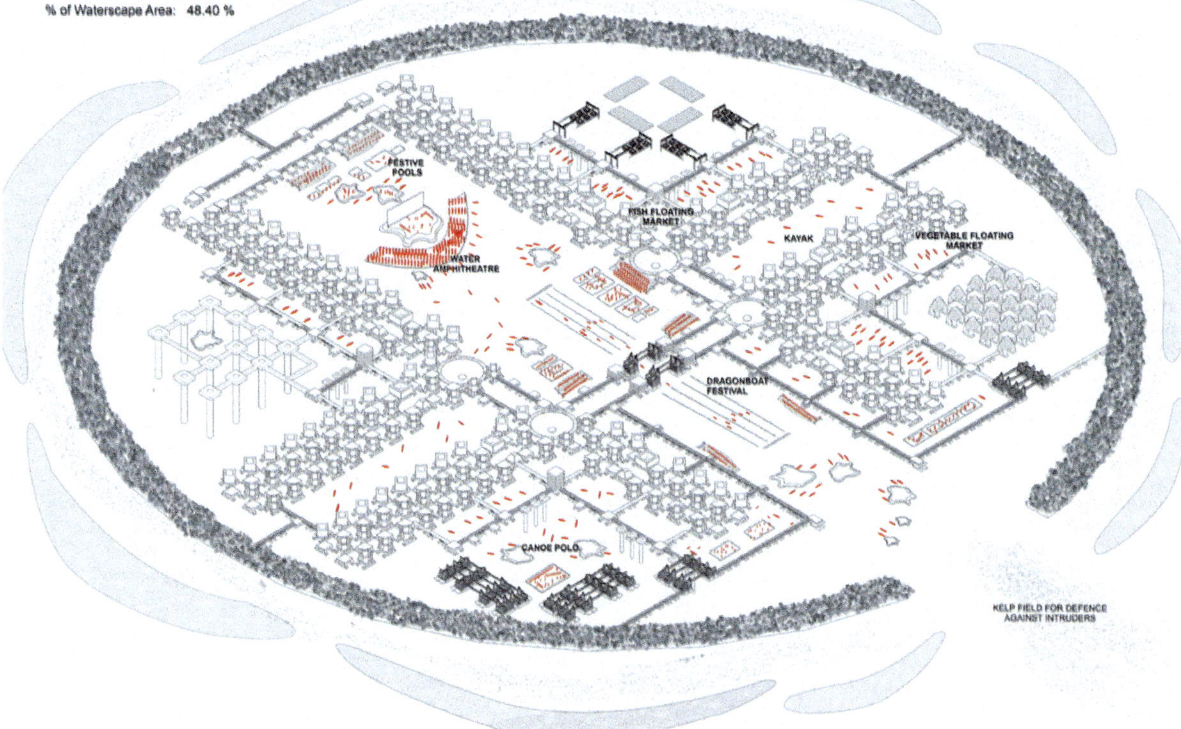

Island Permutation

Production in this masterplan can expand due to the design decision to allocate one quarter of the town plan for it. Expansion might happen to cater to growing needs of people; alternatively, it can also provide exports for other towns.

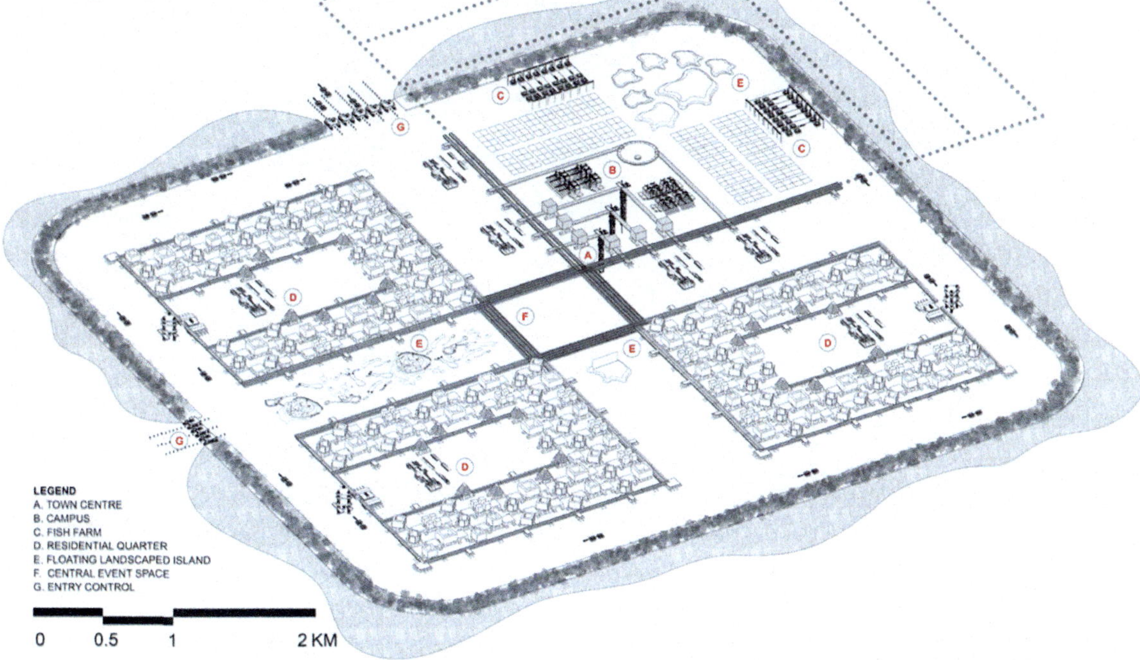

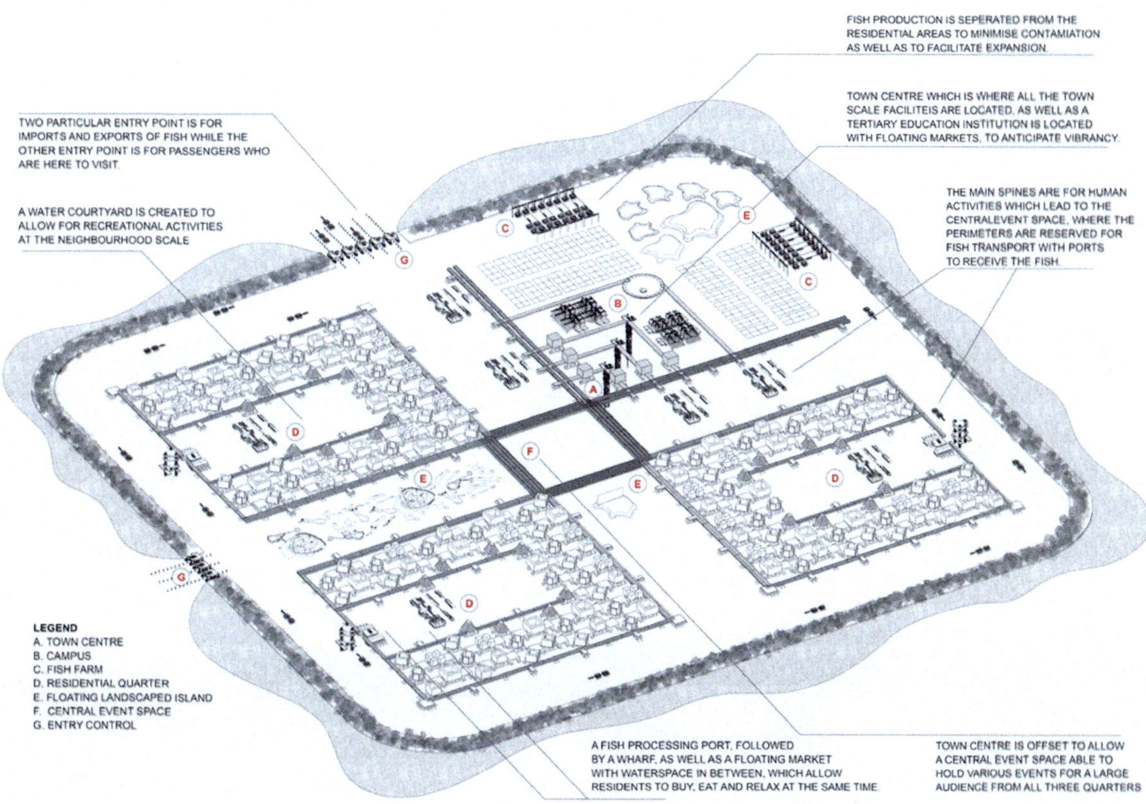

Production Spaces

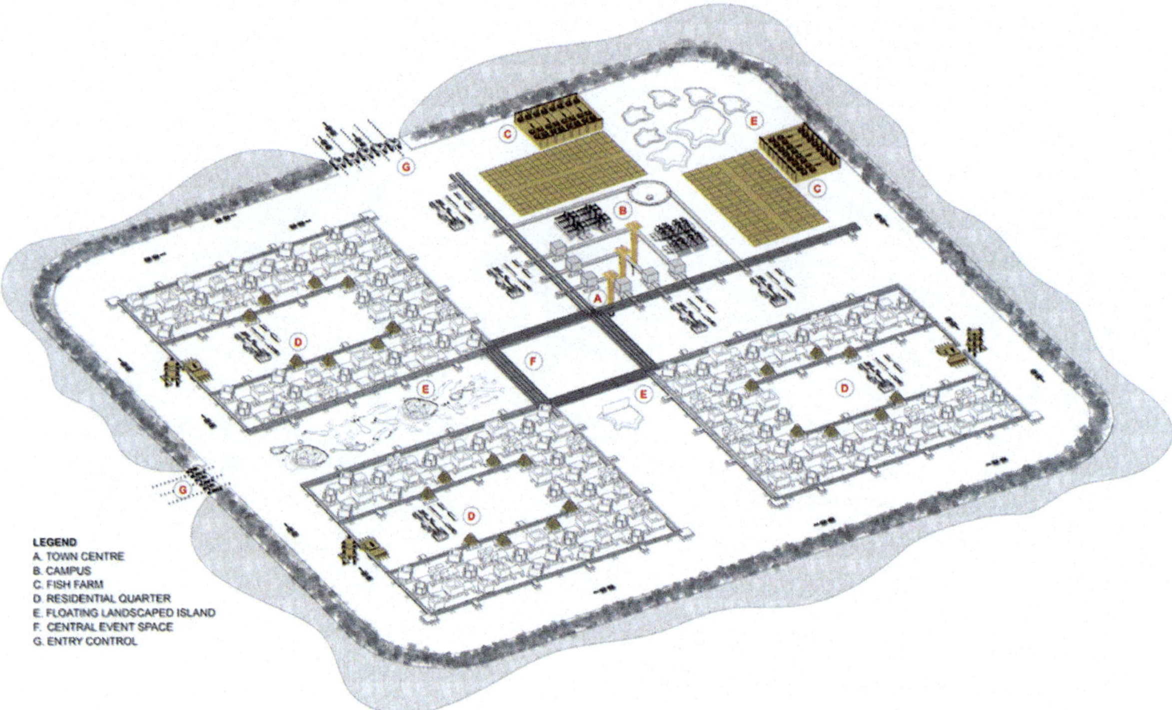

Waterspaces

Total Waterscape Area: 7,057,455 sqm

Percentage of Waterspace Area: 23.7%

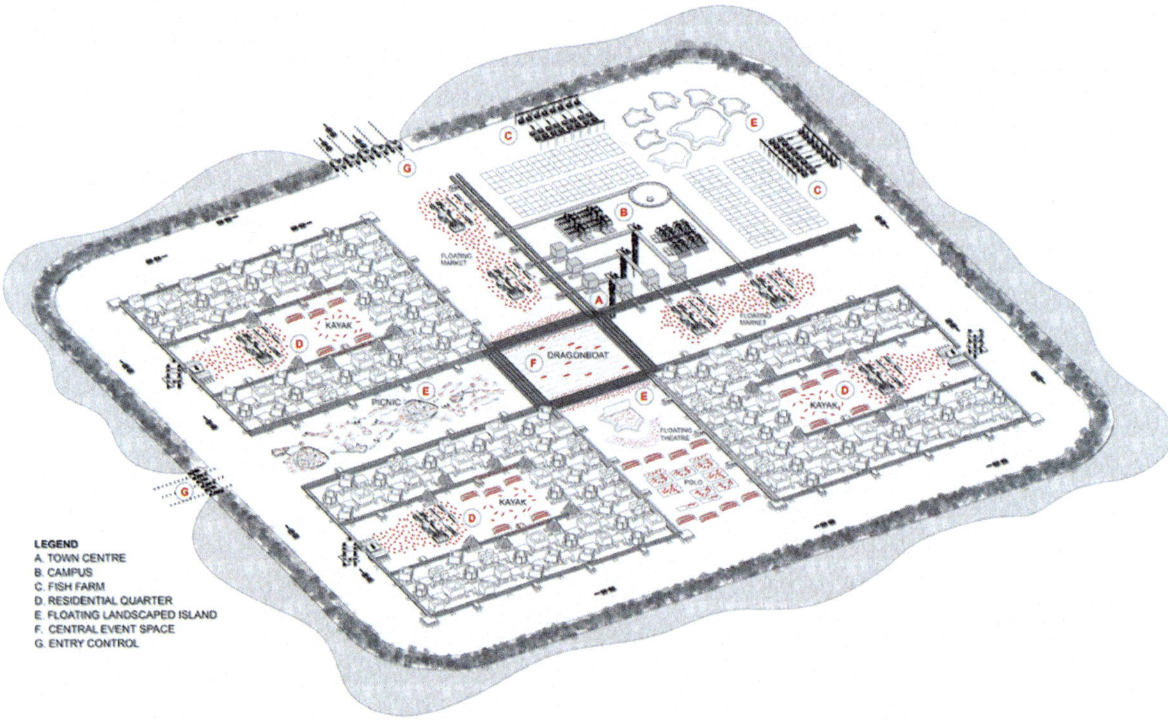

Concentric Square Permutation - Circulation Routes

Water Taxi Ways: 328,000 sqm
Water Taxi Stops: 8
Market Bridge and Walkways: 5,800 m

Water Taxi Stops
Water Service Ways
Market Bridges & Walkways

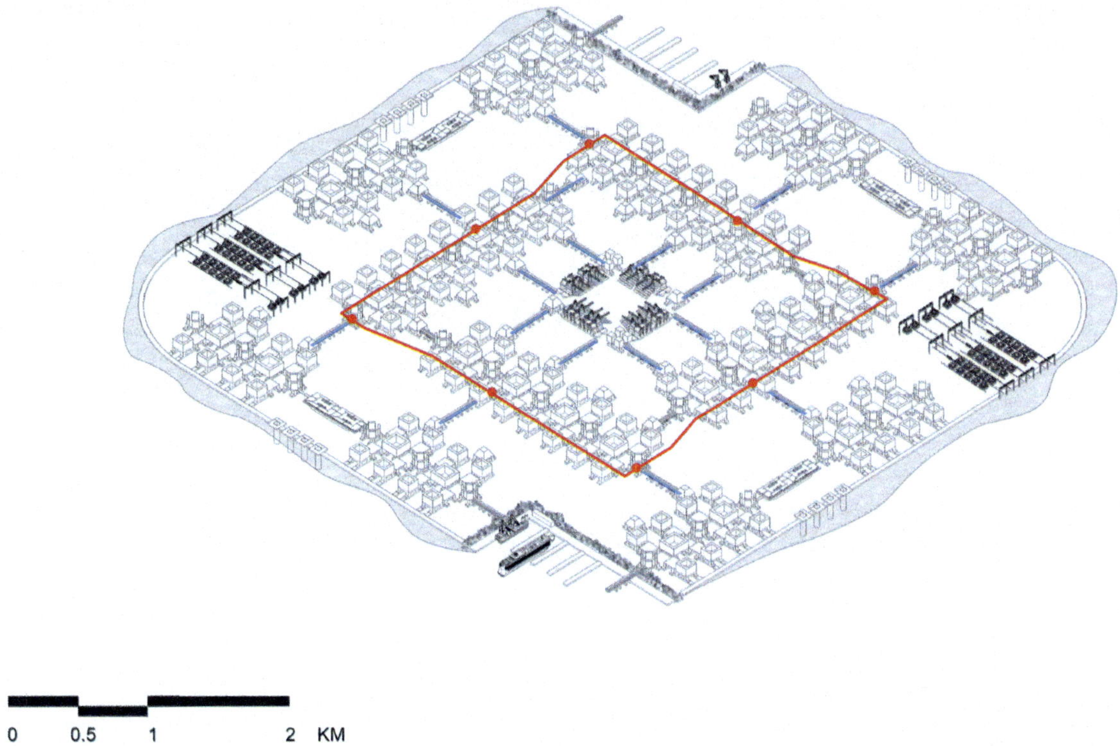

Town Greenery

Total Town Greenery Area: 526,150 sqm
% Town Greenery Area: 4 %

Town Greenery

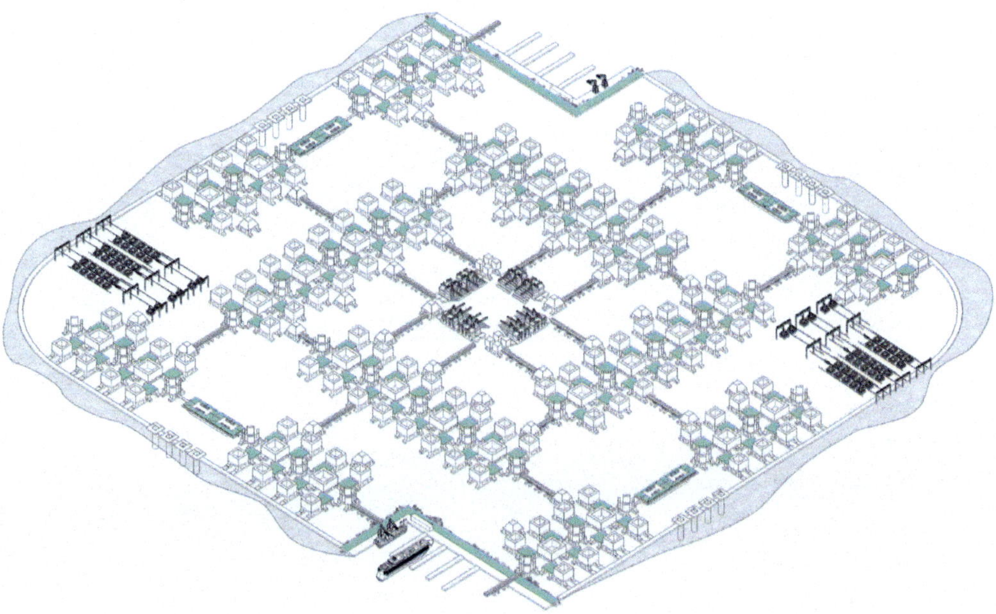

Production Spaces

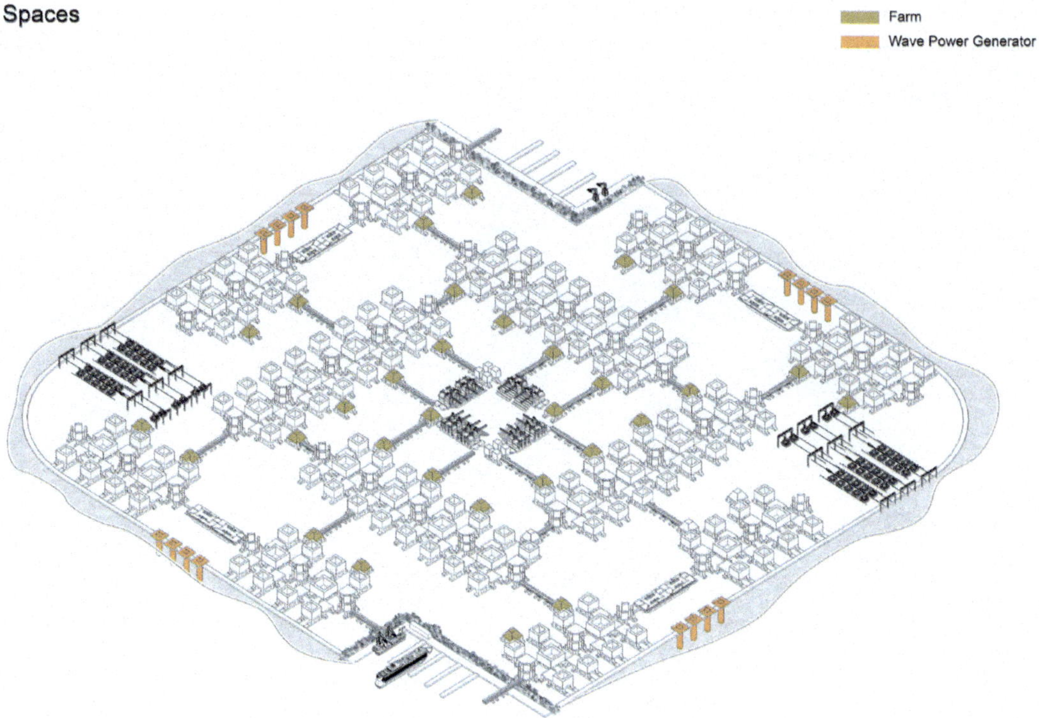

Water Spaces

Total Waterscape Area: 5,304,024 sqm
% of Waterscape Area: 36 %

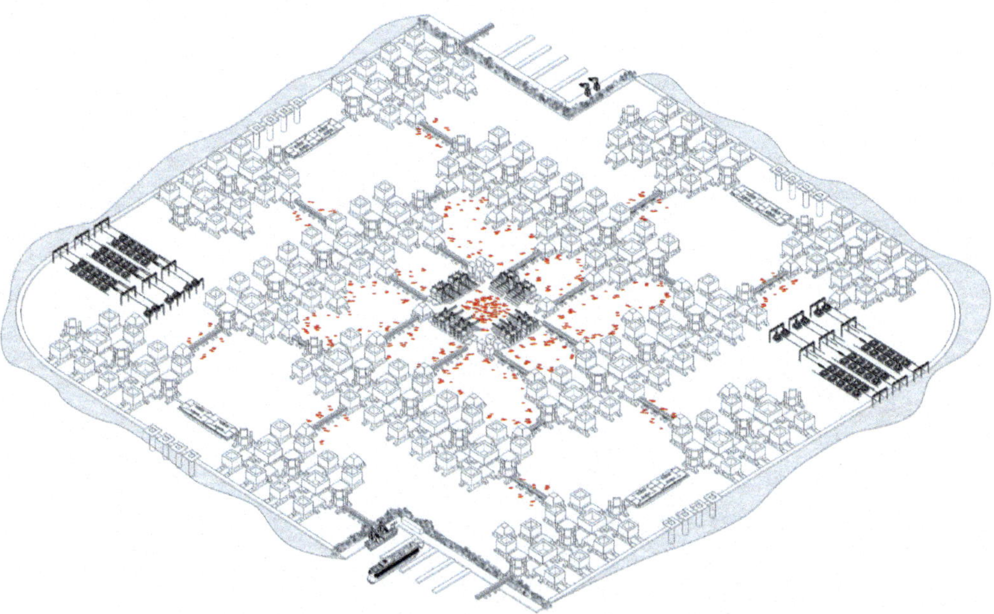

Axial Permutation - Circulation Routes

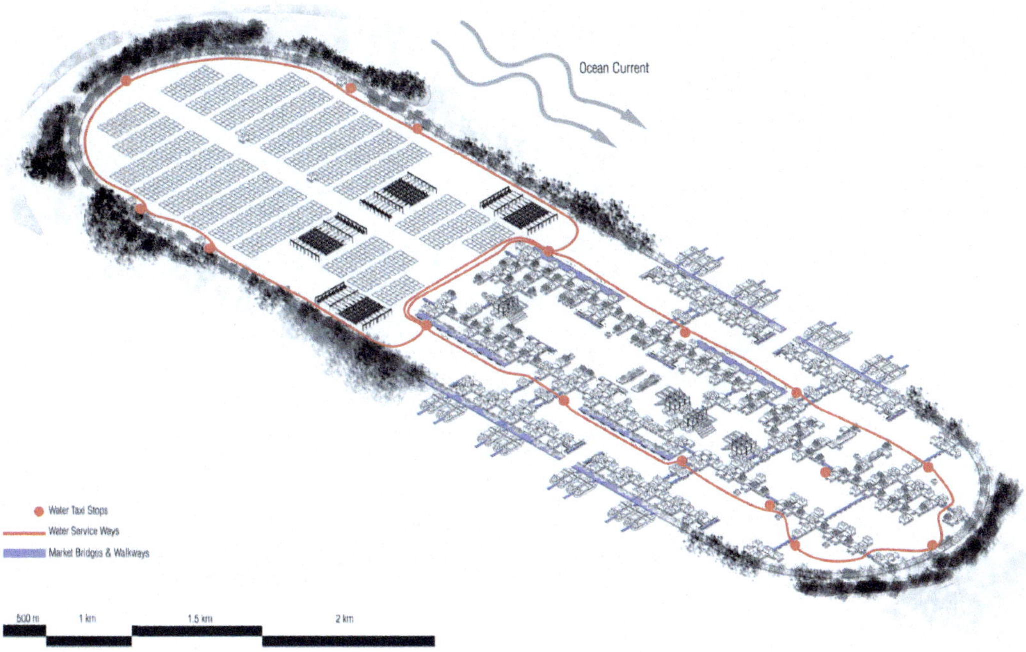

festive spaces

Waterscape Area: 24 km2
Festive spaces : 37%

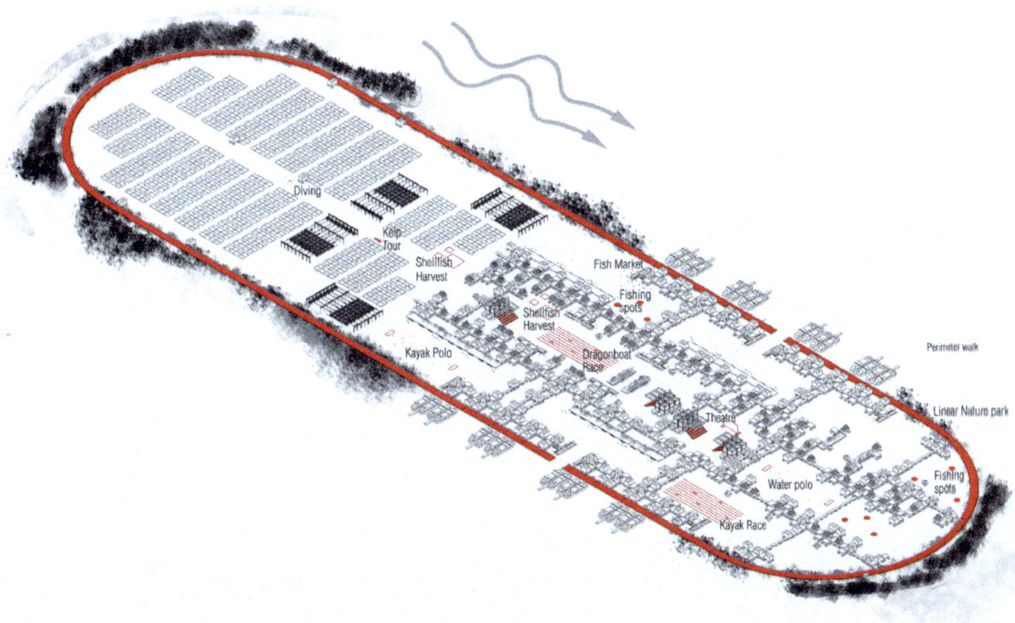

production spaces

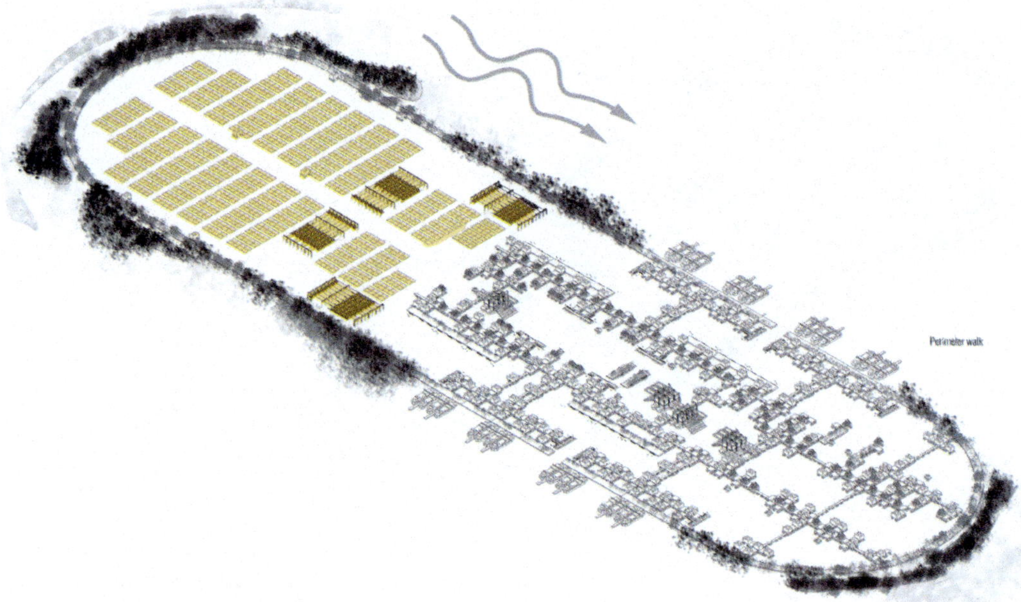

green spaces

250, 600 local residence
82,600 visitor residence

6.3% GREENERY
ENCLOSED AREA: 65 KM2

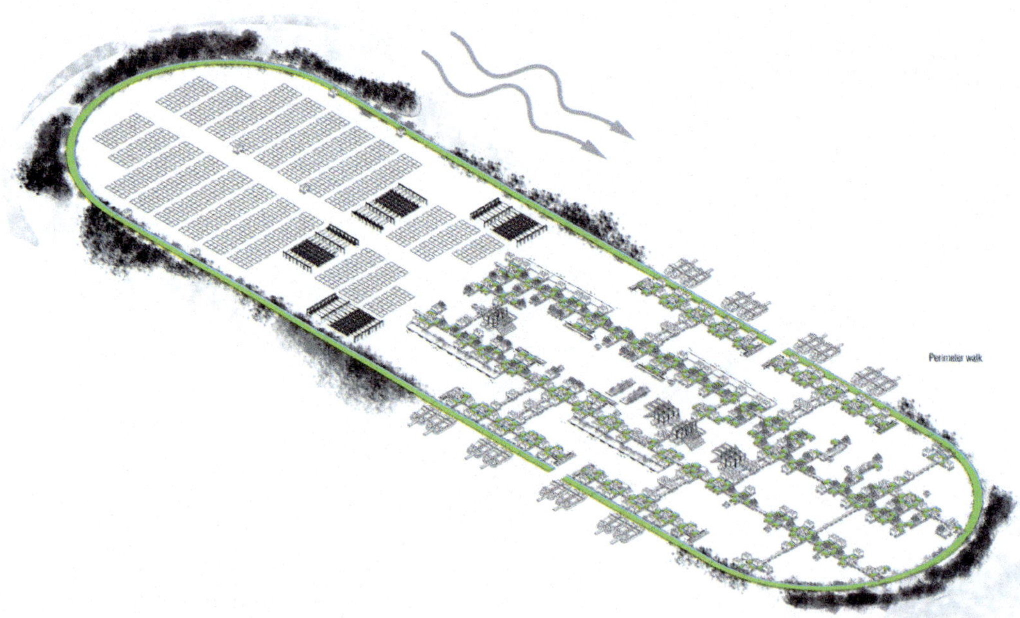

References

1. Capello R, Camagni R (2000) Beyond optimal city size: an evaluation of alternatives urban growth patterns. Urban Stud (Sage Publications) 37(9):1479–1496
2. Moles R et al (2002) Methodologies for the estimation of sustainable settlement size. Environmental Protection Agency, Ireland
3. Oceanix City. https://oceanix.org/
4. Tisdell C (1975) The theory of optimal city-sizes: elementary speculations about analysis and policy. Urban Stud (Sage Publications) 12(1)
5. UNESCO. Technical guidelines for biosphere reserves. https://en.unesco.org/biosphere/guidelines Accessed Dec 2019
6. United Nations, Department of Economic and Social Affairs, Population Division (2019) World population prospects 2019: highlights

Post-disaster Applications for Displaced Populations

Abstract

This chapter generates a separate fleet of floating settlements for post-disaster relief operations. Given an increase in the frequency of natural disasters caused by climate change, there is a corresponding increase with the number of displaced populations. For both climate and political refugees, the timeliness of aid for survival is not the only issue. Beyond the immediate need of restoring damaged infrastructure and providing medical aid; for survivors, is their long-term rehabilitation and recovery. The United Nations (UN) guidelines for post-disaster and humanitarian aid are adopted for response times in providing aid with a flotilla comprising semi-submersibles converted into a 1,500-bed hospital. UN space standards for post-disaster housing are fitted in superbarge accommodation units with additional units for cremation, parks and farm plots to enable survivors to start up activities for self-subsistence. The idea is to create a fully deployable fleet for immediate disaster relief, but also to provide housing, production and communal facilities for establishing and maintaining a settlement in the long term.

1 Global Issues of Human Displacement

The number of disasters worldwide from 1998–2017 totaled 7,255, with flood constituting the largest percentile at 43.4 (Table 1). The intensification of climate variables resulted in the growth of climate related natural disasters such as floods, storms, and droughts. Leaning and Guha-Sapir [4] pointed out that while these intensifying climate variables correlates to the expanding scale of the disasters, the increasing rates of urbanization and environmental degradation play a large part as well.

According to the United Nations Population Fund [8], the most disaster-prone region in the world is Asia and the Pacific. Almost 45% of the word's natural disasters occur in the region, and more than 75% of the population affected by natural disasters live in the region. These countries in Asia have large landmasses, multiple river basins, floodplains and high population densities, making them extremely exposed and vulnerable to the risk of disasters. Flood is Asia's most devastating climate related disaster, often resulting in tens of thousands casualties and hundreds of thousands population displaced. The Indian Ocean Tsunami of 2004 alone left 1.7 million homeless.

In 2018 the UN Refugee Agency reported that 13.6 million of the global population were newly displaced, an increment of 2.3 million people from 2017. As from June 2019, there are 70.8 million refugees forcibly displaced by war, political strife, human rights violations, climate and environmental disasters. This relentless rise in worldwide human displacement, exacerbated by the increasing frequency of disasters and conflicts, is crucial today as governments and not-for-profit organizations struggle to provide timely responses with effective solutions.

1.1 Slow Emergency Response Time

Organizations of humanitarian relief are rife with administrative complications and the time taken to provide medical aid supplies to disaster locations may be prolonged due to inaccessibility of the site. The International Red Cross sets a protocol for providing international emergency aid, detailing a 36-h timeline of the process from notification to on-site assistance. However, such response timeline is not possible in some situations. For example, during Haiti Earthquake in 2010, Haiti's main airport in Port-au-Prince was not equipped to deal with the volume of aid flights arriving, creating a

J. Lim, *Oil Rig and Superbarge Floating Settlements*, Lecture Notes in Civil Engineering 82,
https://doi.org/10.1007/978-981-15-5297-7_4

Table 1 Number and percentage of disasters per type 1998–2017

Flood	3,148	43.4%
Storm	2,049	28.2%
Earthquake	563	7.8%
Extreme temperature	405	5.6%
Landslide	378	5.2%
Drought	347	4.8%
Wildfire	254	3.5%
Volcanic activity	99	1.4%
Mass movement (dry)	12	0.2%

Source CRED, UNISDR, 2018

bottleneck for the provision of food, water and medical aid to the disaster victims. In Hurricane Katrina, search and rescue operations were delayed after communication systems including 911 call centers were damaged, slowing down the emergency response time.

1.2 Lack of Medical Facility and Inadequate Mortuary Space

The 2004 Indian Ocean Tsunami caused a massive strain on the medical infrastructure of the affected countries. Although large populations of victims required emergency treatment, the hospital structures and public health centers of affected countries had sustained structural damage and were no longer serviceable. External medical support was flown in within two weeks to set up mobile field hospitals where deployed emergency healthcare staff distributed medical supplies and vaccine. The handling and identification of dead bodies posed another strain on the medical infrastructures. Lack of refrigerated mortuary space to store and prepare bodies for identification resulted in their rapid decomposition and unsanitary conditions for rescue work.

1.3 Improper Waste Management

In 2010, earthquake-devastated Haiti struggled with water and sanitation until improper waste management resulted in an outbreak of cholera. The epidemic infected large numbers and subsequently killed at least 10,000 Haitians. In post-Tsunami Aceh, sanitation systems were neglected during the building of temporary housing shelters, as open sewage systems along the temporary homes quickly became a dumping ground for settlement waste. Acehnese were exposed to risks of waterborne diseases such as cholera and malaria. Until Aceh authorities could complete repairs to infrastructure systems, survivors were exposed to waterborne diseases such as cholera and malaria over a three year construction period. These however included water, sludge treatment plants and piped water systems, which could only service a fraction of the affected population.

1.4 Poor Housing and Rebuilding Efforts

Countries without effective risk and disaster governance and resources are more susceptible to disaster mortalities. For developing countries, the resources to recover from a post-disaster situation often exceed three to four percent of gross domestic product. Recovery is not a short-term process for developing countries. In the phase of survivor rescue, the provision of medical aid, sanitary conditions, shelter and food is priority and these can be met by foreign aid agencies. But once aid ceases after budgets are consumed, the disaster-stricken country depends on its own resources for medium and long-term recovery. It is often in the medium-term phases of recovery that survivors in temporary accommodation endure long periods of uncertainty before permanent housing is made available. However it is the provision of shelter in temporary settlements that pose environmental and health problems because good sanitary conditions cannot be met if central sewer systems are damaged or when settlement sites are remote from sewer water and electricity lines.

2 Planning Considerations of a Floating Relief Settlement

Displacement posed challenges to relief efforts. Damaged infrastructure slows response times in scouting appropriate sites to set up relief camps. Where resettlement involves large populations in rural sites, masterplan guidelines for land use are necessary to support the livelihoods of survivors through economic development to provide for employment, trade and production. It is proposed that a fleet of aid vessels for ready deployment be developed from semi-submersibles. These aid semi-submersibles would standby onboard medical supplies with air and hovercrafts, operate a 1,500-bed hospital, and provide for wastewater management infrastructure for hydroponic production, which are predicaments in especially challenging site conditions. When mobilized, the logistics units, cremation, mortuary and hospital units arrive on the evening of Day 2 of post-disaster. Additionally, super-barges can provide dwelling accommodation, incinerator facilities, parkways and farm plots. These can be deployed on location and assembled in a few weeks.

2.1 Siting and Deployment Locations for Disaster Relief

As 45% of global natural disasters occur in the Asia and the Pacific regions, the research investigated possible sites in Southeast Asia, to locate on standby, relief aid facilities for deployment in the event of a disaster. Sites in the Gulf of Mexico were also studied as possible locations.

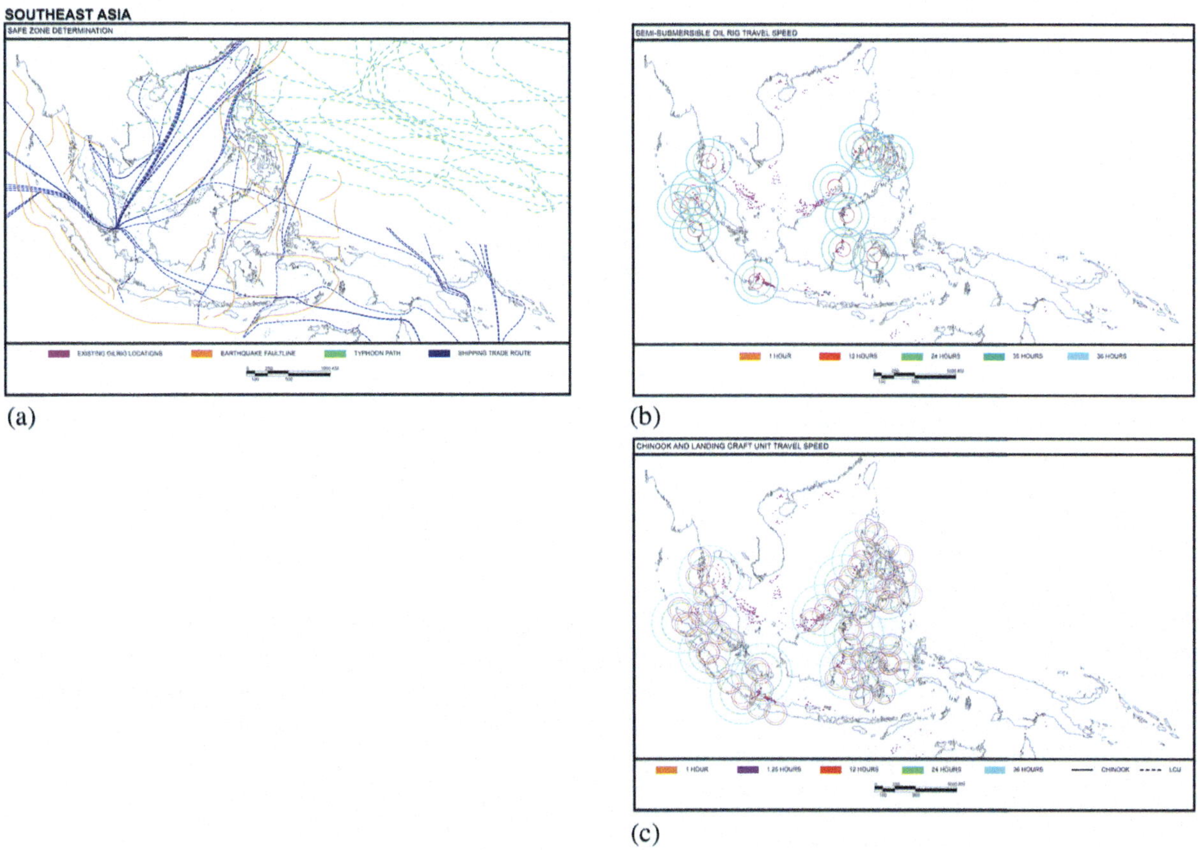

Fig. 1 Siting of relief aid facilities in Southeast Asia. **a** Determining safe zone; **b** overlaying travelling speed of semi-submersible oil rig; **c** overlaying travelling speed of LCU and Chinook

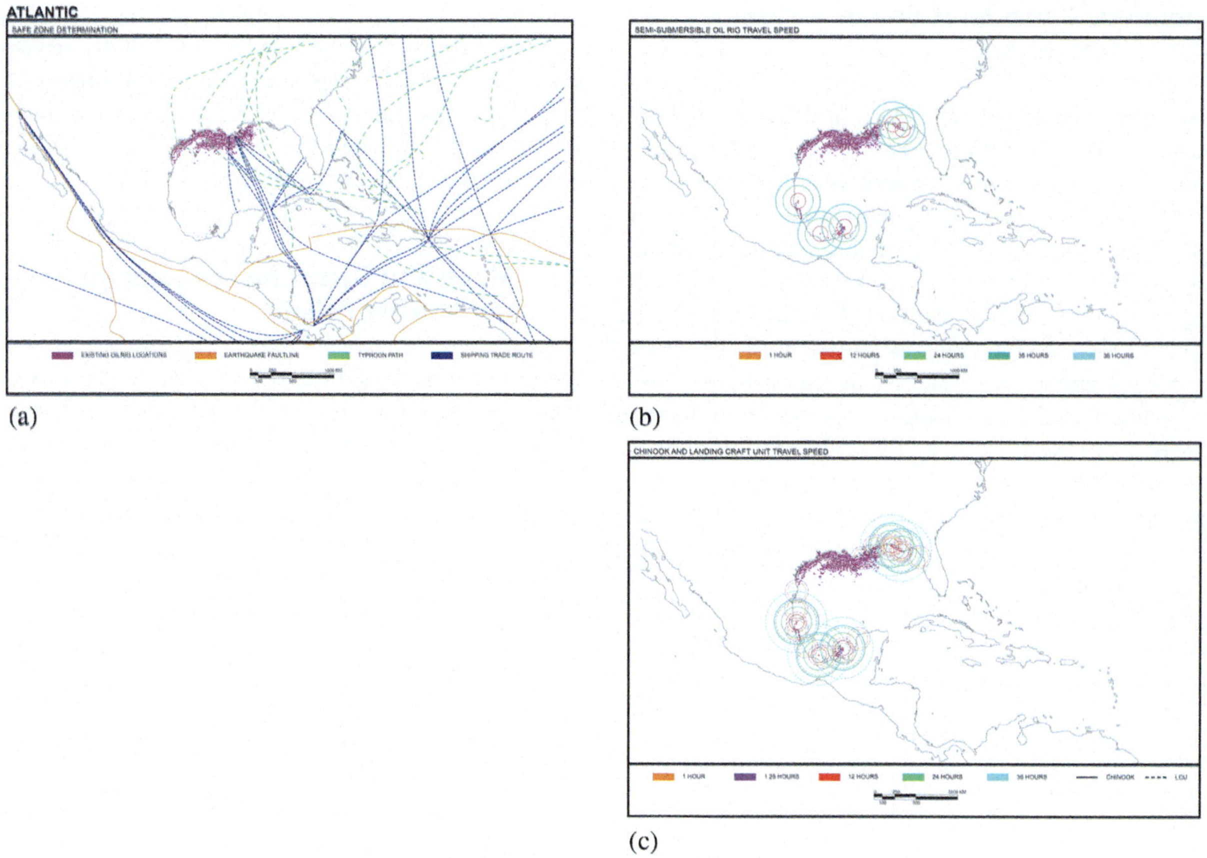

Fig. 2 Siting of relief aid facilities in Atlantic region. **a** Determining safe zone; **b** overlaying travelling speed of semi-submersible oil rig; **c** overlaying travelling speed of LCU and Chinook

Shipping trade routes and potential disaster paths are first mapped out to ensure that the anchorage of post-disaster aid facilities on standby would not obstruct existing port activities nor would they be in the pathway of hurricanes or typhoons and be destroyed by disaster itself. These paths are then cross-referenced with locations of existing oil rigs to determine the safe zone for the placement of the relief facilities.

Upon determining the safe zones to place the relief aid facilities, selection is further narrowed to areas that would allow for faster response during a disaster (Figs. 1 and 2). The response time is calculated by overlaying the travelling time and speed of the corresponding relief vessels, the semi-submersible oil rig, Landing Craft Unit (LCU), and Chinook helicopters used in cargo and rescue operations. This further refined the possible standby placements of post-disaster fleets to be within a 36-hour reach of high-risk zones in Southeast Asia.

In the Gulf of Mexico, most of the existing oil rig locations are located within the high-risk zone, but fall outside the safe zone for the anchorage of post-disaster relief fleets. Hence, relief effort to some high-risk zones near the Gulf may exceed the 36-hour response time.

2.2 Masterplan of a Floating Relief Settlement

Post-Tsunami Aceh was studied as a disaster scenario to model the technical requirements of the floating relief settlement for 400,000 casualties, in one of the deadliest disasters in history. These specialized vessels can also be modified to accommodate different disaster situations and relief operations. As illustrated in Figs. 4 and 5 the floating settlements aim to provide relief aid in five types of specialized vessels i.e. logistics units, hospitals, cremators, hydroponic/wastewater treatment facility and nearshore temporary housing, over a minimum recovery period of five years. Environmental systems between the repurposed rigs and barges are interdependent forming a closed loop system between waste to resource. If effectively maintained, the systems and its settlement may exist over indefinite periods of time.

Situated along or near the coastline of the disaster site, the post-disaster settlement plan was conceived with key considerations in mind. These included the ease with which relief supplies could be delivered and medical aid could be setup in shortest possible time frames to meet the humanitarian needs of a displaced population. While this masterplan is applicable to disaster relief that follows stricter timelines between mobilization and full operation, it is also applicable for the long-term accommodation of any displaced population, and can also be adapted as a construction fleet for larger scale and long-term floating settlements.

The floating relief settlement focused on three distinct groups—the survivors, the wounded, and the casualties. The survivors are directed towards the temporary housing barges, the wounded are brought to the hospital, and the casualties are processed in the mobile crematoriums. Distinguishing these components provides a clear circulation, accommodating operational workflows in parallel processes thereby improving efficiency.

Three different permutations in Fig. 3 illustrate how the masterplan can be adapted to different site contexts and the extent of the disaster along the coastline. A compact masterplan is suitable for concentrated sites, while a linear masterplan is suitable for larger affected sites dispersed along the coastline.

The housing barges are organized to form centralized boat transportation networks. These boat transportation networks streamline the movement of people and resources moving to and fro housing barges. After treatment, survivors are moved from hospital to the temporary housing barges, while families access crematoriums to identify bodies and to mourn the deceased. Treated water, waste material, food and energy provisions flow between the housing barges and hydroponic/wastewater treatment plant, while waste material and energy flow between mobile crematoria barges and the hospital.

Another advantage of the masterplan is its ease of setting up and downstream decommissioning of different relief facilities. The time frame for the provision of these facilities varies according to the different phases of disaster relief and the needs of survivors (Fig. 6). During the first phase of disaster relief, rescue, medical and logistics—food and water—are provided immediately. In the second phase of disaster relief, the focus would be to provide facilities such as long-term shelter, sanitation and healthcare. As such, the masterplan allows for any single relief facility to come in at different times and depart when it has fulfilled the needs of the survivors without affecting the other components. Relief facilities such as the logistics unit, hospital and mobile crematoriums are only needed during the first six-month phase of post-disaster timeline, and are placed further out to sea. The temporary housing barges and wastewater treatment which are placed closer to shore can stay between two to five years in the second phase of the disaster relief while the government concentrate on her rebuilding efforts on land.

In the masterplan, semi-submersible wastewater treatment plants and the housing barges with community spaces, play areas and classrooms are located nearshore in an attempt to lessen the emotional stress of survivors. The hospital and crematoria are placed at more remote ends of the settlement to protect the survivors from extended trauma (Fig. 7).

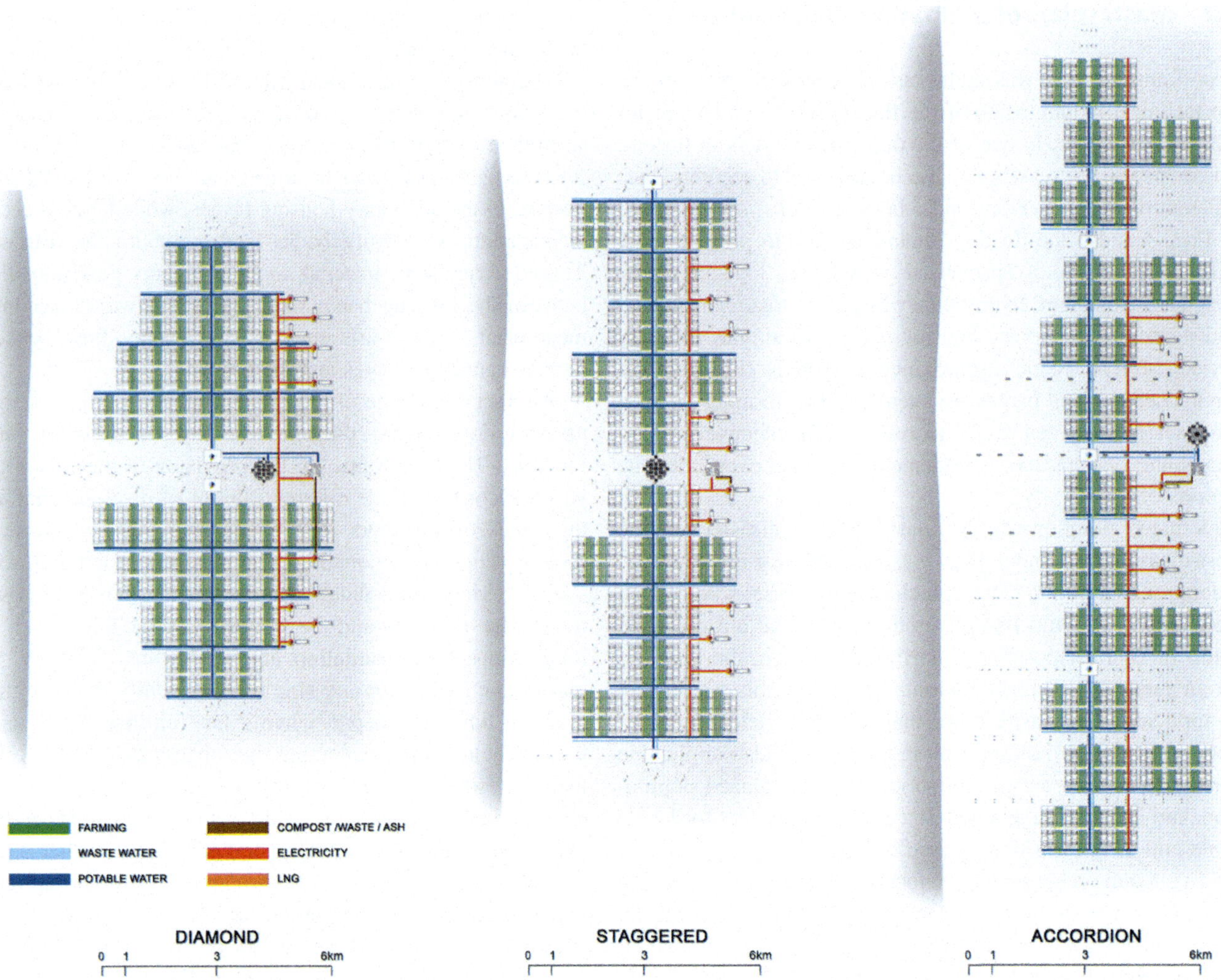

Fig. 3 Layout permutations for a relief aid settlement plan

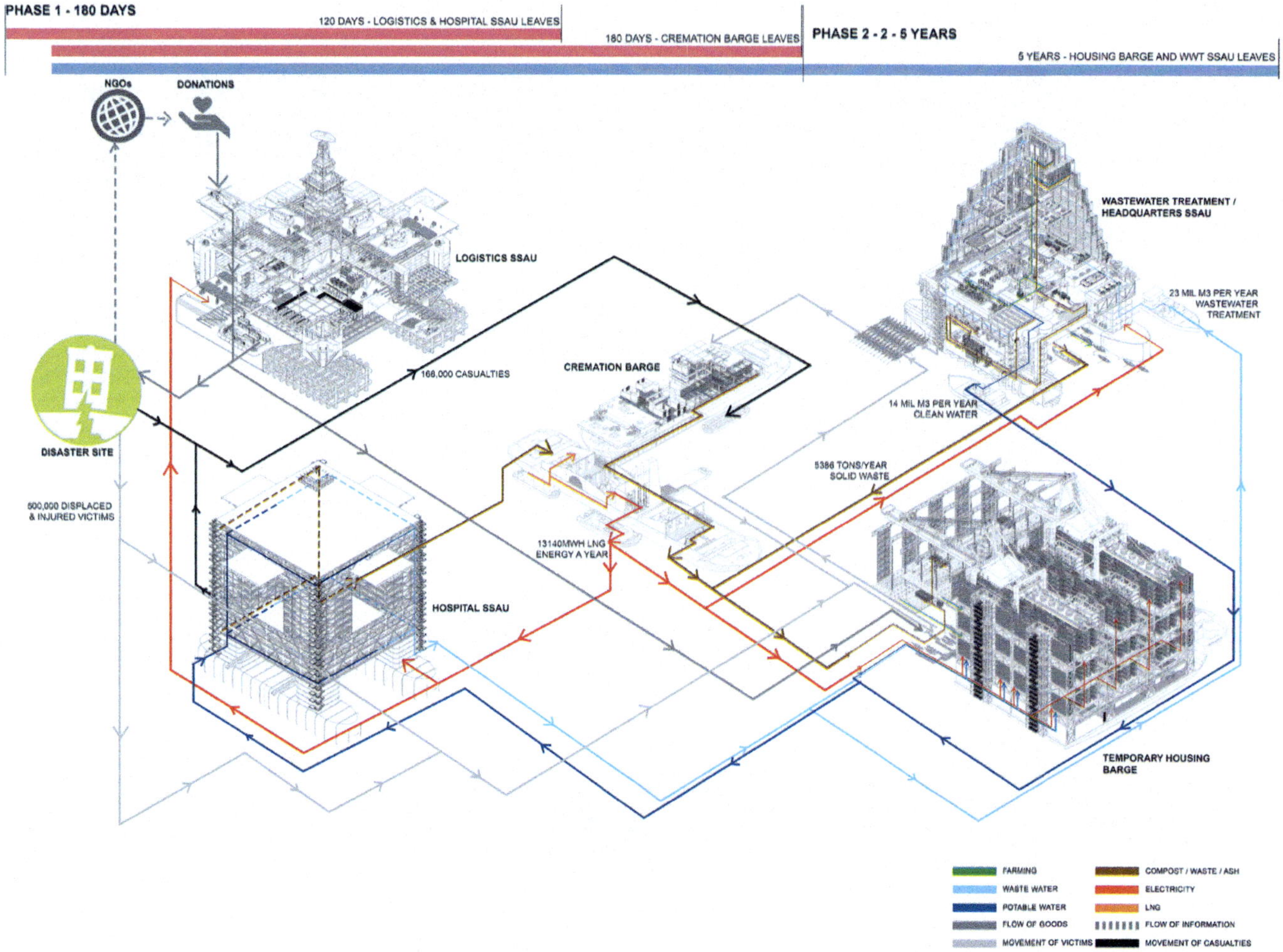

Fig. 4 Floating relief settlement with disaster response and resource loops

Fig. 5 Interdependent specialized vessels forming a closed loop between waste and resource

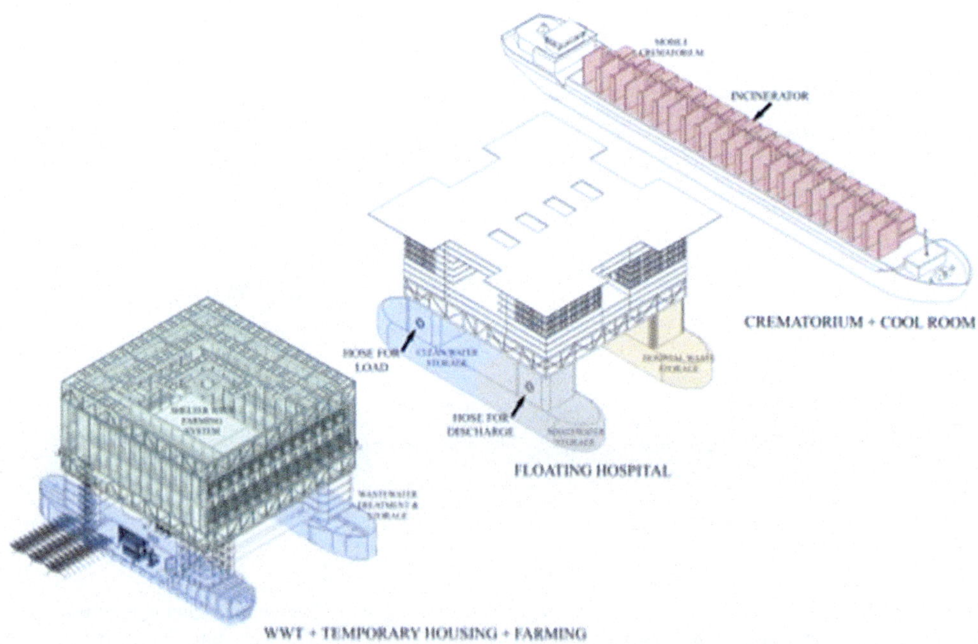

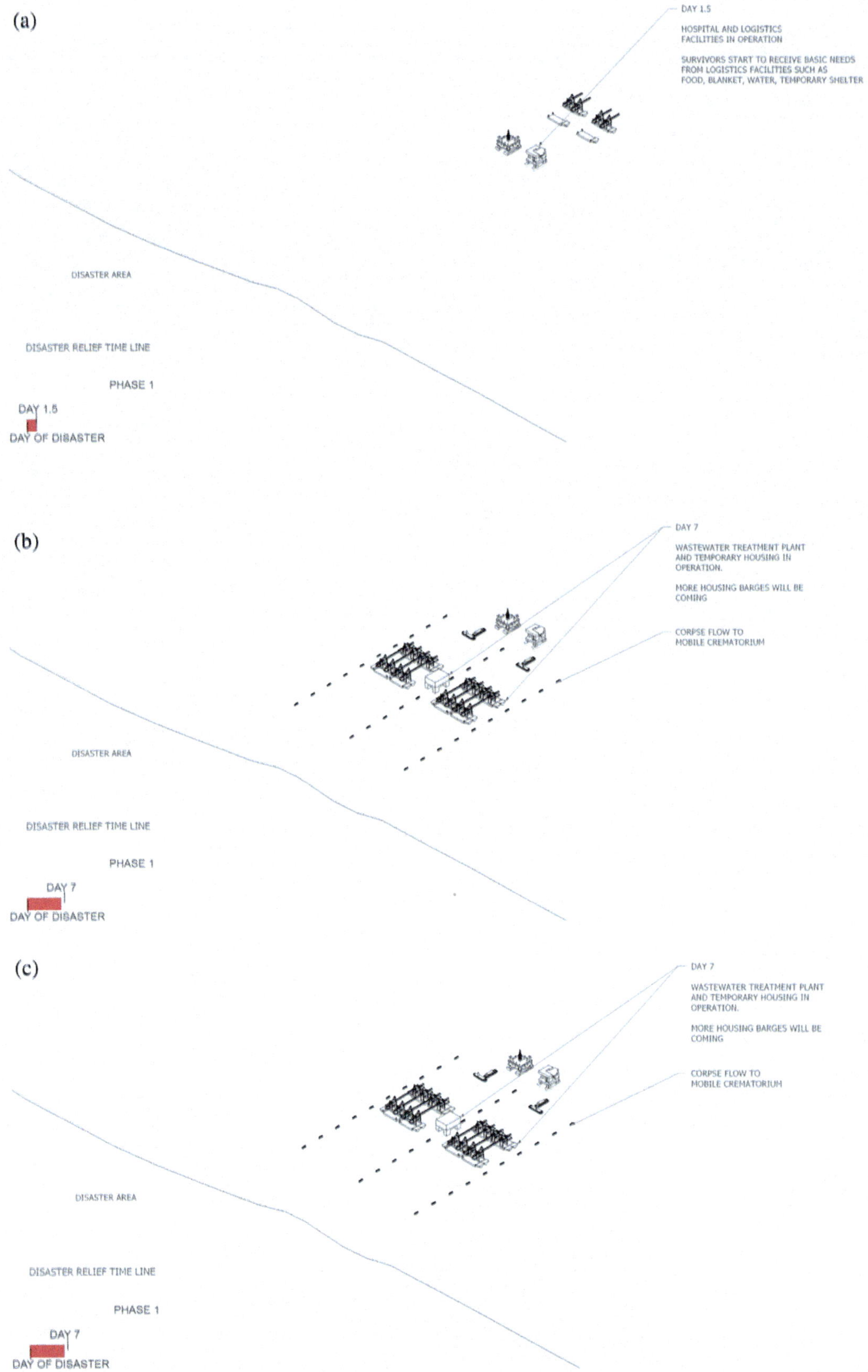

Fig. 6 Timeline of setting up a relief settlement in the event of a disaster. **a** Day 1.5: Hospital and logistics unit in operation; **b** Day 7: Mobile crematoriums, wastewater treatment plant SSAU and first eight temporary housing barges in operation; **c** Day 120: Hospital and logistics unit begin decommissioning process; **d** Day 120–180: Hospital and logistics unit move to the next disaster site, while remaining relief components continue in operation; **e** Day 180: Mobile crematoriums leave while housing barges, wastewater treatment SSAU and LNG barges remain for Phase 2 of disaster relief efforts; **f** Two years and beyond: relief floating settlement remaining self-sustaining until permanent infrastructure is rebuilt on land

Fig. 6 (continued)

(d)

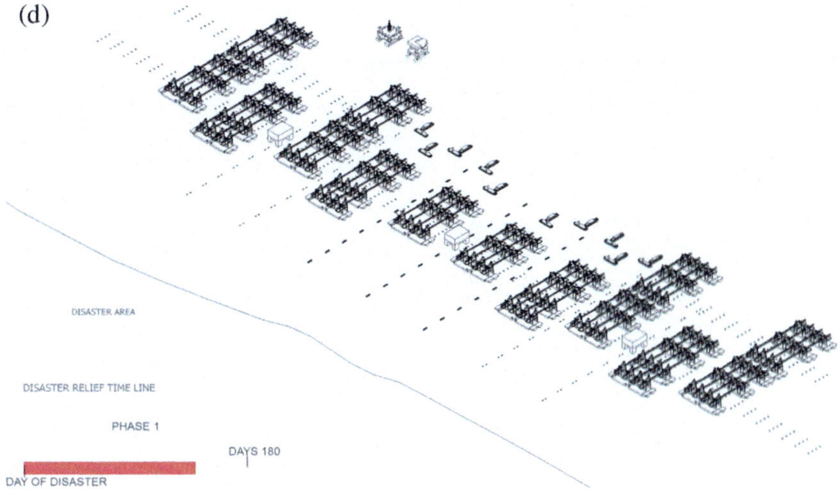

(e)

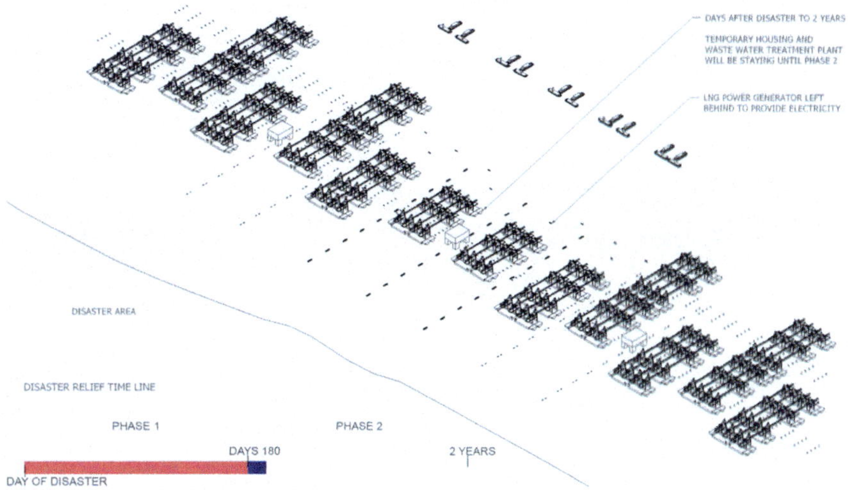

(f)

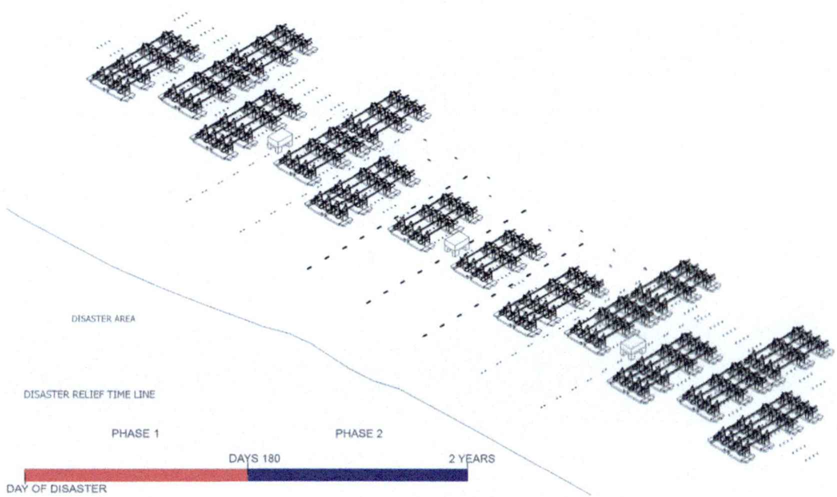

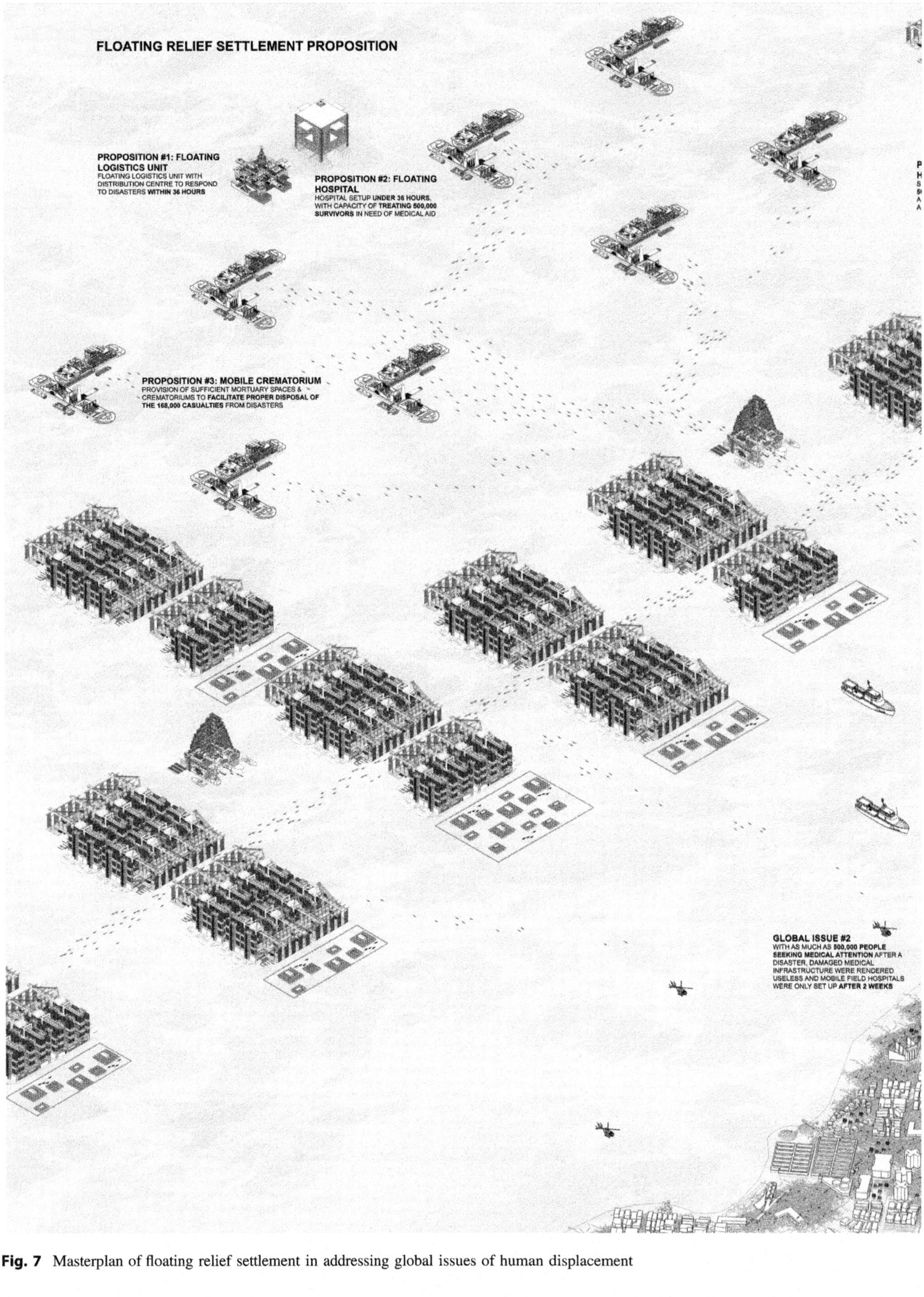

Fig. 7 Masterplan of floating relief settlement in addressing global issues of human displacement

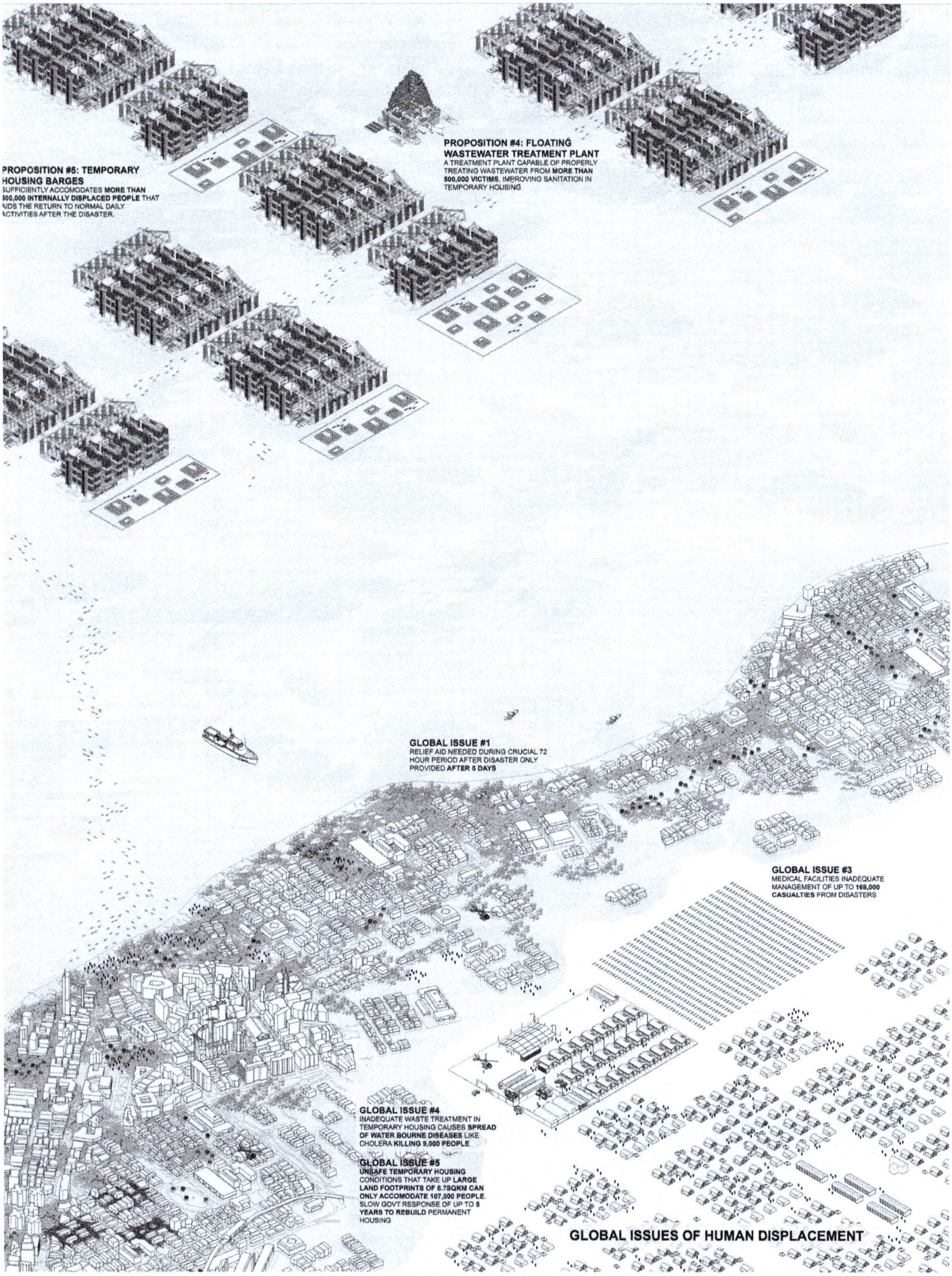

Fig. 7 (continued)

3 Semi-submersible Logistics Unit

A semi-submersible rig is repurposed into a logistics unit and defensible outpost, fitted with warehouses storing aid supplies and equipment. The outpost is repurposed to include landing and hangar decks to receive aid aircrafts and docking area to accommodate sea vessels. In the immediate aftermath of a disaster, response can reach the site by deploying the nearest semi-submersible on standby in the closest safe zone. Figures 8, 9, 10, 11 and 12 illustrates in detail, the capacities and operations of the logistic unit.

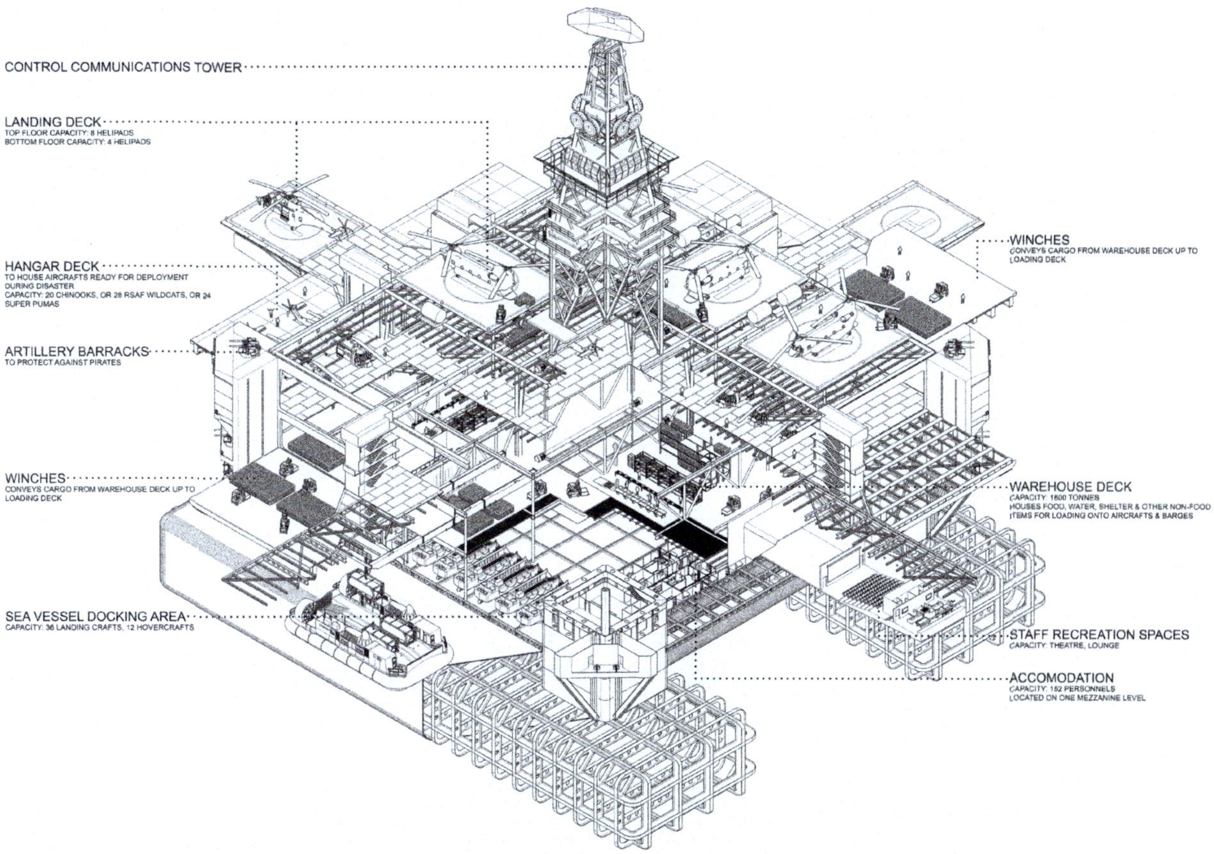

Fig. 8 Axonometric illustrating the components and workflow of a semi-submersible logistics unit. Aid supplies are hoisted up to the aircraft deck from the warehouse via external air lifts. Similarly, the existing moon pool is fitted with a lift to move aid supplies to the sea vessel decks for loading onto smaller landing crafts stowed on board

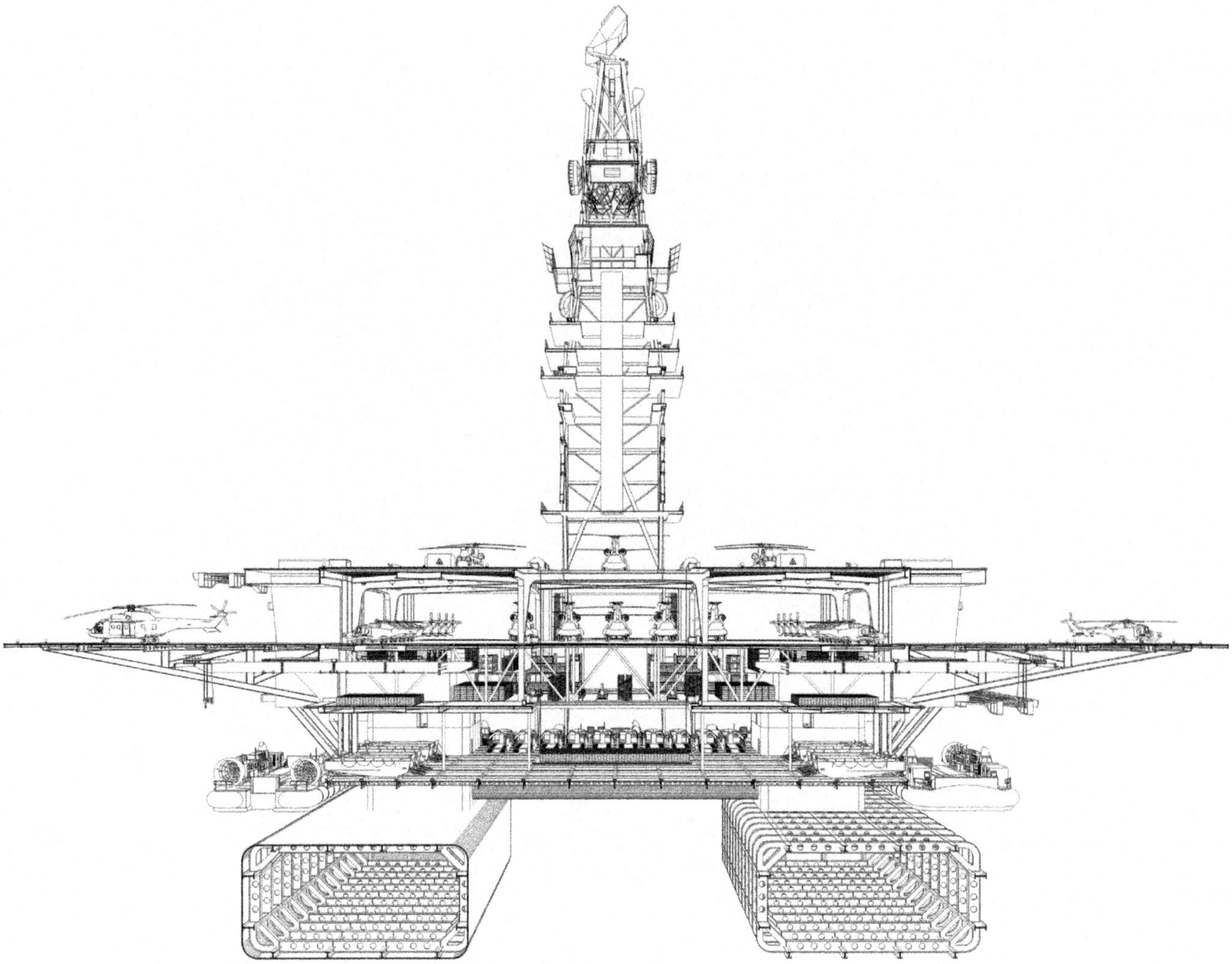

Fig. 9 Sectional perspective

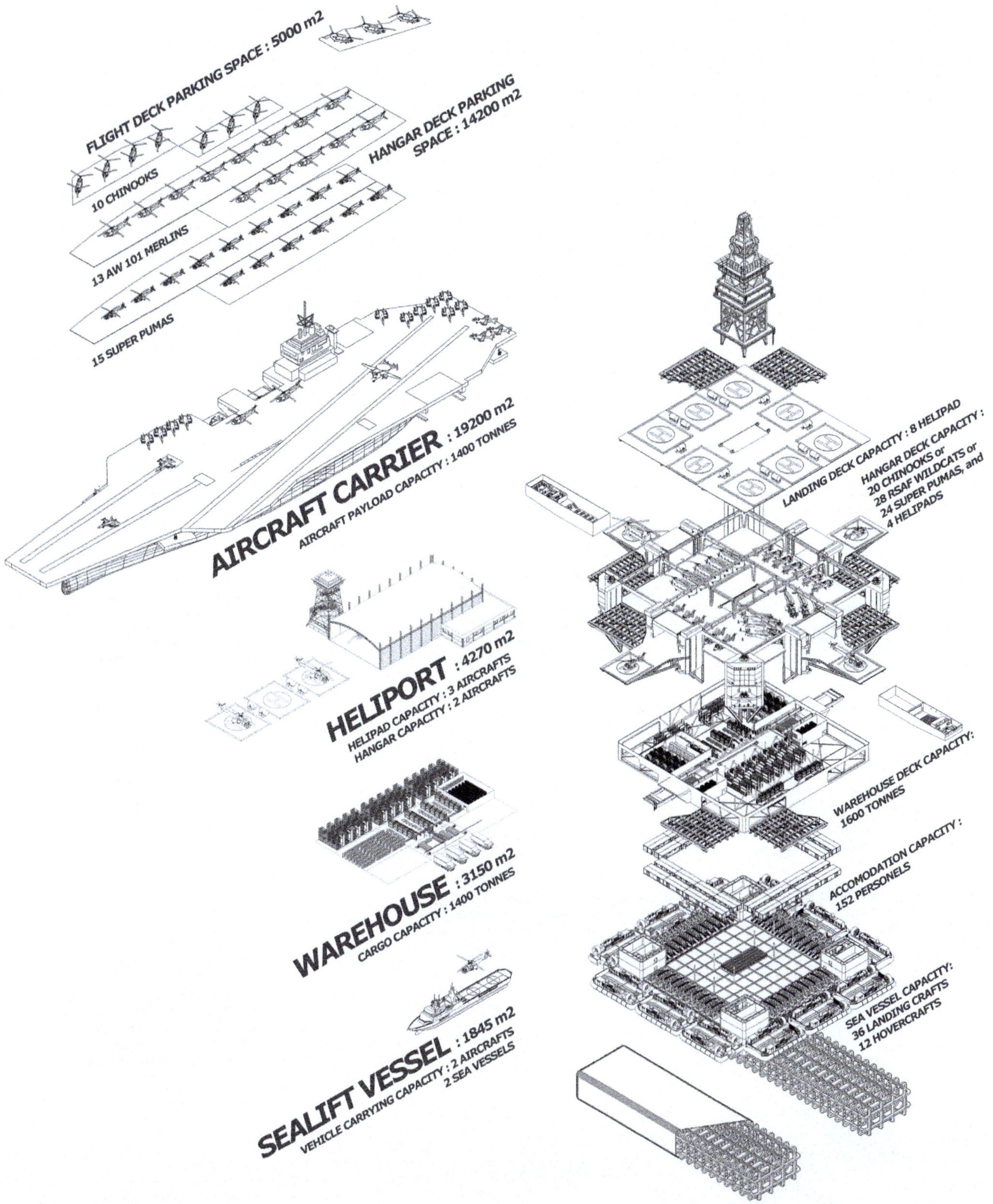

Fig. 10 Comparing loading capacity of single-purpose facility into a semi-submersible logistics unit and outpost. The design considered the specifications of rescue crafts: LCAC-4 air cushioned landing craft (cargo delivery) and Boeing CH-47D Chinook (medical evacuation and cargo delivery)

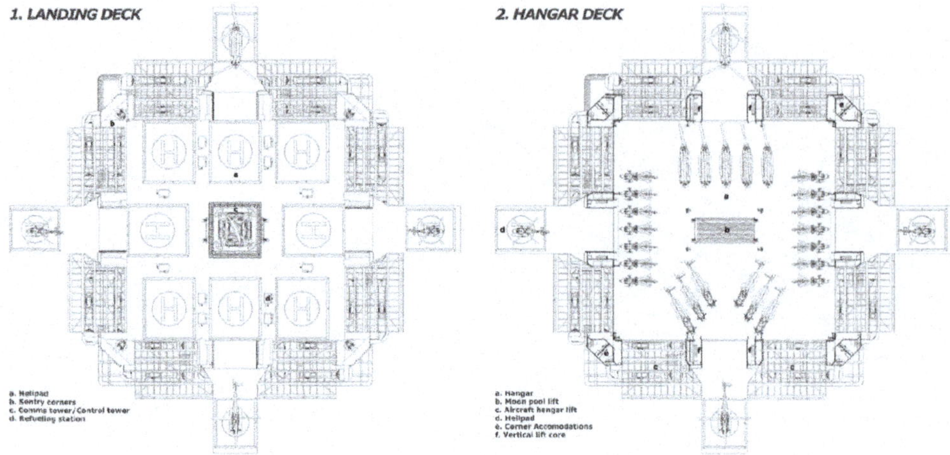

Fig. 11 Floor plans

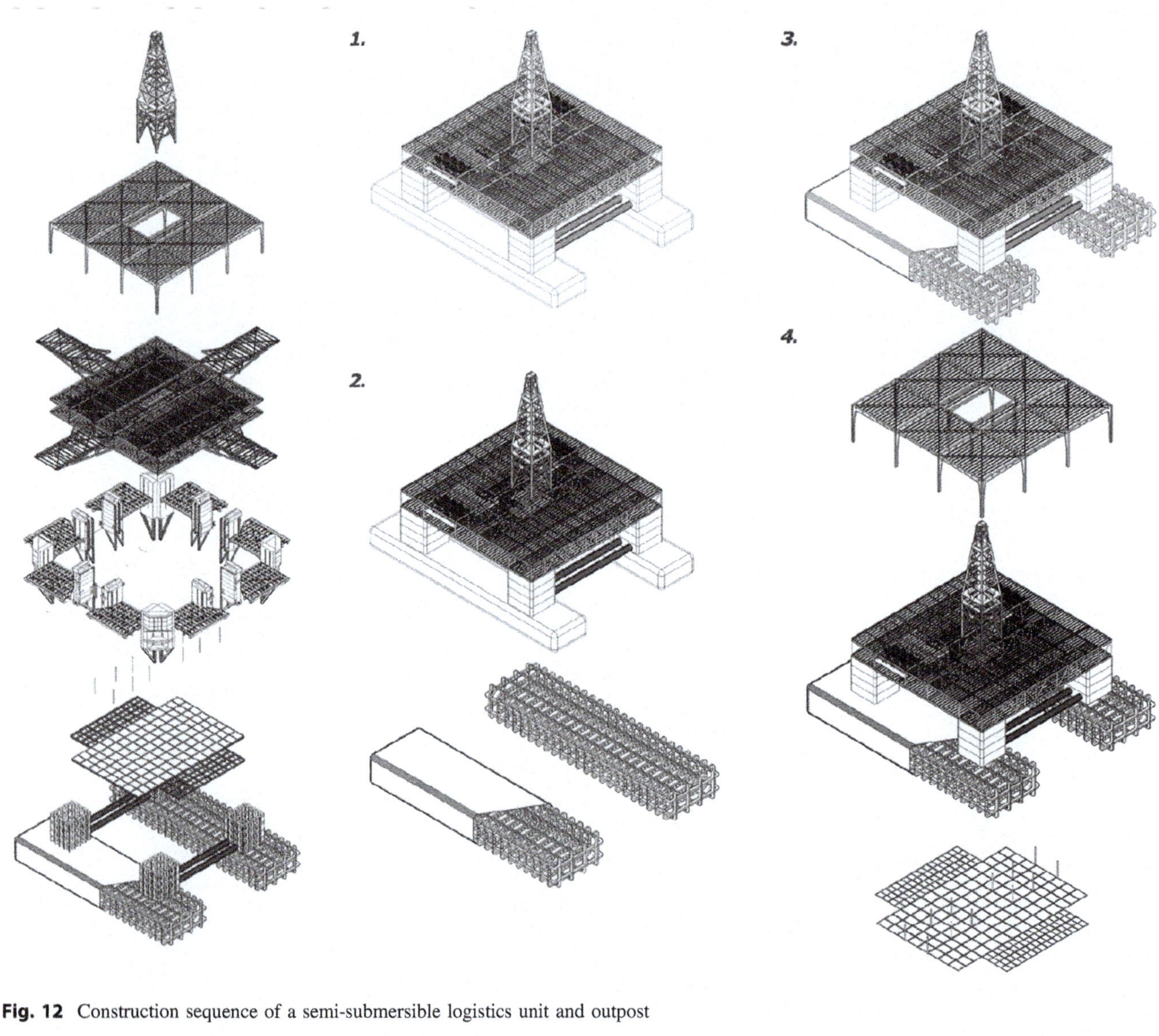

Fig. 12 Construction sequence of a semi-submersible logistics unit and outpost

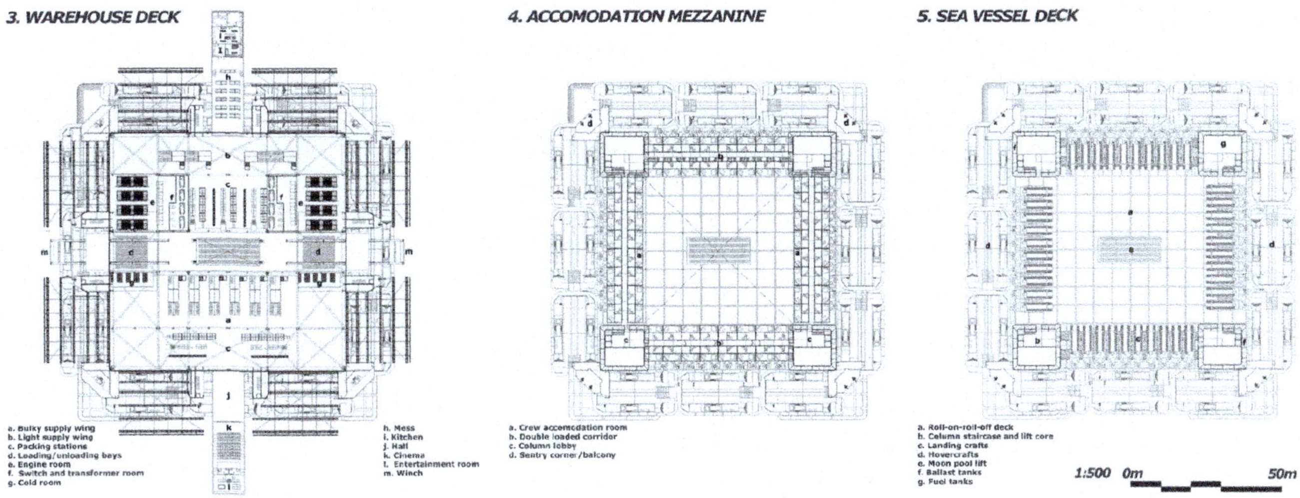

Fig. 11 (continued)

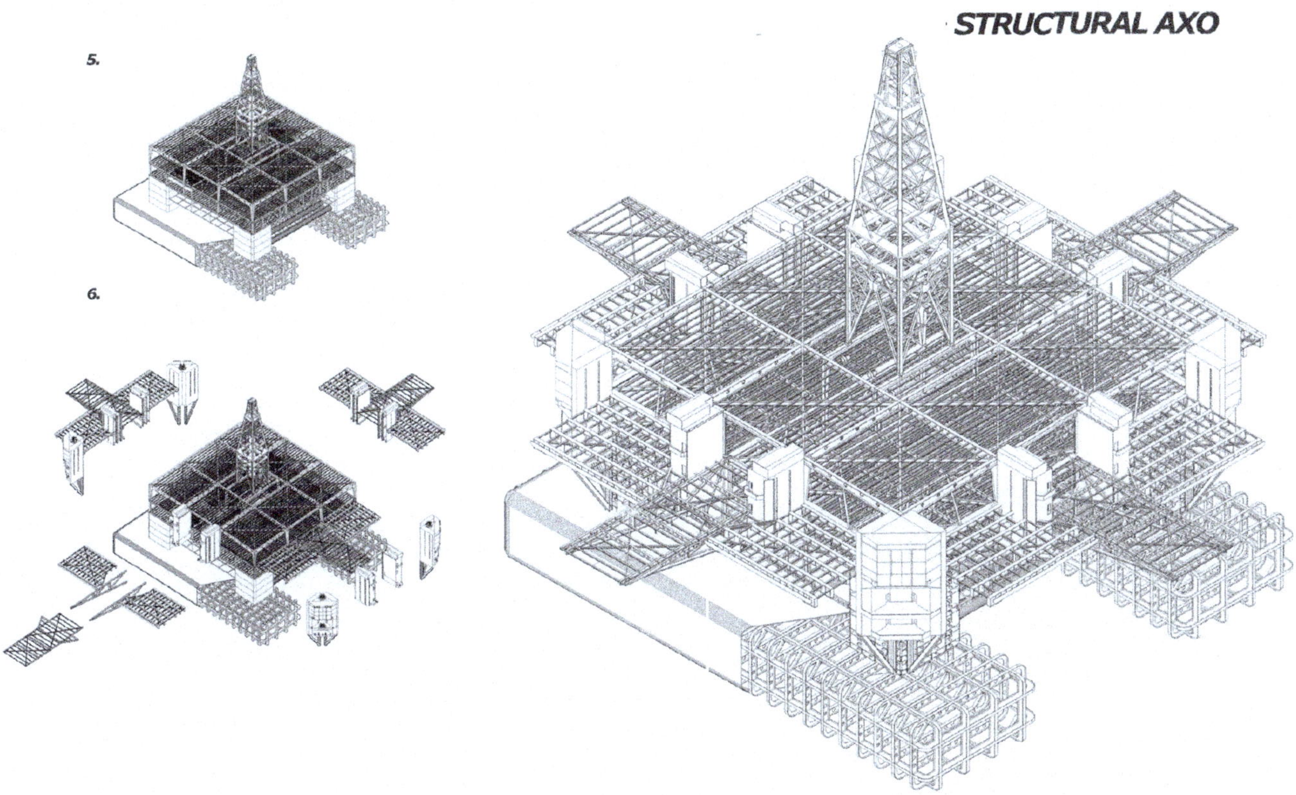

Fig. 12 (continued)

4 Semi-submersible Hospital

A semi-submersible hospital can be deployed to a disaster stricken coastline in 36 h and remain stable during harsh weather conditions. United Nations Disaster Response tools and services in large-scale emergencies are between 12 and 72 h [9]. Conversely, a field hospital requires three days to set-up, excluding transportation period that is dependent on the road conditions post-disaster. Field hospitals comprise lightweight tents, which are easy and quick to set-up, but their capacity depends on large land plots, which may not be available on disaster sites [6].

A semi-submersible hospital, however, is a standby facility anchored in disaster prone areas, with vertically stacked provisions and which can accommodate large usable areas within a small footprint at sea, equivalent to the pontoon dimensions of the semi-submersible. It is less energy intensive, as 80% of the hospital bed spaces are naturally ventilated. The hospital can receive 96 and 210 casualties from both the Chinook and LCU respectively. Not limited to space constraints of field hospitals, the semi-submersible hospital can perform up to 140 operations a day. So triage and treatment areas need to occupy substantial areas of column free space when receiving many casualties. Figures 13 and 14 illustrate, the design and workflow of a semi-submersible hospital; and Table 2 details its operational capacity.

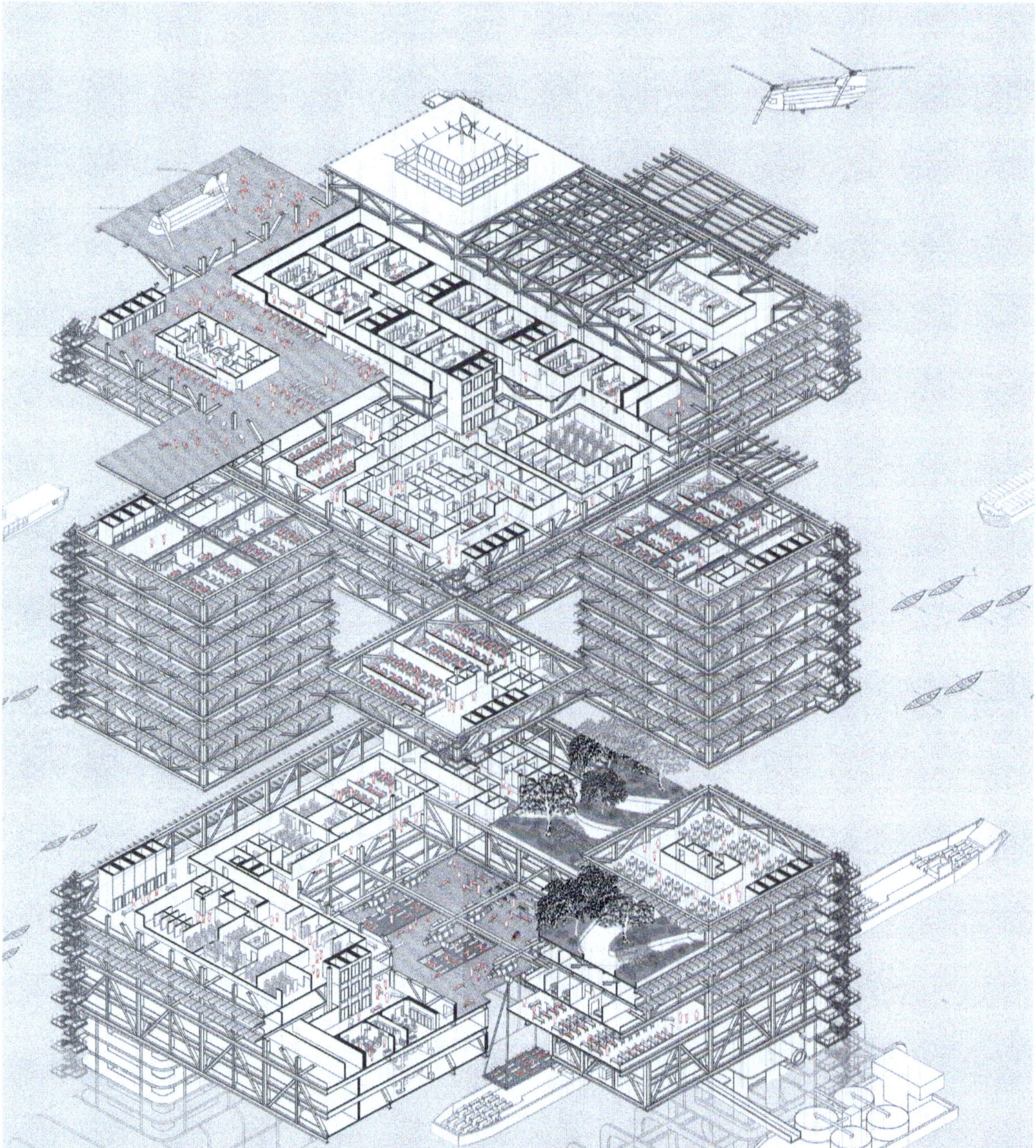

Fig. 13 Axonometric of a semi-submersible hospital with 1,500 beds

Fig. 14 Sectional perspective and floor plans of a semi-submersible hospital with 1,500 beds

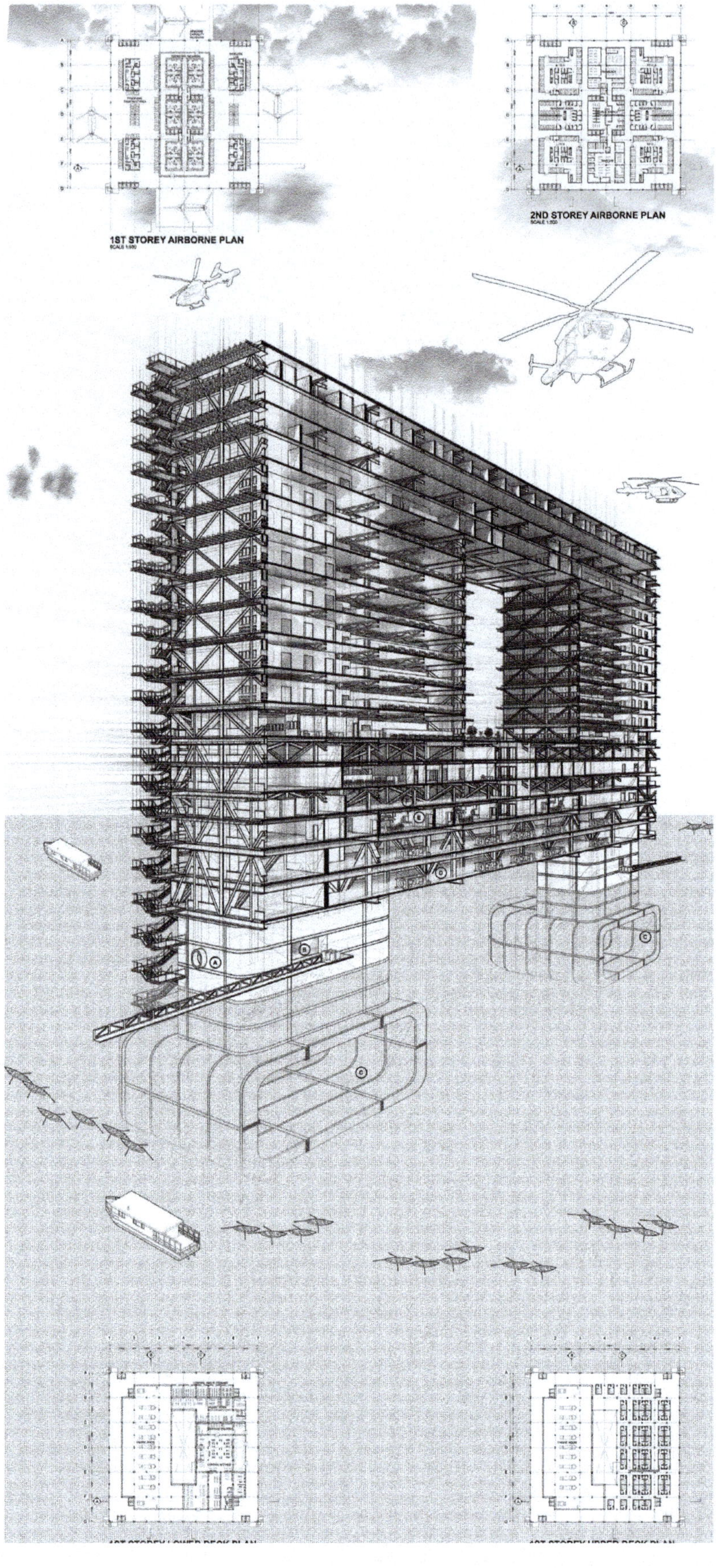

Fig. 14 (continued)

Table 2 Operational capacity of a semi-submersible hospital

Rig type	Semi-submersible	*Area comparison*		
No. of chinook helipads	4			
No. of landing craft unit	3	Red Cross field hospital	Medecins Sans Frontieres inflatable hospital	SS Hospital
No. of incoming rescued persons in one landing	96 by air 210 by sea	32,400 m^2	14,000 m^2	6,162 m^2
Area of each triage	3,363 m^2 for air rescue 1,425 m^2 for sea rescue	Waste type and treatment strategies	Hospital waste sent to mobile crematorium for incineration Wastewater sent to treatment plant	
No. of ward beds [19,960 m^2 total area]	1,424 beds [28 wards]			
No. of ICUs	144 [3,153 m^2]	No. of rigs needed for world population?	8,810 [provision guideline: 2 beds per 1000 persons 1680 beds per SS hospital can support 840,000 people 14,800,000 beds or 8,810 SS hospitals can support 7.4 billion people]	
No. X-ray rooms	18 [841 m^2]			
No. of surgical rooms	23 [3,452 m^2]			
No. of recovery beds	98 [1,288 m^2]			
Medical crew accommodation	3,465 m^2 can house 400 crews			
Canteen	1,426 m^2			
Central kitchen	1,237 m^2			
Recreational facilities	1,426 m^2			

5 Mobile Crematorium Barge

A barge is repurposed into a crematorium and while it is comparable to traditional methods of handling deceased on land, this all-in-one facility houses morgues, crematoriums and wake halls. The mobile crematorium barge provides several key advantages over the traditional methods of handling dead bodies in the event of a disaster.

Large areas of land are needed for temporary body storages, identification of bodies, and burials. This further diminishes available land to set up medical facilities and to house survivors. Crematorium barges alleviate land scarcity issues by shifting these operations out to sea. Cremation is a preferred means of disposing bodies when poorly managed mass burials risk disease outbreaks.

In the mobile crematorium, (Fig. 15) hulls of barges become chilled morgues that can temporarily hold 2,106 bodies. Each barge houses 20 wake halls for families of the deceased to perform last rites before cremation; and comprises 12 cremators which process 144 bodies a day. Ten such mobile crematoriums would only occupy 8,000 m^2 of the sea.

LNG is proposed as a fuel source to run the cremations. Safety precautions for bunkering would be planned for in the operational layout. The waste heat from cremations can be harnessed to power steam turbines for electricity. On average, 250 kWh of energy can be recovered from each cremation. This translates to each crematorium barge contributing to 36,000 kWh of energy a day to sustain its operations.

The waste heat generated from cremation can be captured to power steam turbines, generating electricity for the survivors. Ashes left over from cremation can also become a resource in the form of compost or a substrate for construction works.

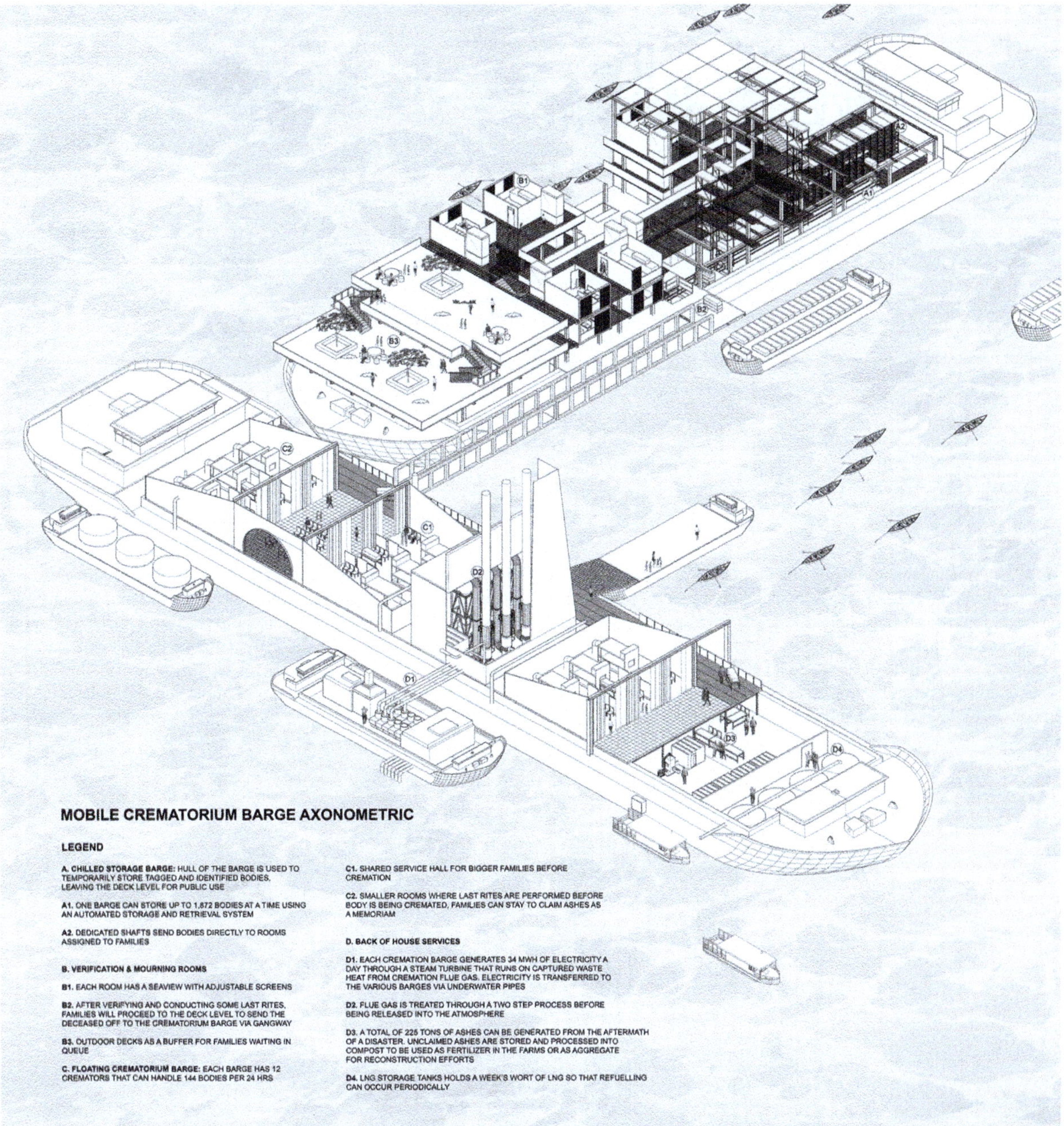

Fig. 15 Mobile crematorium barge equipped with morgues and wake halls

6 Waste Management and Resource Loop

In preventing contamination or an epidemic onboard a semi-submersible hospital, proper management of generated waste generated is crucial. Studies show that 85% of the waste produced by health-care providers is comparable to domestic waste, which can be classified as non-hazardous. The remaining 15% comprising chemical/radioactive/infectious waste are regarded as hazardous and pose environmental and health risks [1]. Hazardous waste has to be treated before discharging into the sea.

Waste management practices for an inland hospital [5] and a USNS Mercy [10] vessel—offshore/nearshore hospital—are illustrated in Figs. 16 and 17. Treated waste from Singapore's inland hospitals are converted to energy while treated wastes from USNS Mercy vessel are disposed at inland facilities or discharged into ocean without any form of energy recovery. A Mercy vessel comprising 1,000 patient beds and 1,430 staff generates an estimated 395

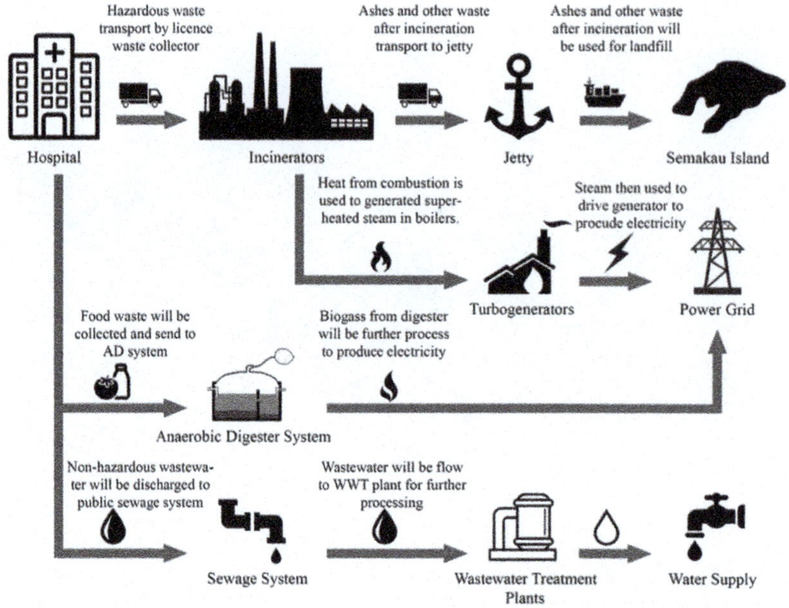

Fig. 16 Waste management in an inland hospital. Under National Environment Agency (N.E.A), waste management guideline stipulates that non-hazardous waste from inland hospital such as food waste and wastewater can be disposed as general household waste by general waste contractors at public waste disposal facilities. Hazardous waste on the mainland requires collection by a licensed waste collector with a registered truck to the incineration plant on a daily basis. Heat produced during incineration is converted to steam to heat up the boiler. The steam from the boiler then drives the generator and produce electricity, which is then looped back into the power grid. After incineration, ash reduced from waste weighs only 10% of its original volume, and is transported to Semakau Island for landfill purposes. In 2002, ash was used successfully on a 150 m section of Jalan Buroh, Singapore [3]

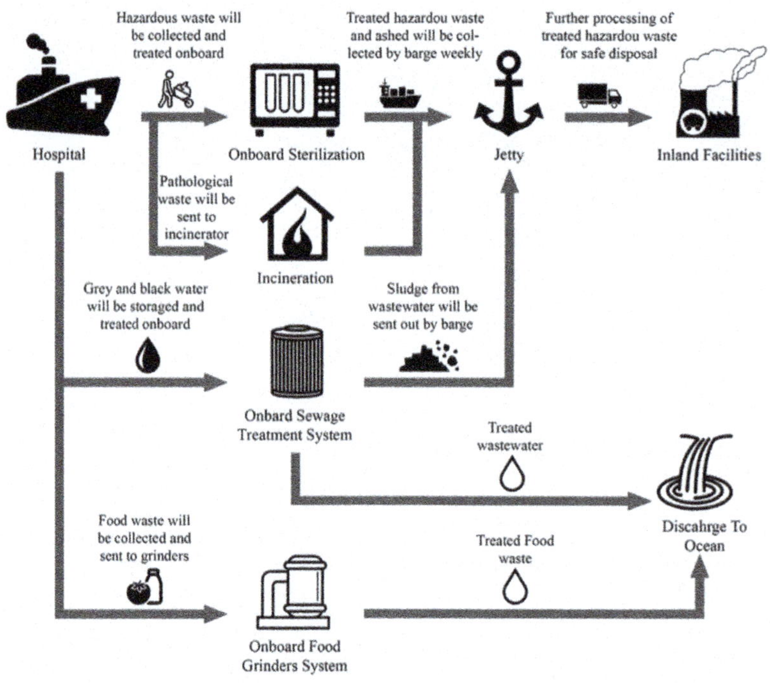

Fig. 17 Waste management of a USNS Mercy vessel. World Health Organization's (W.H.O) guide to ship sanitation allows food waste and wastewater to be discharged into the ocean provided it is 5.5 km away from the coastline. One reason is this waste can be decomposed by the microorganism in the ocean or eaten by the fishes. Second, it can obviate the need for medical waste storage. Third, it reduces the frequency of barges needed to transport the waste to inland facilities for further processing. Most of the medical waste—excluding pathological waste, which is incinerated—is sterilized by microwave irradiation and collected weekly by a barge that is ferried to an inland facility for safe disposal

tonnes of food waste, 304,045 m^3 of wastewater and 183 tonnes of hazardous waste a year. In comparison, Table 3 shows an estimated amount of waste and energy that can be generated from an inland hospital with 1,250 beds.

With this energy potential, a semi-submersible hospital enables a waste treatment loop with a crematorium barge and a semi-submersible wastewater treatment facility during a disaster relief operation. The hazardous waste from the hospital is incinerated at the crematorium; and waste heat generated from the incineration can be harnessed to power steam turbines generating electricity to sustain the operations of the other relief facilities (Fig. 18).

In selecting energy production systems, the systems using wave energy or LNG were considered due to the energy loads and material flows required for food and water production. The control of wastewater discharge quality for the

Table 3 Waste to energy for an inland hospital with 1,250 beds

	Patient	Staff
Total number of beds/staff on duty	1250 beds	3966 persons per shift
Food waste (include food waste during food processing and leftover food)	0.9 kg/bed/day = 410 tonnes/year	0.126 kg/person/day = 182 tonnes/year
Waste to energy conversion	**592 tonnes/year can generate -> 302 m^3/year fertilizer (by-product) -> 53.28 MWh/year biogas energy**	
Wastewater (include equipment washing, shower, human waste)	0.69 m^3/bed/day = 314,630 m^3/year	0.1 m^3/person/day = 144,905 m^3/year
Waste to energy conversion	**459,535 m^3/year can generate -> 275,721 m^3/year potable and non-potable water -> 17.8 MWh/year biogas energy**	
Hazardous waste	0.5 kg/bed/day = 228 tonnes/year	N.A
Waste to energy conversion	**228 tonnes/year can generate -> 22.8 tonnes/year ash (by-product that can be used for landfills or road-base)**	

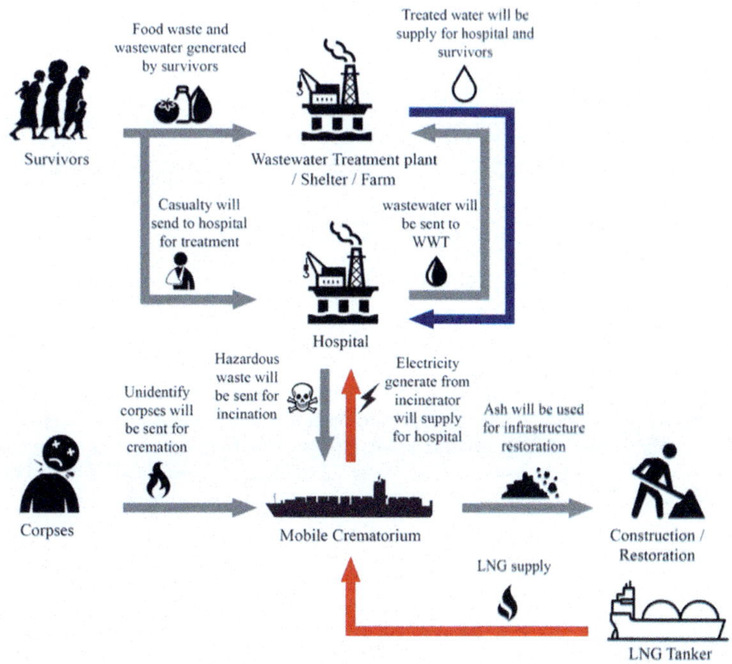

Fig. 18 Proposed waste management of a floating relief settlement

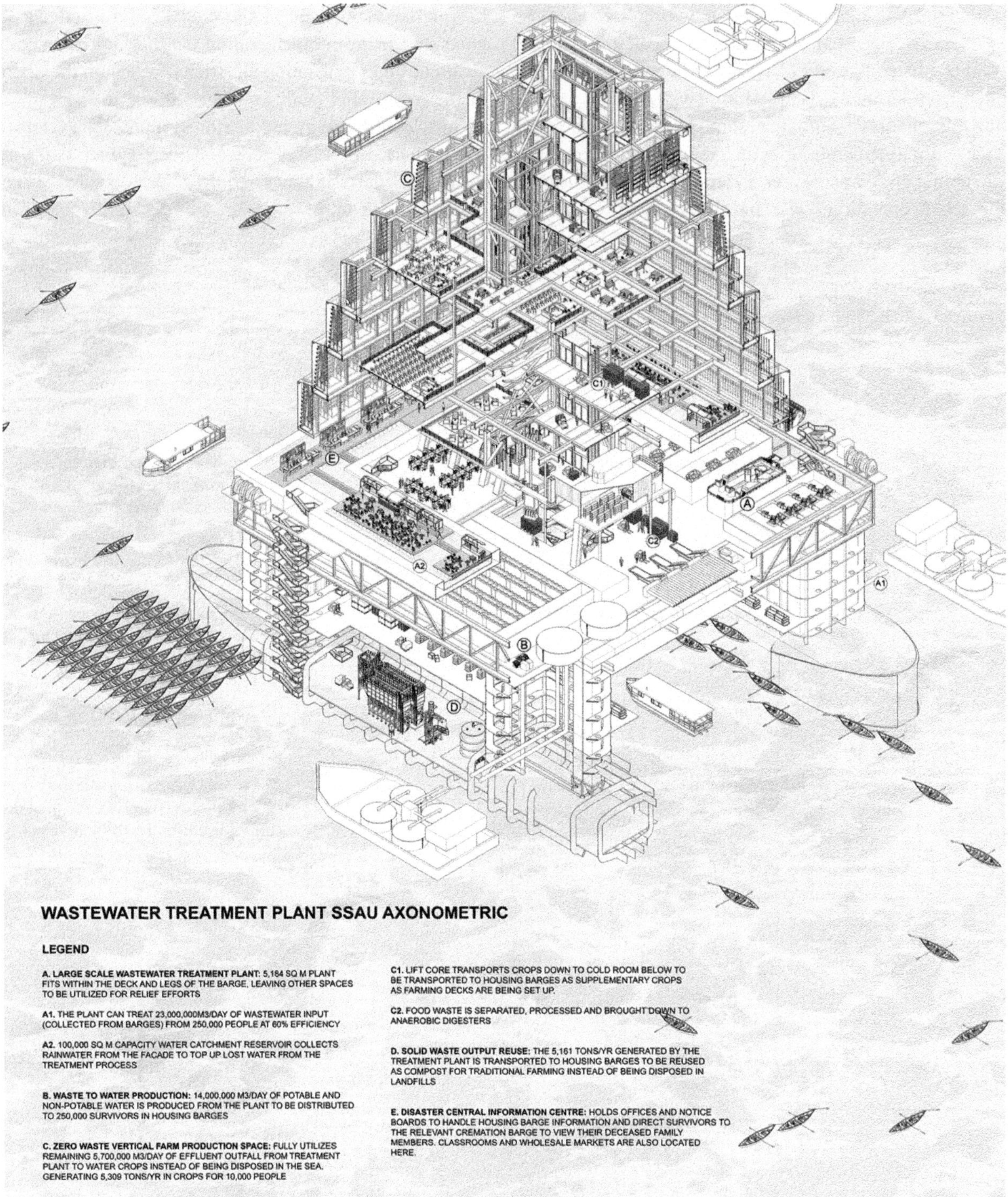

Fig. 19 This repurposed semi-submersible wastewater treatment facility is kitted with compact systems to meet the demand and supply of potable water during disaster relief efforts. The footprint of this specialized vessel is 6,400 m^2. It treats 23 million m^3 per year of wastewater for 250,000 people. Rainwater collection is estimated at 100,000 m^3 a year. One semi-submersible wastewater treatment facility provides 14 million m^3 of treated water and 5,309 tonnes of hydroponic vegetables a year. As 37,000 tonnes of vegetables and fruits a year (at five helpings a day) are needed for 250,000 people [2], the quantum per semi-submersible is only one-seventh of demand needed

protection of marine eco-systems is imperative for floating settlements. Residential, food production and consumption wastes are known types generated by settlement occupancy. While footprint sizes of treatment plants are unlikely to be standardized (e.g. sludge management facilities vary), systems design incorporating compact technologies with stackable configurations are preferred.

Compact technologies such as membrane batch reactors (MBR) use more energy than conventional activated sludge because of greater aeration requirements. Research into reducing the energy consumption of MBR processes considers the relationship between effluent quality and aeration rate [7]. The MBR recycling wastewater rates meet township demand capacities in compact sizes suitable for onboard integration with semi-submersible structures.

The population equivalents fed by compact wastewater treatment plants (chapter "Optimal Settlement Size and Masterplan Strategies", Table 1) on a single semi-submersible would form the basic unit of a settlement plan. There are three scales: specialized vessel, flotilla and fleet. One such specialized vessel combines hydroponic production with wastewater treatment in a semi-submersible rig as illustrated in Fig. 19. The wastewater treatment facility can provide 14 million cu m of treated water and 5,309 tonnes of hydroponic vegetables a year.

Wastewater from the housing barges is brought to the semi-submersible wastewater treatment facility via barges to be treated into potable water, effluent waste and solid waste. The potable water is reticulated to the housing barges, providing occupants with clean, fresh water without the threat of contamination, unlike groundwater from bored wells during post-tsunami Aceh, which was exposed to contaminants from the sea and from pathogens discharged by temporary settlements with improper or absent waste treatment systems.

Instead of being disposed into the sea, or taking up space onboard a vessel waiting to be transferred to landfill, the treated effluent waste can be effectively used in agriculture to sustain the livelihood of inhabitants. The additional decks on the semi-submersible wastewater treatment facility are used to integrate agricultural landscape, as effluent water is fed into the vertical hydroponic farms. The crops generated from these hydroponic farms are supplementary to the food generated from the suspended farming decks, which utilizes the solid waste as compost to grow vegetables on the superbarge housing.

In this respect, the semi-submersible wastewater treatment facility manages waste efficiently, providing potable water during disaster relief efforts and supporting the growth of supplementary crops on the housing barges. Also, biogas extracted from waste treated onboard can partially meet the energy needs of the hydroponic production integrated with the wastewater treatment process.

7 Superbarge Settlement

Post-disaster housing is fraught with delays in organization, dubious build quality and the absence of sanitary infrastructure and water supply. Superbarges by virtue of mobility and large displacement capacity can be refurbished into floating housing relief for post-disaster situations. The plight of refugee camps such as Kenya's Dadaab exist today since it was first established in 1992 for refugees of the Somalian war. The largest Dadaab camp, Hagadera has a capacity of 106,926 people within a space standard of 1.25 m^2 per person.

By comparison, one module of the superbarge housing as illustrated in Fig. 20 accommodates 5,184 people in a space standard of 2.5 m^2 per person. It does so with an ingenious tartan grid layout of container box housing. Food is grown on cantilevered farming decks with direct access to elevated housing. The cantilevered farm decks do not increase the footprint of the housing barges on water.

There are ample sanitary provisions on these housing barges. Communal bathrooms of three shower cubicles and two latrines are shared between 20 people. The wastes from these bathrooms are brought to the nearby wastewater treatment facility for treatment, reducing the risk of contamination and pollution from improper waste disposal. Spaces for drying laundry are also adequately provided to prevent mold growth, which can fester within the housing communities. Communal spaces such as playgrounds, classrooms, religious spaces, cooking and dining areas are designed to encourage the displaced community to gradually return to their normal daily activities (Fig. 21).

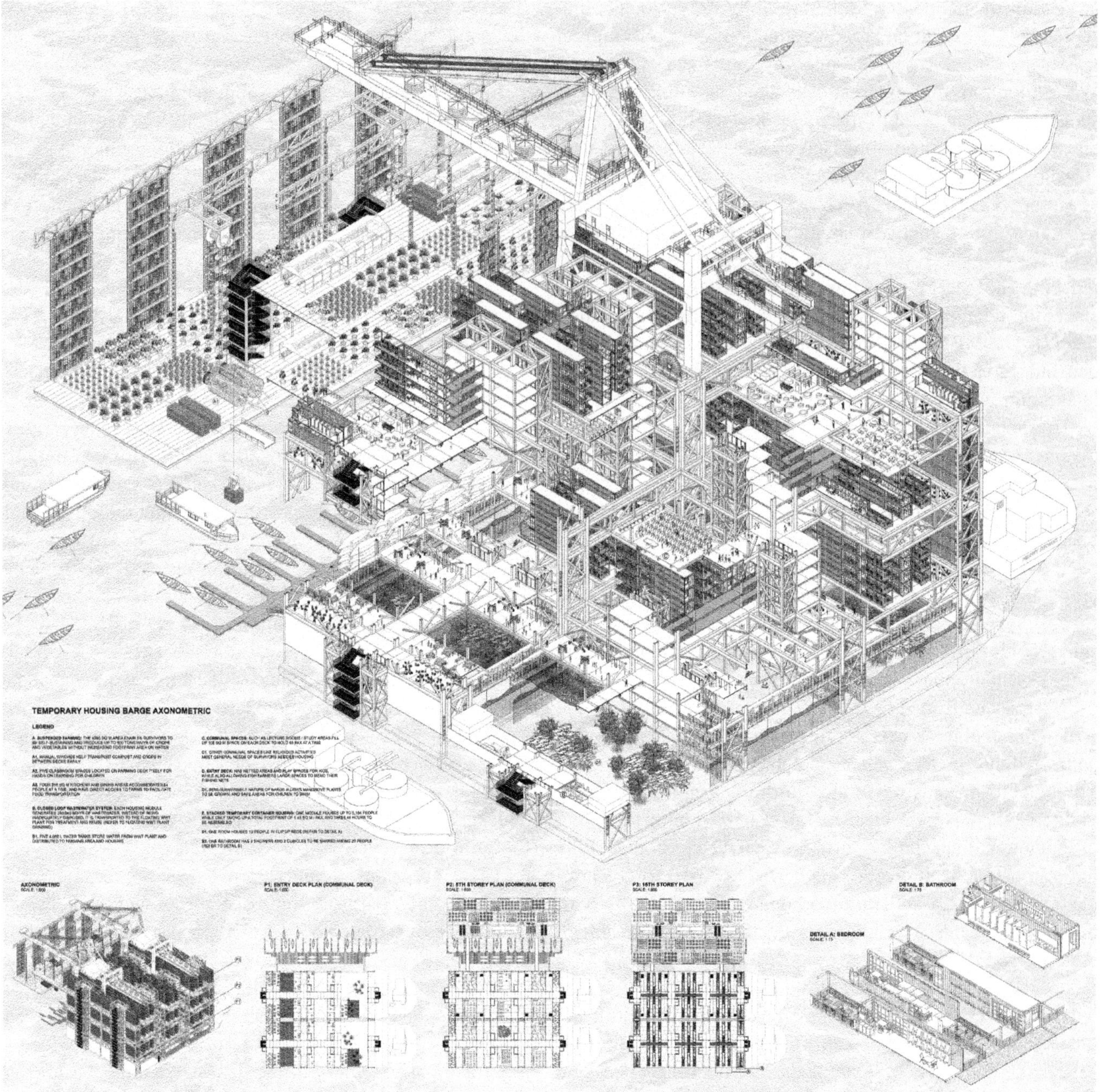

Fig. 20 One module of housing barge

Fig. 21 Communal atria within housing barge

Fig. 22 A superbarge settlement accommodating 107,712 people

Post-disaster settlement A post-disaster settlement comprising one semi-submersible logistics unit, one semi-submersible hospital, ten mobile crematorium barges, 106 temporary housing barges, and three semi-submersible wastewater treatment plants can provide relief with closed-loop resources for 549,504 people. This relief aid would have sufficiently provided for the 500,000 injured and displaced, and 168,000 deceased in Aceh during the Indian Ocean Tsunami disaster of 2004.

The relief settlement masterplan that has an immediate response time of 36 h and a total set-up time of 180 days can be fully operable from two to five years. Longer serviceable time frames are possible with proper maintenance or renewal of plant equipment. It takes 48 h to set up four housing barges, with on-board cranes. The housing units can be disassembled upon recovery of the disaster site, and can also be redeployed for subsequent disasters.

The housing barges have the advantage of long term accommodation with minimal food and water aid from organizations when permanent housing for the displaced people is extensively delayed. The modular nature of the housing barges allows for scaling upwards or downwards according to the requirements of the displaced population. Figures 22 and 23 illustrate the construction of a floating relief settlement, which can accommodate 107,712 people on 22 barge-housing modules. Table 4 details the operation capacity of a superbarge settlement.

Fig. 22 (continued)

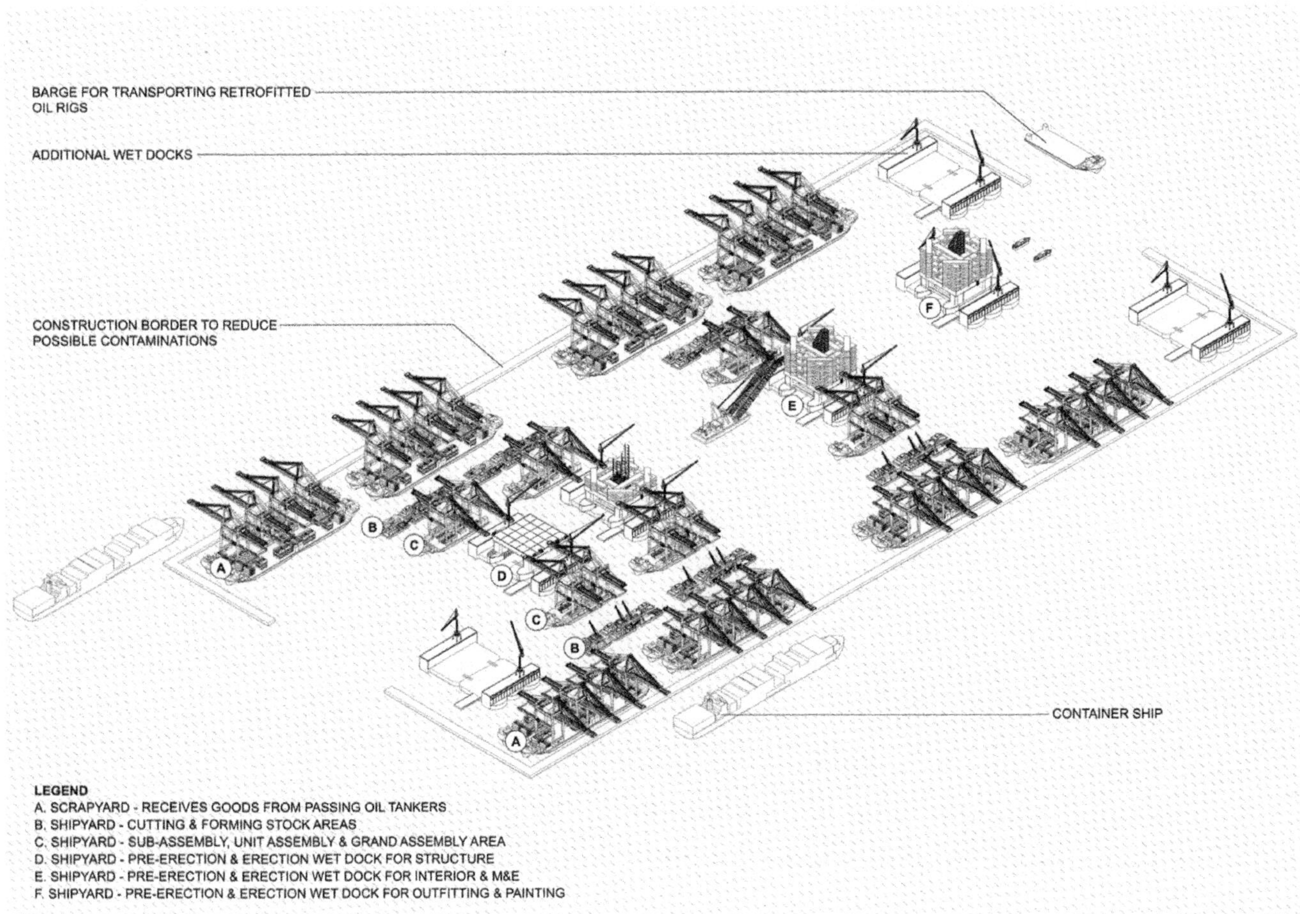

Fig. 23 Construction of a floating relief settlement

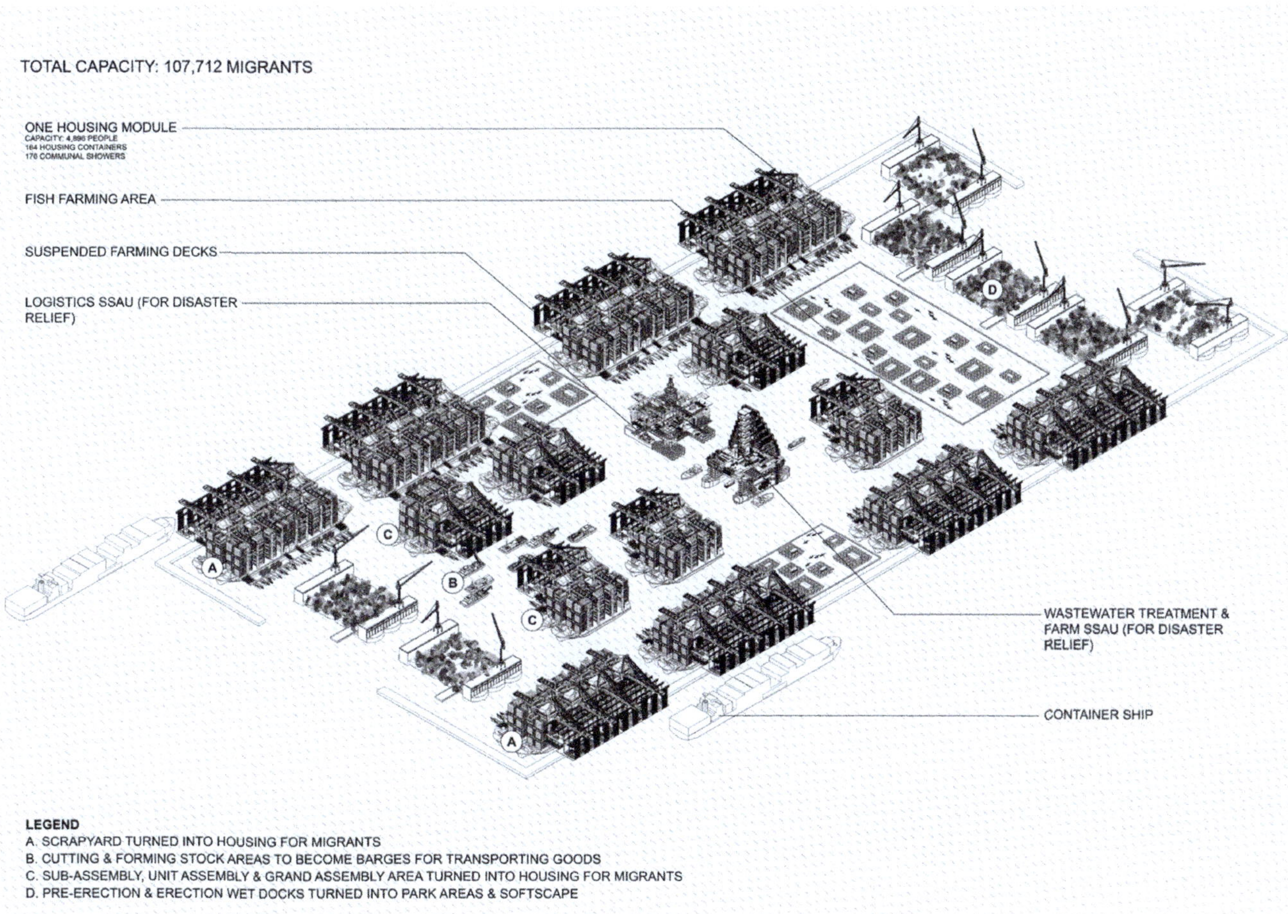

Fig. 23 (continued)

Table 4 Operational capacity of a superbarge settlement

Vessel type	Two superbarge modules	*Area comparison*		
Structural additions	• Two panamax ship-to-shore gantry canes • Steel frame structure to support container stacks		Hagadera refugee camp (largest in the world)	Superbarge settlement
		Footprint (m^2/person)	81.3	1.45
		No. of people	1,06,926	549,504 [106 modules]
		Per housing unit	10 m^2 per 8 people	30 m^2 per 12 people
Total settlement area	12,000 m^2	No. of superbarges needed for future projection?	106 [to accommodate 549,504 displaced people in a disaster of the scale as 2008 Aceh tsunami (4th largest in the world)]	
No. of people per superbarge module	5,184			
No. of days to assemble each module	Cargo arrives at disaster site in 7 days and assembled within 48 h			
No. of containers	624 × 40 ft containers [housing, bathroom] 648 DNV crash container frames [verandahs, corridors]	Energy source	LNG from cremation barges	
No. of people per housing unit	12	Waste type and treatment strategies	per superbarge module 477,358 m^3 wastewater per year treated at three WWT SSAUs [treating capacity of 69 million m^3 wastewater per year]	
Total bathroom area	2,880 m^2			
Cooking and food preparation	432 m^2			
Dining area	432 m^2			
Farming yield capacity	575 tonnes/year [6,393 m^2]	Water supply	42 million m^3 potable and non-potable water produced by three WWT SSAUs per superbarge module consumes 285,716 m^3 water per day	
Learning spaces	594 m^2			
Play and recreational area	1,296 m^2			

References

1. Chartier Y (ed) (2014) Safe management of wastes from health-care activities. World Health Organization
2. Gabbatis J (2018) Independent. Environment. Fruit and vegetable waste from farms 'could feed population of Birmingham or Manchester for a year,' says environmental charity
3. Goh CL (2010) Straits Times. NEA studying feasibility of setting up a plant to recycle Pulau Semakau's burnt waste
4. Leaning J, Guha-Sapir D (2013) Natural disasters, armed conflict and public health. N Engl J Med
5. National Environment Agency. Waste-to-energy incinerator plants waste management infrastructure. https://www.nea.gov.sg/our-services/waste-management/overview
6. Red Cross. Surgical Field Hospital. https://www.redcross.fi/about-red-cross/our-work-throughout-world/surgical-field-hospital
7. Sun J et al (2016) Reducing aeration energy consumption in a large-scale membrane bioreactor: process simulation and engineering application. Water Res. Elsevier
8. UNFPA (2018) Regional overview: Asia and the Pacific
9. UN Office for the Coordination of Humanitarian Affairs (OCHA) (2013) Disaster response in Asia and the Pacific
10. World Health Organization (2011) Guide to ship sanitation. Waste management and disposal, 3rd edn. Geneva

Other Forms of Repurposing

Abstract

This chapter explores six specific applications when repurposing oil rigs into specialized vessels. The proposals are premised on nearshore/offshore prototypes which improved by design, the challenging issues of inefficient land plot requirements (e.g. prisons, funeral facilities, resorts) or unsustainable business models (Olympic stadia) or which improved industrial workflow (LNG bunkering). This leads to a streamlining of processes saving time and land resources. A jack-up funeral complex integrates wake halls, chapels, funeral parlors and crematorium in a single nearshore facility. A prison on a semi-submersible can be a solution to unsanitary conditions of many prisons worldwide that are operating in severe overcapacity. A fleet of semi-submersibles are converted into floating stadia capable of staging international sporting competitions involving games, courts and fields and which can be mobilized and reused at sea. Jack-up oil rigs arranged radially to integrate LNG bunkering with wharves in a nearshore facility can reduce time taken for loading and unloading cargo separately from time needed for bunkering. With jack-up rigs, recreational parks requiring large amounts of water can be placed nearshore so as to free up coastal land for strategic and essential developments. The propostion of offshore resort to mitigate environmental impact is explored with a semi-submersible rig and a jack-up rig of 5,000 m^2 deck area, to accommodate the number of hotel rooms and a range of facilities recommended in the hospitality industry.

1 Jack-up Funeral Facility

1.1 Unsustainable Burial Practices

The notion of population growth does not just translate into housing issues; it extends to land needed for burial. But more than finding pragmatic and utilitarian solutions to deal with the end of individual lives, planning considerations for burial or cremation involves considerations of religious practices and family preference. While cremation is appropriate in locations of land scarcity, it is not for groups who prefer burial as religious practices. Only under exigencies such as post-disaster situations or extreme remoteness and costs of moving the corpse is burial tradition broken. The ceremonial and undertaker processes performed in morgues, cemeteries, wake halls, funeral parlors, churches, crematoria and columbaria practiced in different locations are compounded by land scarcity for burial.

Funeral parlors in Singapore are located in the industrial part of housing estates such as Ang Mo Kio, in city fringe Lavender Street, at peri-urban Lim Chu Kang and in Mount Vernon near housing estates. The latter is being redeveloped into a new complex on a 1.1 hA plot, one-seventh of its original land size comprising twelve funeral parlor halls. The remaining site is currently being developed into premium public housing. Crematoria are located in Mandai Bright Hill for Buddhists and; in Lim Chu Kang and Kranji for all religions

The increase in global population and improved life expectancy leads to a growing aged population. Number of persons aged 65 years or over, is projected to triple for Africa, Asia, Latin America and Oceania regions (Table 1) by 2050. Singapore is the second country in the world with the largest percentage point increase in the share of older persons between 2019 and 2050 as seen in Fig. 1 [28]. While more resources are being put into healthcare and prolonging lifespan of populations, the projected number of burials or cremations (Fig. 2) poses a logistic challenge for the future demand on resources.

The lack of crematoria in Japan has spurred the funeral industry to use corpse hotels as a temporary holding area for the deceased and for their family to mourn [23]. In South Korea, there were 49 crematoria nationwide in 2010, and government's plans to build more crematoria are often met with opposition with nearby residents [10]. As such, Bucheon City with a population of about 900,000 does not

J. Lim, *Oil Rig and Superbarge Floating Settlements*, Lecture Notes in Civil Engineering 82,
https://doi.org/10.1007/978-981-15-5297-7_5

Table 1 Number of persons aged 65 years or over by geographic region, 2019 and 2050

Region	Number of persons aged 65 or over in 2019 (millions)	Number of persons aged 65 or over in 2050 (millions)	Percentage change between 2019 and 2050
World	702.9	1548.9	120
Sub-Saharan Africa	31.9	101.4	218
Northern Africa and Western Asia	29.4	95.8	226
Central and Southern Asia	119.0	328.1	176
Eastern and South-Eastern Asia	260.6	572.5	120
Latin America and the Caribbean	56.4	144.6	156
Australia and New Zealand	4.8	8.8	84
Oceania, excluding Australia and New Zealand	0.5	1.5	190
Europe and Northern America	200.4	296.2	48

Source United Nations, Department of Economic and Social Affairs, Population Division [28], World Population Prospects (2019)

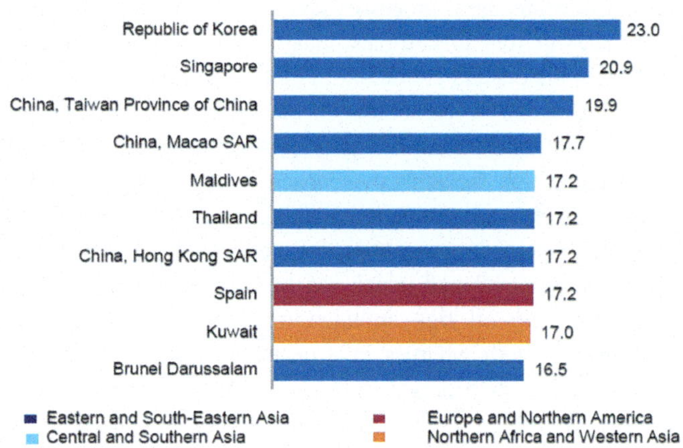

Fig. 1 Countries with the largest percentage point increase in the share of older persons aged 65 years or over between 2019 and 2050 (*Source* United Nations [28])

have its own crematorium [11]. Similarly in Hong Kong—one of the densest cities in the world—the cremation and burial process takes months longer than most other places. The lack of crematoria facilities and its construction delays are also attributed to public dissent and objection.

1.2 Urban Strategies

Urban burial grounds were conceived in response to problems with hygiene and overcapacity in older graveyards [16]. However, competition in land use meant that existing plots could not be easily expanded to keep up with the rising demands. These urban funeral facilities are in peri-urban or urban locations in order to be accessible to the public (Fig. 3).

In Israel and in Brazil, coffins are stacked in several floors/shelves for space efficiency in vertical cemeteries. Catholic cemeteries in New Orleans are towering tombs because burial is difficult in the city built on swampland. Necropole Ecumenica in Santos Brazil first built in 1983; is a tower and podium block 32 storeys and 108 m high, accommodating 25,000 storing units or tombs on leasehold. It has wake rooms, crypts and mausoleums with every floor having 150 mechanically ventilated tombs that can accommodate six bodies each. Similar proposals are being made in Oslo, Verona, Mexico City, Mumbai and Paris, as a result of future land constraints.

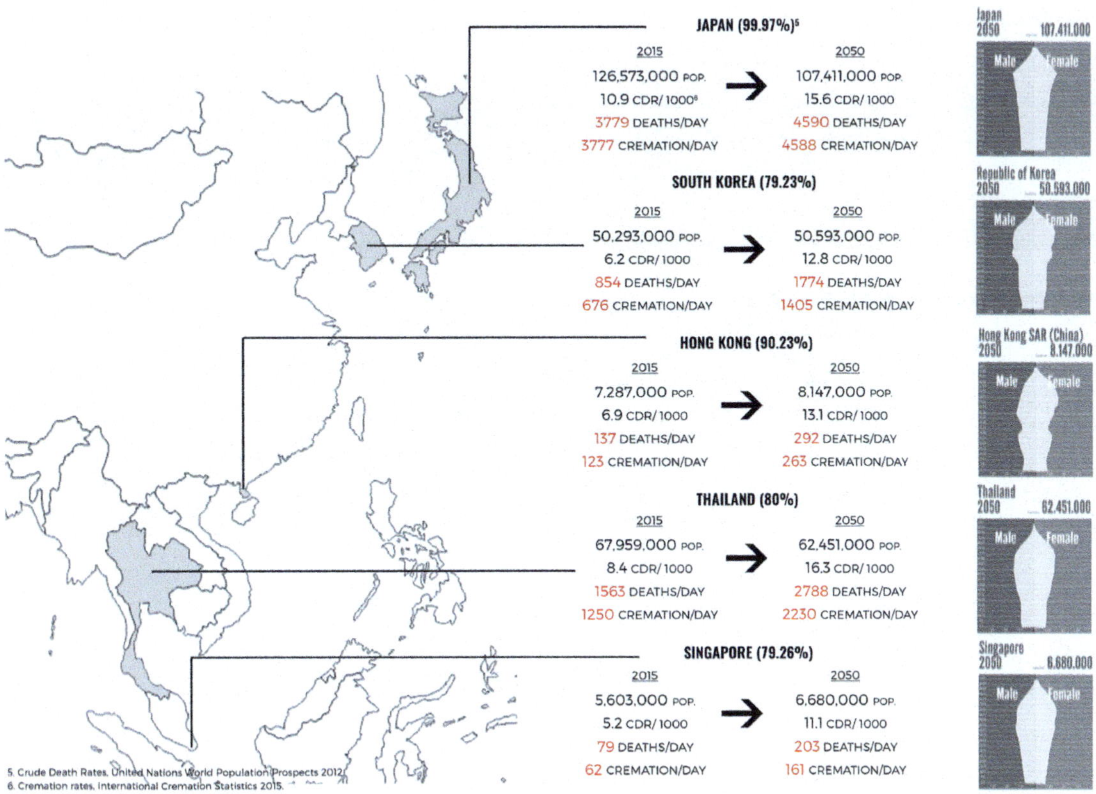

Fig. 2 Regional cremation and projected crude death rates for 2050 (*Source* United Nations World Population Prospects 2012, International Cremation Statistics 2015)

Large-scale viable solutions are needed worldwide when an additional 6,500 km^2 of burial land is anticipated in 2050. This is equivalent to 9.2 times the land area of Singapore. Cities are finding it difficult to sustain urban cemeteries as landscaped parks in order to portray the idea of civility in a dedicated zone outside the urban environment of a city [35]. But the integration of funeral facilities within the urban fabric would require a paradigm shift in planning as well as in the perceptions of environmental and health concerns with respect to the increased proximity of incinerating processes in cremation. Apart from existing and proposed vertical cemeteries, a funeral complex may be planned as a new civic space to allow for public engagement with the topic of death and remembrance in a city community [25].

Alternative solution The funeral industry is besieged by land scarcity and environmental concerns including technical and operational regulations governing location and siting, waste management and structural considerations. The land intensive processes of death-related facilities can be replaced by a nearshore funeral facility, which integrates crematoria practices in a repurposed oil rig. Figure 4 illustrates three versions of offshore funeral facilities. Where a floating settlement is envisioned as an agglomeration of specialized vessels, the funeral facility—as one of these vessels—is contextualized within a neighborhood precinct with possible resource loops integrated with orchards, parks and farmland. There is the possibility of acquiring electrical power through a heat recovery process in cremation (Fig. 5).

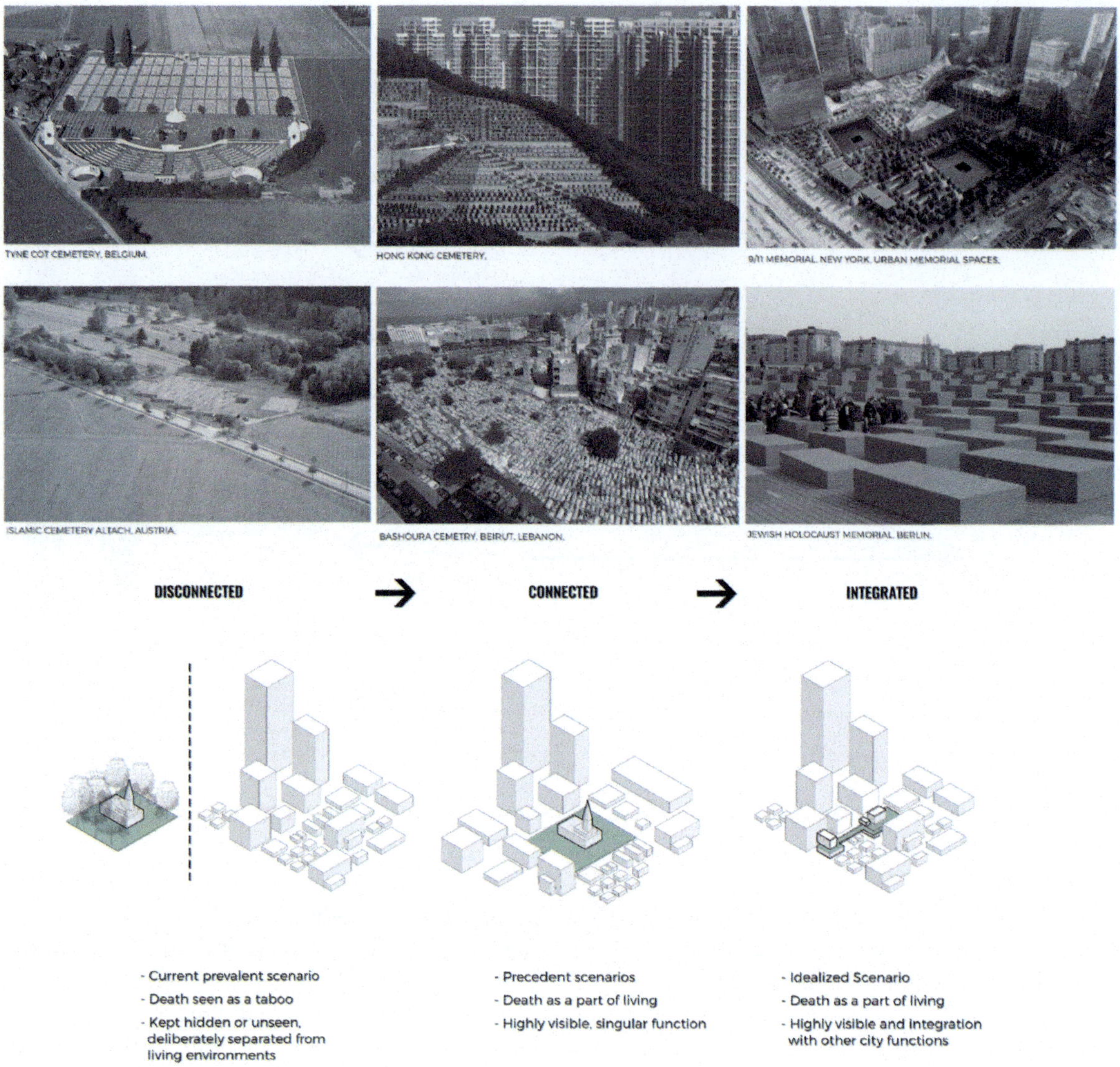

Fig. 3 Urban strategies integrating burial grounds with city

A mobile crematorium in a form of a barge can be designed as part of a response team in a post-disaster scenario where cremation facilities enable quick disposal of decomposing bodies before sanitary environments can be restored (see also Sect. 5 in chapter "Post-disaster Applications for Displaced Populations").

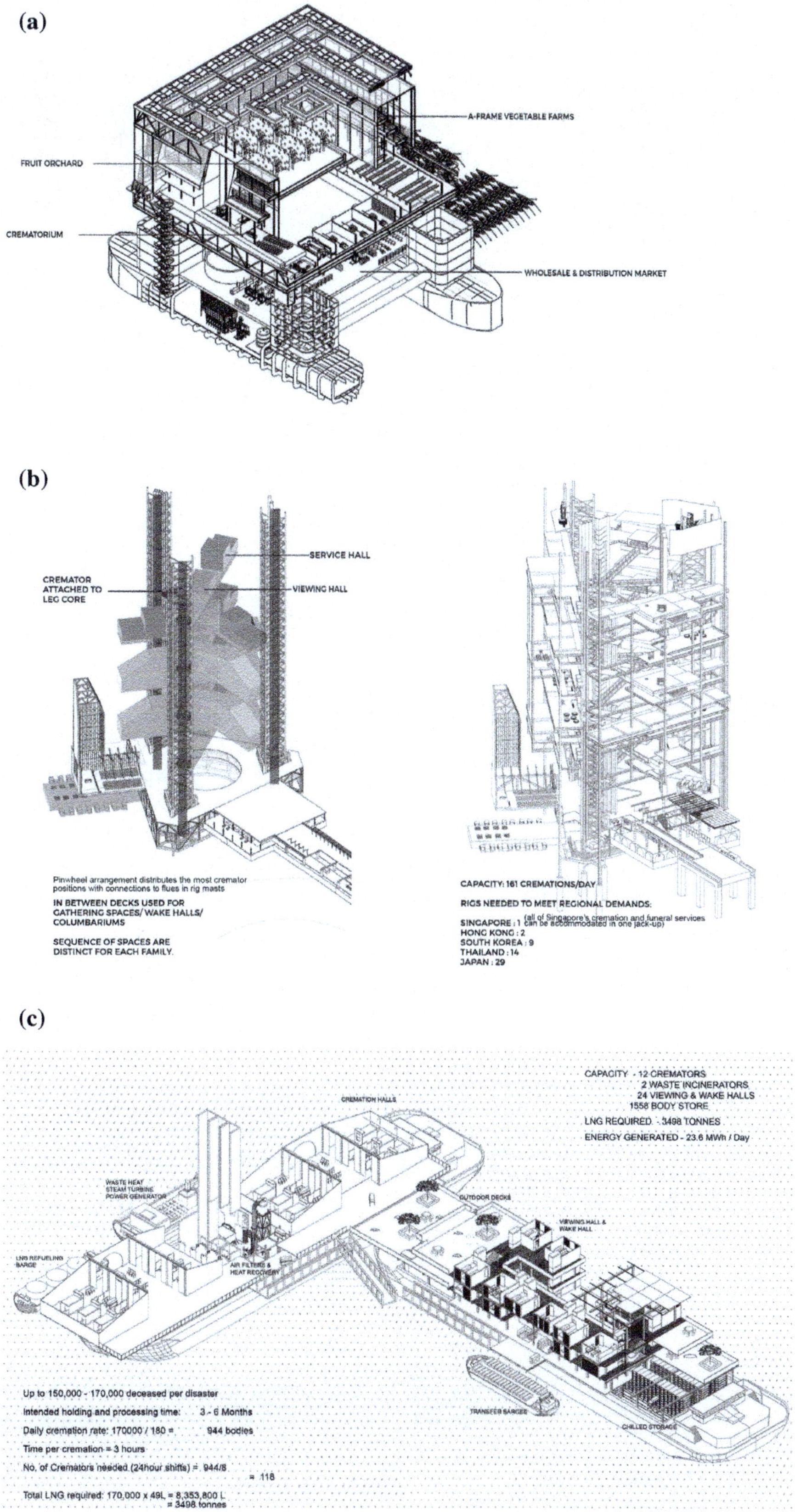

Fig. 4 Three versions of funeral and crematoria facilities on semi-submersibles, jack-ups and barges. **a** Floating settlement scale crematoria on a semi-submersible, **b** national scale funeral facility on a jack-up (Singapore population of 5.5 million), **c** mobile crematorium barge for post-disasters

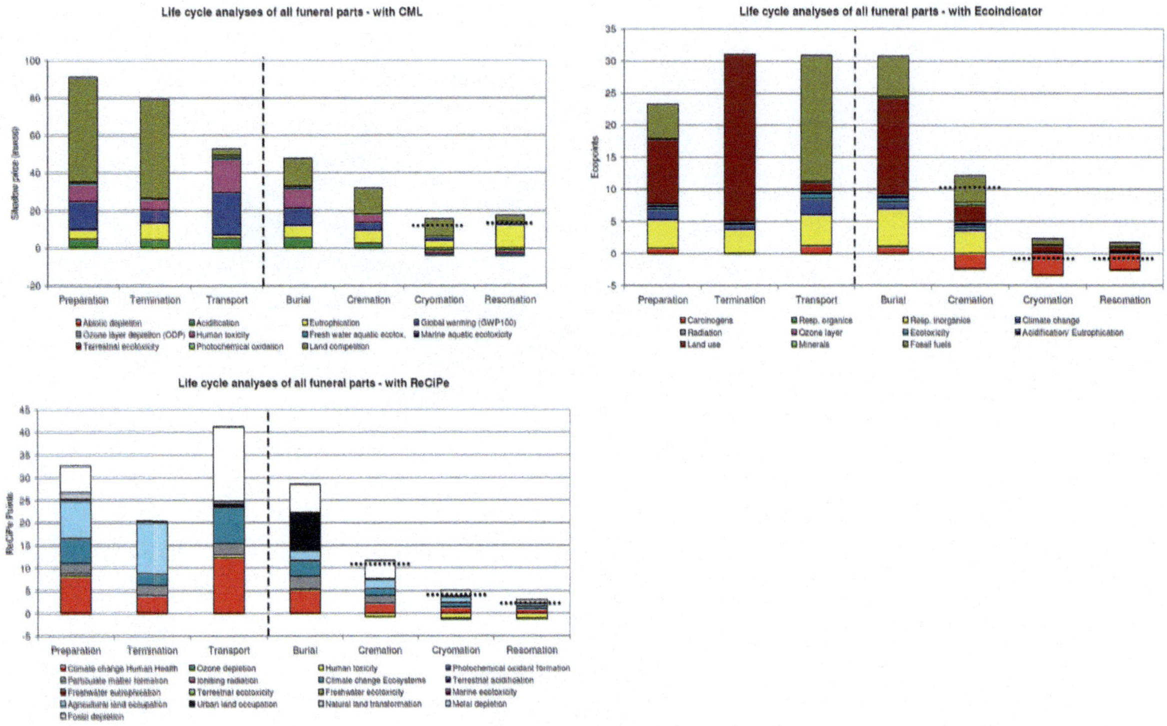

Fig. 5 Energy consumption in stages of funeral processes (*Source* Keijzer [9])

1.3 Funeral Processes

Workflow In Mandai crematorium, user zones and back-of-house are kept on separate levels. The floor plan is organized in horizontal and linear spatial distributions. User experience is homogenized, as there is little privacy between groups of mourning families. While way finding is clear in linear horizontal circulation, it however occupies a larger land area compared to a vertical organization of functions. Figures 6 and 7 detail the workflow and layout.

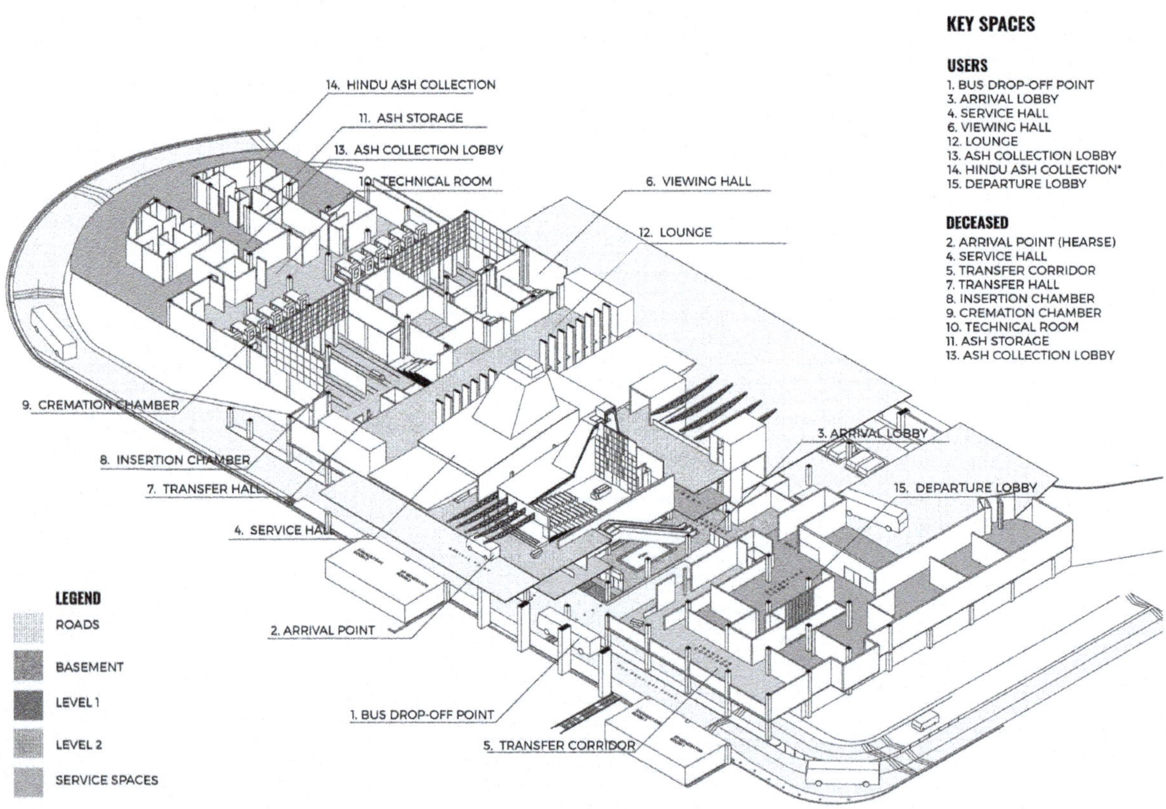

Fig. 6 Mandai crematorium workflow

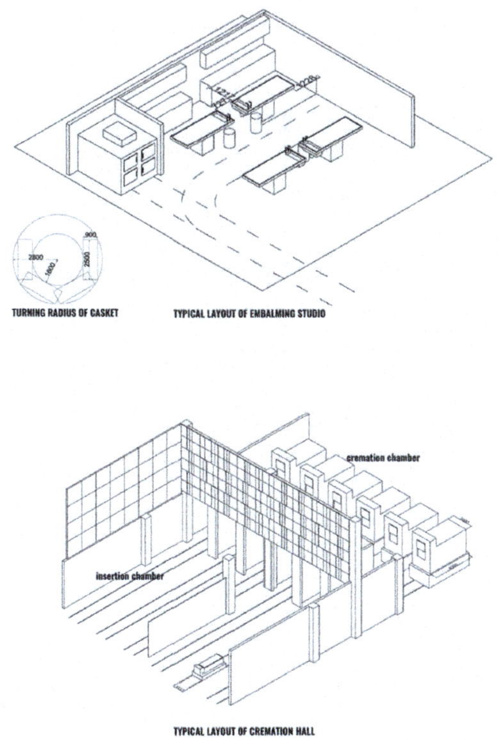

Fig. 7 **a** Typical layouts of an embalming studio and a cremation hall. Embalming tables should be a minimum of 2.1 m by 0.9 m. There should be at least 0.6 m clear space on all sides of the table. Biohazard wastes are collected in a biohazard wastebin during the embalming process. No blood waste should be discharged into the sewers/seawater. Biohazard wastes are incinerated. **b** Cremation equipment or cremators can weigh up to 14.5 tons and is housed in the cremation chamber [5]. A concrete plinth is required for the equipment to be placed onto. A length of track is provided to guide the insertion machine towards the cremator. The insertion chamber is the point of farewell for family members and the deceased. Each cremation cycle takes up to 90 min. Waste heat energy from flue gas can be harnessed through a boiler and used to generate electricity using a turbine. Flue gas will also need to be filtered before discharged

Resources and Energy Keijzer [9] identified three types of analysis models to determine the impacts of the funeral processes (Fig. 5). First, resource consumption is heaviest in the preparation, termination and transport phases of the funeral when preparation and termination locations are far apart in distance. Second, there is land competition when funeral facilities occupy peri-urban sites adjacent to nature reserve and water catchment areas. Third, decomposition methods such as Cryomation and Resomation may save land, but chemical waste disposal and their effect on environment need to be assessed. Cryomation processes use liquid nitrogen to freeze-dry the corpse before breaking it into disposable fragments. Resomation processes dispose of a corpse by submerging them in a chemical compound under high pressure and temperature to dissolve it. Table 2 summarizes the resource and energy consumption in these processes.

1.4 Cremation

Cremation Abatement Technologies The abatement technologies for crematoriums include the heat recovery and waste gas treatment. The flue gas has to adhere to local regulations on air emissions. For Singapore, the emissions are to comply with the schedule for standards of concentration of air impurities. The primary concern from crematorium flue gas is the mercury vapor that comes from incinerating dental amalgam. Mercury is itself a neurotoxin and thus could cause severe health impacts if left unchecked. Figures 8 and 9 show the technical systems and sequence of a cremation.

Determining the number of cremators on a jack-up funeral facility In the context of Singapore, the projected death rates in 2050 is multiplied by the national cremation rate of 79%, giving an estimate of 161 cremations a day. Assuming, one cremator is in operation 8 h a day and one cremation is 90 min duration, only five bodies can be processed in a day. Thus, 30 cremators are estimated per jack-up funeral facility.

Heat Recovery The temperature of flue gas from the cremators ranges from 800 to 1,000 °C. This gas has to be cooled before the cyclone and wet bed absorber for fine particles and mercury vapors can treat it. This falls under heat sources with a high temperature range, similar to industrial smelting furnaces. The most viable option to recover high temperature heat sources would be to use the traditional steam cycle, because of its low capital cost to energy production ratio [3].

A cremator capacity of 170 kg produces 1,798 m^3 per hour of flue gas (Fu Shen Crematorium EIA). If Singapore

Table 2 Resource and energy consumption of disposal methods

	Resources		Energy (kWh)	Carbon footprint (kg)
Resomation	15 l	Potassium hydroxide	90	10
	300 l	Water		
Cryomation	1 kg	Corn starch	130	15
	83 l	Liquid nitrogen		
Cremation	50 l	Natural gas	15	150
Embalmed burial	35 kg	Wood	1	160
	5 l	Embalming fluid		
	150 kg	Concrete		

Source Keijzer [9]

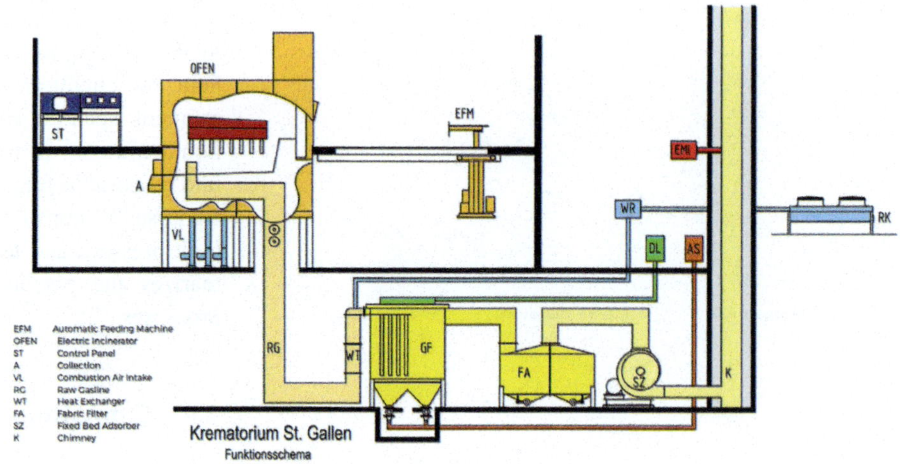

Fig. 8 Cremation technical systems (*Source* TREMA Verfahrenstechnik GmbH)

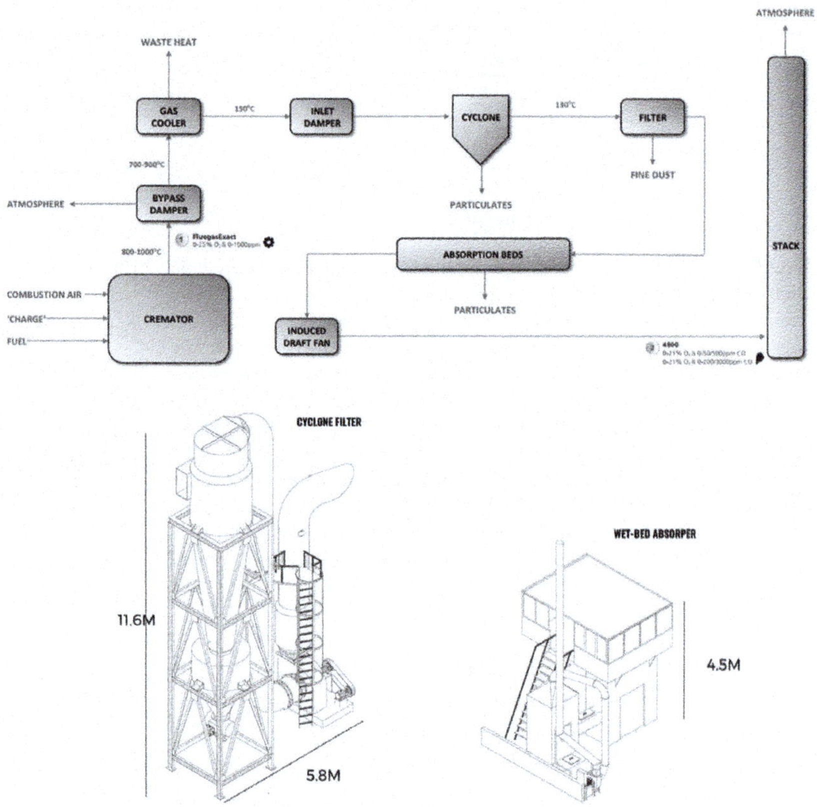

Fig. 9 Cremation workflow visualization (*Source* Servomax Crematoria Monitoring)

requires 30 such cremators to match the demand in 2050, the flue gas output is calculated at 53,940 m^3. This translates to three units of steam boilers, each with maximum volume of flue gas at 23,500 m^3 [2]. Potentially, each cremation in Singapore could recover up to 250 kW of energy and with 161 cremations per day in 2050, a total of 40.23 MW of energy can be recovered.

Exhaust Management The Singapore Urban Redevelopment Authority (URA) guide to air pollution control under "Place of Worship and Institutional Developments" require four compliances:

(1) "… Exhaust gases from the pollution control equipment shall be emitted into the atmosphere through a discharge stack of a height approved by PCD to ensure safe dispersion."

(2) "… within 100 m of any residential building shall use gaseous fuel or diesel with sulfur content of not more than 0.05% by weight."

(3) "… The minimum height shall be at least 3 m above roof level of the factory building or 15 m measured from ground level whichever is the higher."

(4) "... Monitoring equipment shall be provided at the discharge stacks and chimneys to monitor air impurities emitted. The monitoring equipment shall be installed in accordance with the technical specifications of the equipment supplier to give accurate readings."

As such, flue gas chimneys are conveniently located in the jack-up mast structures and are connected to the cremators. This forms a basis for distributing wake halls and ceremonial rooms in a pinwheel plan form, allowing for visual contact with sea surroundings in all spaces where external view is appropriate. The URA guideline is used in sizing the chimney dimension.

Sulphur content of LPG $< 0.1\%$, assume 0.05%

LNG usage = 12 kg/h/cremator

Cremators needed to meet Singapore's demand = 30 units

Chimney stack height $H = 14(Q)^{0.3}$

(where H is the height of chimney, Q is (quantity of fuel (kg/h) × sulfur content (%) × 2/100

$H = 14(12 \times 0.05 \times 2/100)^{0.3}$

= 7.4 m

Chimney exit diameter @ 15 m/s efflux velocity

$= \text{volumetric gas flow rate}/(3600 \times 15)(4/\pi))^{0.5}$

$= 20770/(3600 \times 15)(4/\pi))^{0.5}$

= 0.7 m

1.5 Siting and Location

Currently, crematoria are zoned under cemeteries with a built-in buffer from other forms of land use. To integrate funeral functions into urban programs, the crematorium would have to comply with guidelines on its waste output and environmental impact on its immediate vicinity. Taking into account the technical processes involved in cremation, crematoriums would need to fit definitions of a 'special industry' or a 'scheduled premise' in the URA Masterplan.

Under clause 2.4a of 'special industry' and 'scheduled premise': on which there is erected any boiler of steam generating capacity of 2,300 kg or more, there shall be a buffer distance of at least 500 m between the boundaries of a factory and the nearest residential building. Figure 10 shows a possible location of a funeral facility nearshore.

1.6 Ashes and Cremated Remains

The cremated remains of an average person weigh about 1.8–2.7 kg [33]. There are monetary and environmental implications to the retention or disposal of cremated remains when: (1) placed in an urn and interred in a columbarium, (2) dispersed at sea, (3) converted into compost/fertilizers for plants, and (4) used as aggregate for artificial coral reefs.

Case for Columbaria Columbaria are popular with Buddhists, Hindus and Christians as a means of remembrance. Public columbaria in Singapore offer 'standard' and 'family' niche spaces for S$500 and $900 respectively [17] However, the allocation of niche space is random and no reservations are allowed. Private columbaria—for example, Kong Men San Columbarium—allow specific niche spaces that range from S $2880 to S$9880 for a 60-year lease period.

The inherent problem for columbaria lies in its limited usage. Apart from special occasions like the Qing Ming Festival (All Souls Day equivalent), columbaria are often underutilized. The niche spaces are finite and would eventually reach its maximum capacity. Thus, ancestral archives and libraries in meditative sanctuaries may eventually replace physical niches.

Dispersal at Sea Dispersal of ash at sea is gaining popularity as a cost effective way to dispose of cremated remains instead of paying for columbaria niches and also to obviate the maintenance of graves on an annual basis [27]. Singapore has a designated site for the dispersal of ash at sea, located 2.6 km south of Pulau Semakau. A new burial facility built in 2019, is located along the shoreline in Tanah Merah, with a boardwalk that extends into the sea [24]. The U.S. Environment Protection Agency however requires cremated remains to be scattered at a minimum distance of three nautical miles away from land.

Ashes as compost/fertilizers The practice, in recent years, has been to plant cremated remains together with trees or plants in parkland or private gardens as a symbol of a life returning to nature. Products like Bio Urn allow cremated remains to be stored in a special capsule mixed with nutrients that will grow into a plant of family's choosing. The concept of greening the world even after death is a poetic and ceremonial gesture [1].

However, cremated human remains have a high pH and sodium level, which can be toxic to many plants. But once neutralized and treated, the cremated remains are suitable for greening and may provide an alternative way to pay tribute to the deceased, in a resource scarce future. Several plant species such as the wheatgrass, wild rye, barley, hawthorn and rose have a range of survivability in alkaline and sodic soil conditions [18]. These could be cultivated for grain, fodder, fruits and flower and simultaneously, contribute to habitat for wildlife.

In Singapore, an ash scattering facility, 'Garden of Peace' located at Choa Chu Kang cemetery is scheduled for completion in 2020. The garden is designed as a secular facility and would be open to people of all faiths. NEA said Singapore has been studying the option since 2014 alongside consultations with religious groups, after-death care service providers and the public. The service will also be implemented in the Mandai Crematorium Complex in 2021 [20].

Fig. 10 Singapore map showing the existing cemetery areas and possible location of a nearshore funeral facility

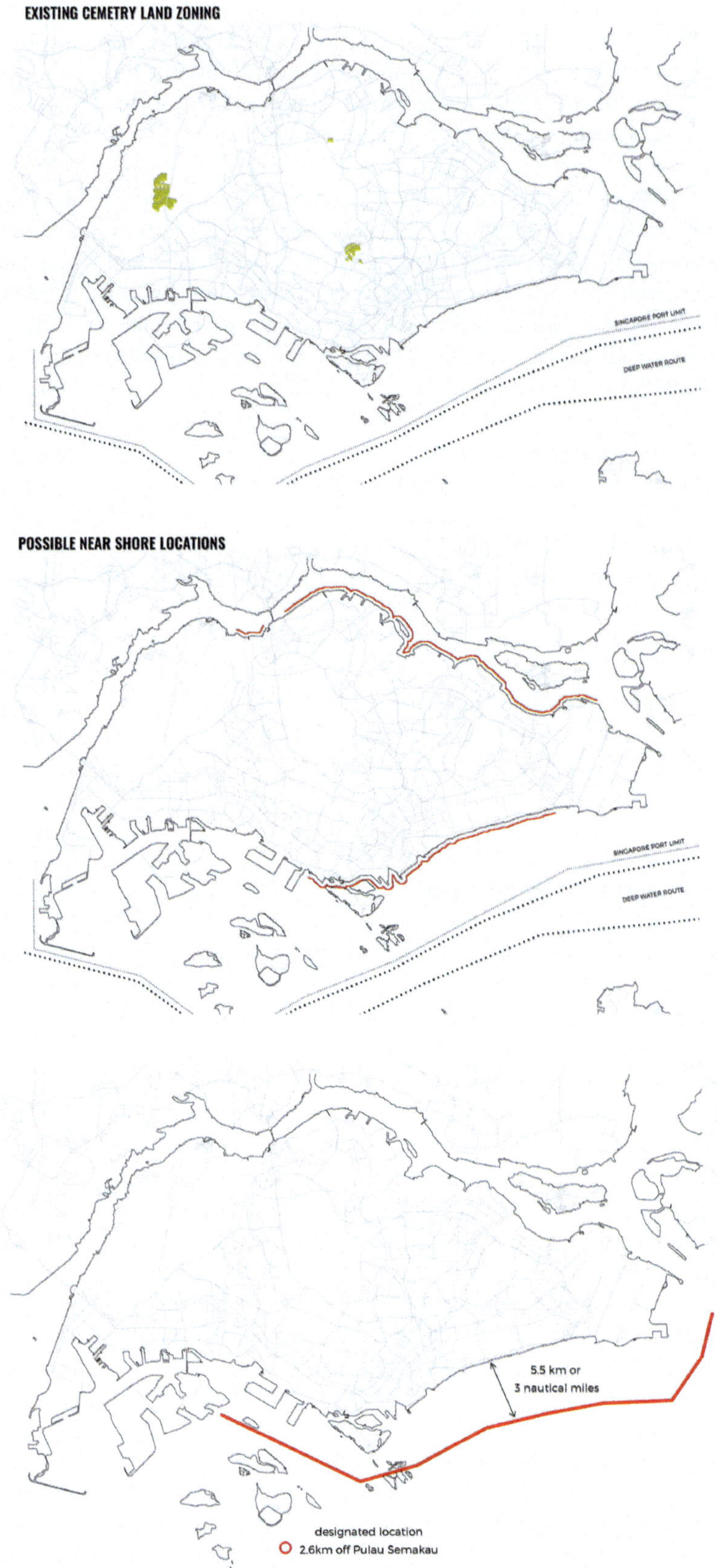

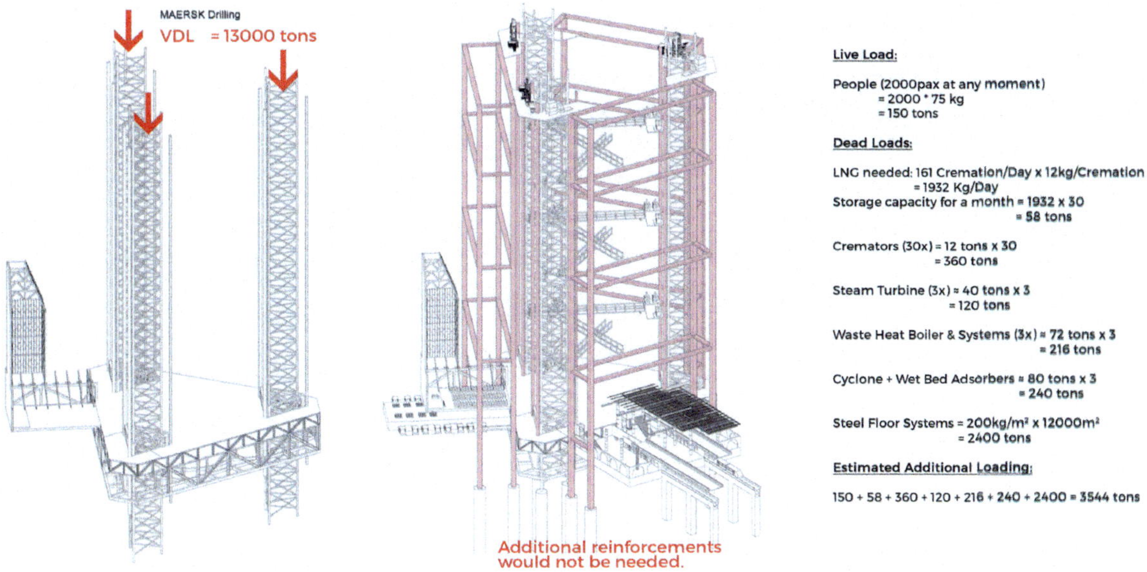

Fig. 11 Structural considerations. An estimate of additional live and dead loads of the proposed funeral programs is calculated and found to be within the capacity of the existing variable deck load of a three-legged jack-up

Fig. 12 Key cremation and waste management processes organized along masts of jack-up

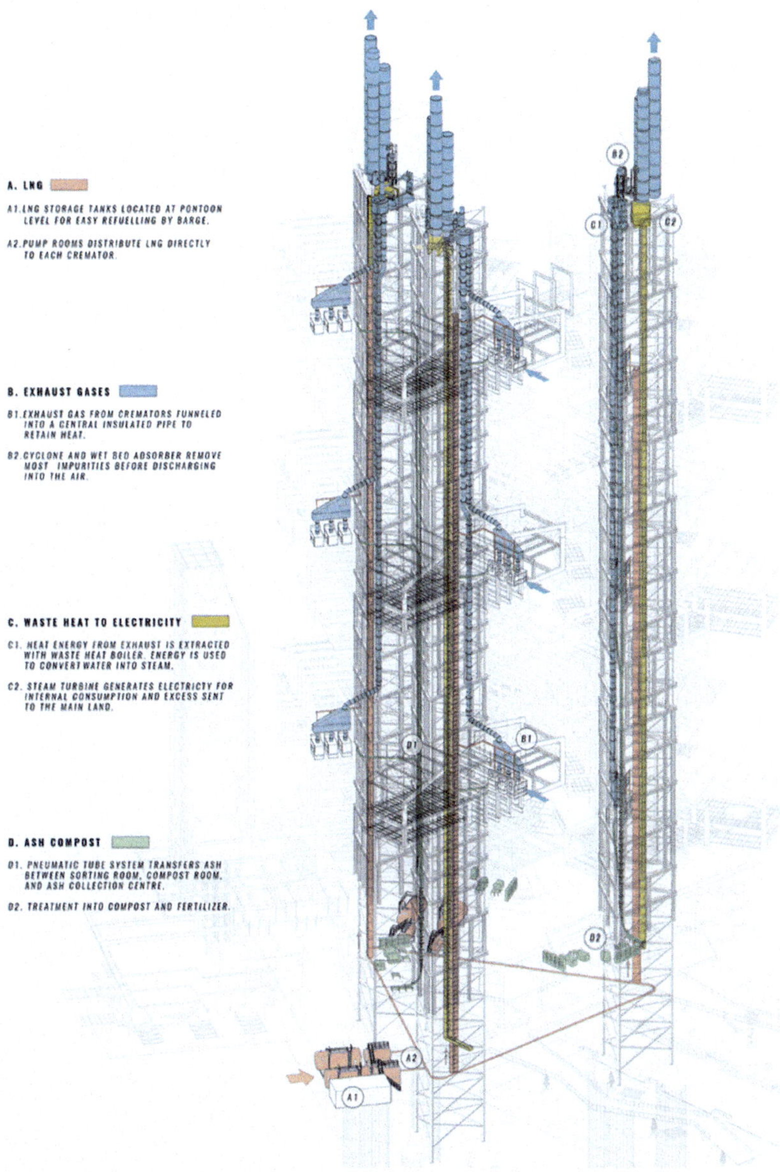

1.7 Design in Detail

Figures 11, 12, 13 and 14 illustrate the spatial sequence, structural and construction considerations of a jack-up funeral facility. Its capacity and waste reclamation capability is summarized in Table 3.

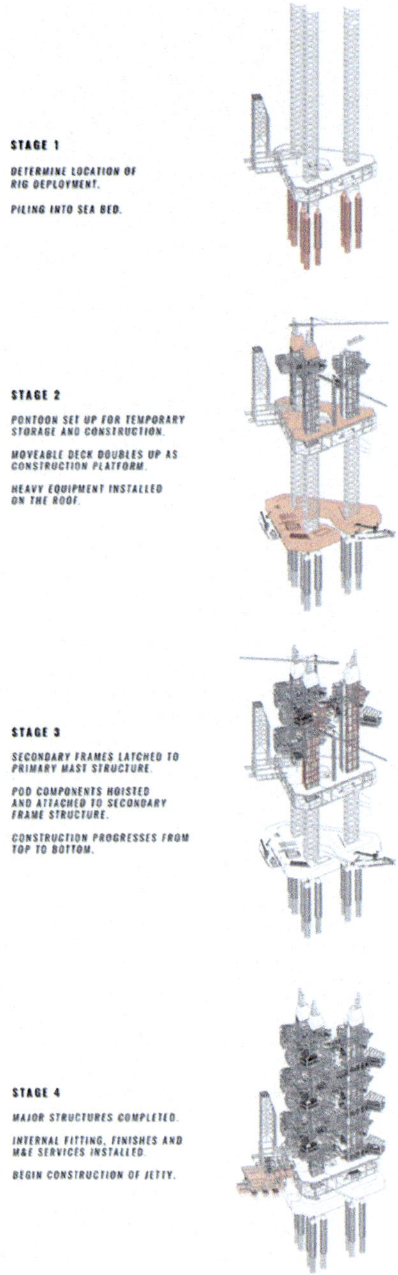

Fig. 13 Construction sequence

Fig. 14 Spatial sequence. **A. Arrival Boardwalk** Families arrive via a scenic boardwalk overlooking the sea. The linear water body set into the boardwalk is for 'the festival of lights' and for Hindu ash dispersal. Vehicular access is limited to passenger drop-off, hearses emergency and goods delivery. **B. Central Atrium** A large atrium, the full height of the oil rig structure is the central space, which orientates visitors before approaching elevator lobbies. **C. Service Hall** Service halls are used in 45-min intervals where the last rituals are performed. Here the casket is lowered to the insertion chamber. Family members then proceed along the viewing corridor parallel to the insertion chamber. **D. Wake Hall** Wake halls are used for up to a week for families and friends to gather, mourn and to pray. On cremation day, family

members accompany the deceased across a bridge toward the insertion chambers. **E. Insertion Chamber** Caskets approach the cremators with an insertion machine. Families witness the final send-off at the viewing corridor parallel to the chamber before proceeding to the garden pods for a reception. **F. Ash Collection Center** Families can collect the ashes on the same day or the next day in private rooms with a sea view. Unclaimed ashes are sent to the hull to be processed into compost. **G. Back-of-House Services** Embalming, ash composting, equipment storage, offices and staff accommodations are located at the lowest hull level. Ash sorting takes place at the ends of insertion chambers. Heat recovery, power generation and exhaust filters are located at the top of each mast structure. **H. Waiting Deck and Jetty** Families wait on decks with cafes with a panoramic sea view before boarding a boat to disperse ashes further out at sea

Table 3 Operational capacity of a jack-up funeral facility

Rig type	3-legged jackup	Energy source	LNG
No. of wake halls	9		
No. of service halls	9		
No. of reception garden decks	18	Waste reclamation capability	126 tonnes/year of ash compost which can grow 61,300 trees/year or 12,678 tonnes/year of tomato fruit or 40 MWh electricity daily
No. of cremators	27		
Area of public rituals	2,000 m^2		
No. of rigs to accommodate world cremation figure in 2050?	2,165 [10% of land-based footprint]		

2 Semi-submersible Prison

2.1 Global Problems

With increasing population, land and resource scarcity, conventional land-based prisons are not sustainable. Although prison complexes are land-intensive globally, many are overcapacity in cities with high conviction rates. More than 10.74 million people are held in penal institutions worldwide with the U.S. having the highest prison population rate at 655 per 100,000 people totaling 2.1 million in prison. The world prison population grew by 24% since 2000, proportionately to global population increase [32].

After the 19th century hulks, the last British prison was *HM Prison Weare* berthed at Portland Port. It was bought by the prison service to ease overcrowding in British jails. It could only hold 400 inmates in claustrophobic conditions in cells without a sea view, lacked adequate exercise space for inmates and had insufficient daylight conditions. Its permit expired in May 2006 [15].

Oil rigs can be repurposed into offshore prisons that support rehabilitative programs in a paradigm shift from those of incarceration. Rehabilitative activities involve food, energy and water production in excess of inmate demand for the benefit of settlement populations. Prison complexes attempt to reduce operational expenditure attributed to energy and manpower costs of wardens, medical and technical crew. Case studies in prison operations and workflow informed the planning of space in the proposed offshore semi-submersible prison. In addition, energy and wastewater treatment plant enabled closed resource to waste loops, to minimize pollution to the marine environment.

Overcrowding, Poor Quality of Prison Spaces and Recidivism Prisons worldwide face the problem of overcrowding. In 2016, nearly a million inmates were confined in extremely inadequate living spaces [34]. Overcrowding results in poor ventilation of prison spaces that can lead to epidemics. Prisoners are five times more likely to be HIV positive and tuberculosis notification rates are 11 and 81 times higher in European prisons than the general population. Data from 2016 suggests that at least 115 countries exceeded their official prison capacities with 51 countries in extreme overcrowding at above capacity in excess of 150% [22].

Where cell space is underprovided, it is unlikely that spaces for skills, training and rehabilitation would be possible. The outcome is that inmates cannot develop useful skills to reintegrate into society after their sentence. Because of this, many inmates tend to revert back to a life of crime upon release.

High Cost of Incarceration and Shortage of Prison Wardens Incarceration is expensive. In U.S. and Singapore, it costs tens of thousands of dollars to keep an inmate in prison for a year. This sum pays for food, water and the salaries of the prison wardens. Every dollar spent on incarceration is a dollar away from schools and public infrastructure. Further, due to poor quality of workspaces and a high occurrence of riots, prisons often experience chronic staff shortage (Fig. 16). Less staff meant longer work hours and inadequate manpower for emergency situations such as riots and outbreak scenarios (Figs. 15 and 17).

Unmet Needs for Inmates with Special Conditions

Many prisons are ill-equipped to accommodate the special needs of juveniles, pregnant inmates [29] and the elderly, as these require specially trained staff and facilities resulting in higher costs not in conventional prison budgets. Without proper facilities to help special needs inmates cope with rehabilitation, their reintegration with society would also be difficult.

2.2 Comparing Existing Prisons

Existing prison models were studied to identify program components and workflow (Table 4). This determined relationships between space and layout for repurposing the oil rig structure. Design strategies to overcome the limitations of existing prison layouts related to sightlines in the configuration of the building/oil rig section and the provision of adequate exercise yards and space standards for incarceration and communal recreation using above and below deck spaces.

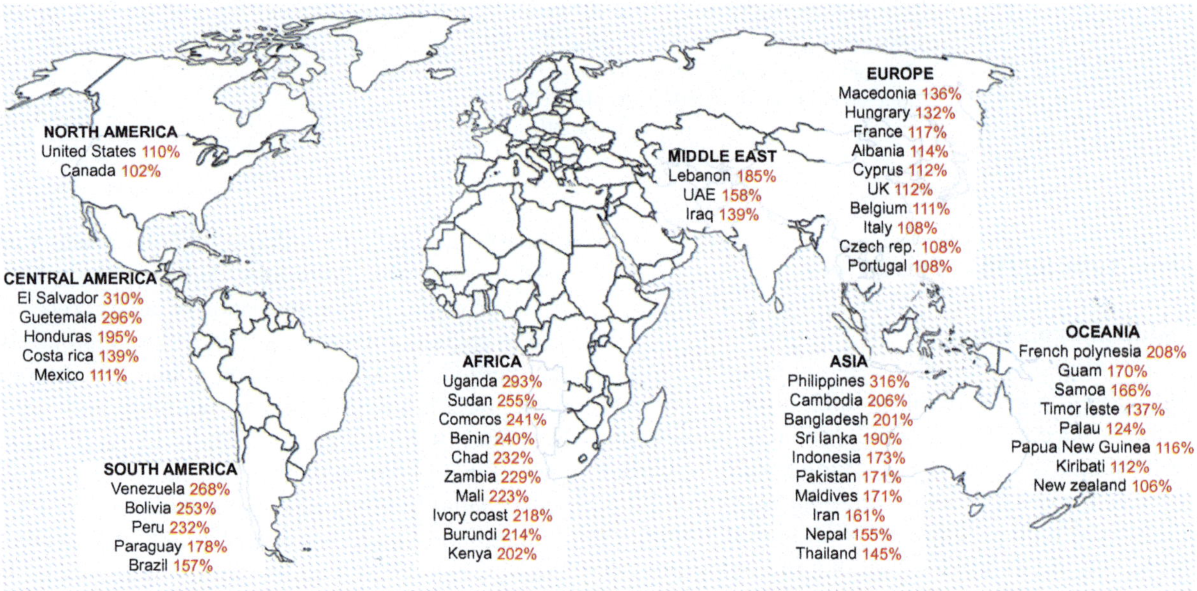

Fig. 15 Overcrowding percentage of prisons in the world. Prison capacity for 2016 was 955,146

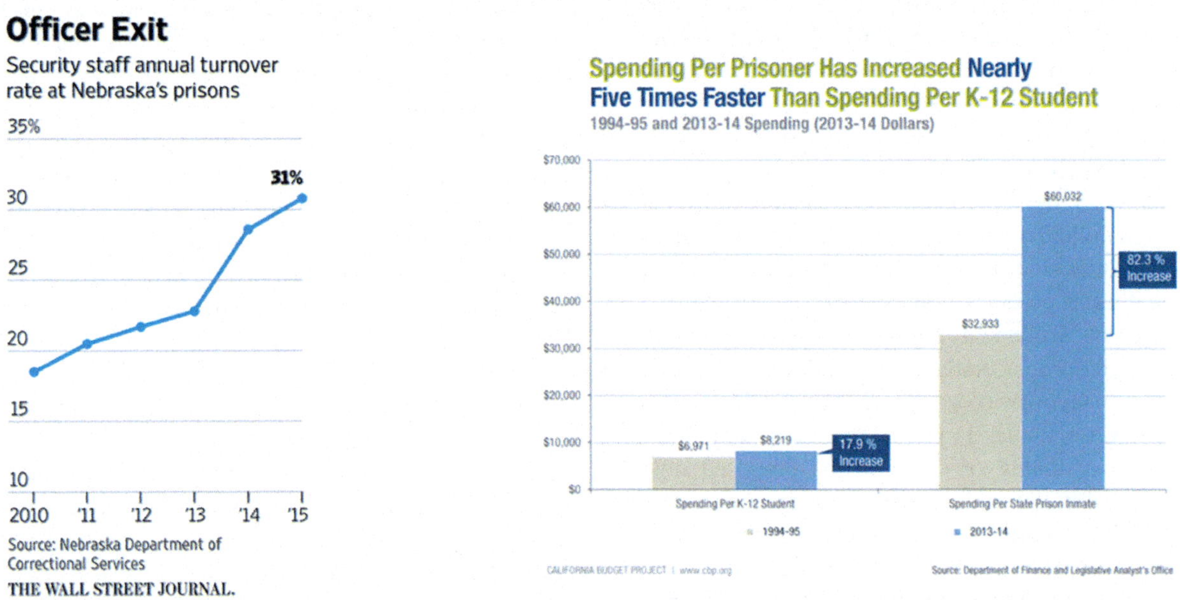

Fig. 16 Graphic trends on declining number of prison staff and rising prison spending

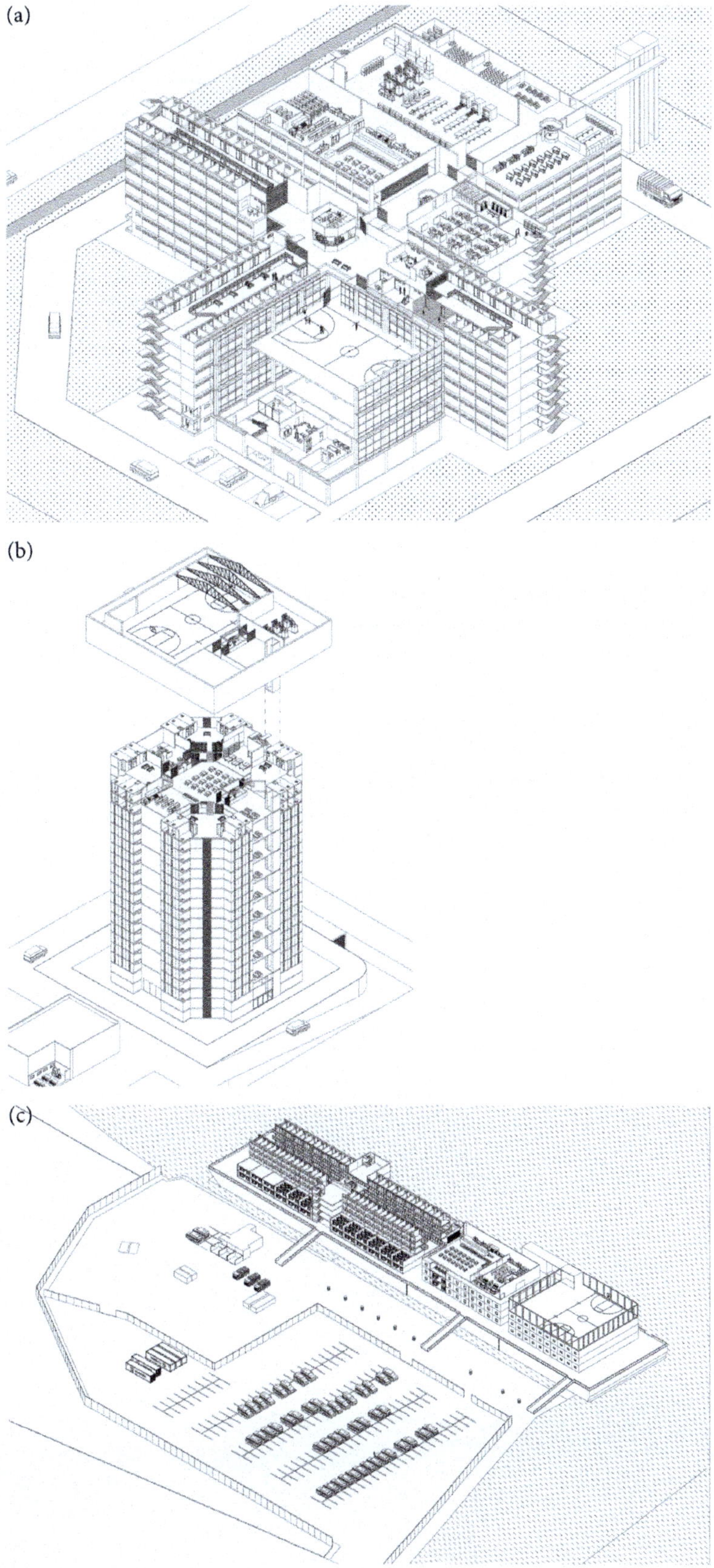

Fig. 17 Precedent prison and correctional centers (dimensions in meters). **a** Changi prison 106 × 112 × 25.6H, **b** Metropolitan Correctional Center 40 × 40 × 60H, **c** Vernon C. Bain Correctional Center 38.1 × 190 × 15H

Table 4 Comparing three prisons

	Changi Prison (Singapore)	Metropolitan Correctional Center (San Diego)	Vernon C. Bain Correctional Center (New York)
Typology	High-rise	High-rise	Barge vessel
No. of cells	768	432	16 dormitories and 100 cells
Dimension (m)	106 × 112 × 25.6H	40 × 40 × 60H	38.1 × 190 × 15H
Cells per 1000 m^3	2.5	4.5	–
Pros	Substantial provision for training spaces and medical services	More dense and compact than Changi Prison complex at 4.5 cells per 100 m^3	Built on a barge, it is able to sail to where there is a greater demand for prison
Cons	Complex takes up a large land footprint	• Not self-sufficient • Located next to residential • Limited medical services • Limited training/rehabilitation spaces	• Not self-sufficient • Far from city meant long travel times for deliveries and visits

Case Study: Singapore Changi Prison

Clear sightlines to each cellblock The three sides of the control room look into inmate cellblocks that are two storeys in height. The control room is strategically located in the center to achieve a clear sight of every cellblock (Fig. 18).

Controlled movement of inmates During transfer of inmates from cellblock to another location, the inmate must always move through the center to allow for warden to keep track of inmates. Between the control and other spaces, a sally port is provided to act as a buffer in case of a riot where lockdown is necessary (Fig. 19d).

Execution workflow The execution chamber is adjacent to the surgery and organ harvesting rooms. These lead directly to an ambulance on the ground floor. The colocation facilitates speedy extraction and preservation of the organs (Fig. 19a).

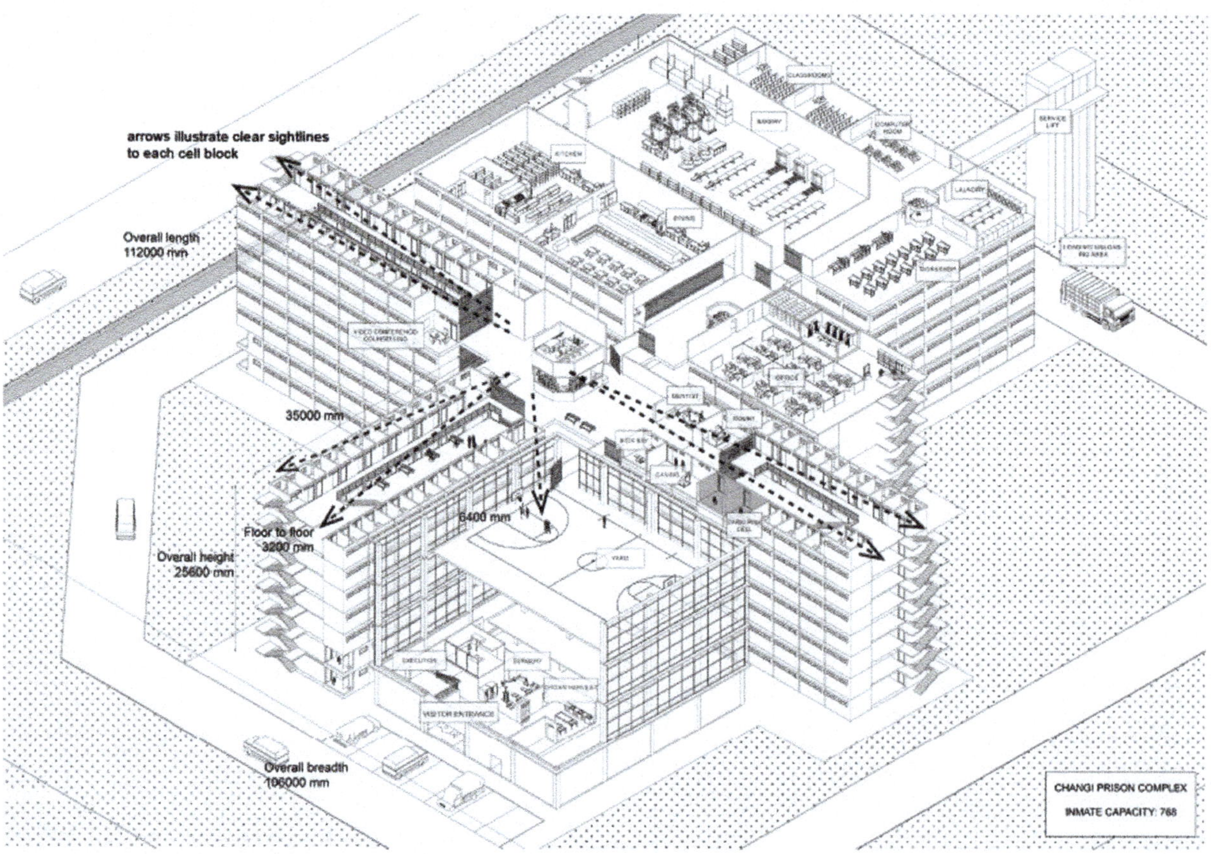

Fig. 18 Clear sightlines in Changi prison

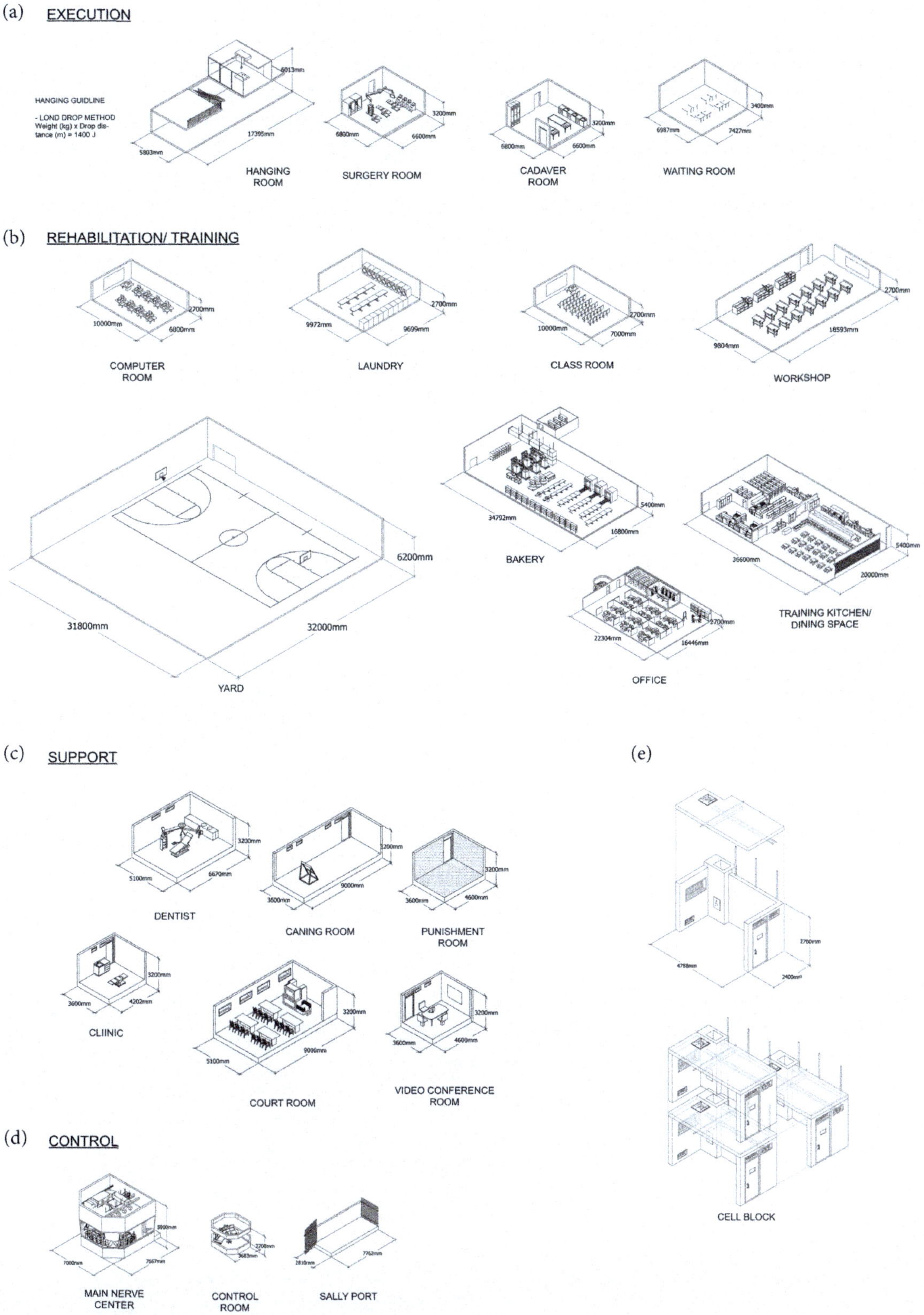

Fig. 19 Program and spatial components study of Changi prison

Support Rooms for caning are placed next to the sick bay, as inmates have to be screened by a doctor before, during and after punishment/caning is administered, in order to safely monitor the health of the inmate. A courtroom is also provided sentencing by the judiciary (Fig. 19c).

Rehabilitation/training spaces A variety of workspaces and programs are provided for inmates to be productive during incarceration while upgrading skills for future vocation. Inmates participate in the kitchen, bakeries and laundry, thereby minimizing external manpower needed to run the prison (Fig. 19b).

Cell construction The cells are designed in PPVC. Walls and floors are composed of concrete. The sanitary and light fittings, doors and windows are cast securely into the walls to prevent tampering (Fig. 19e).

2.3 Sectional Design of a Semi-submersible Cellblock

The Singapore Changi prison sits on a land plot of 11,972 m^2. To double the capacity of the Singapore prison on a semi-submersible deck area of 11,660 m^2, a stacking strategy is used. Here, four blocks of 20 storeys each occupied the semi-submersible deck. In each block, there are four sections—of cells and exercise void deck—stacked on top of each other. Each five-storey cellblock has one exercise deck to allow the daily routine of moving between decks to communal spaces and learning facilities.

Of great advantage is the waterbody sheltered beneath the deck and between the pontoons of the semi-submersible. This waterbody can be used for recreation, with surrounding spaces overlooking it in an arena-like configuration. The concentric water body to the outer edge of the pontoons serves as fish-rearing areas with a further outer moat of plankton serving as motion detector to prevent escape. The outermost layer is a ring of wave energy harnessers demarcating the boundary of the semi-submersible prison facility.

Figures 20, 21, 22, 23, 24, 25, 26, 27 and 28 illustrates in detail the design and security considerations in planning the internal spaces and circulation. Special provisions for juvenile, women and elderly inmates are also incorporated. Table 5 summarized the operational capacity of a repurposed rig prison.

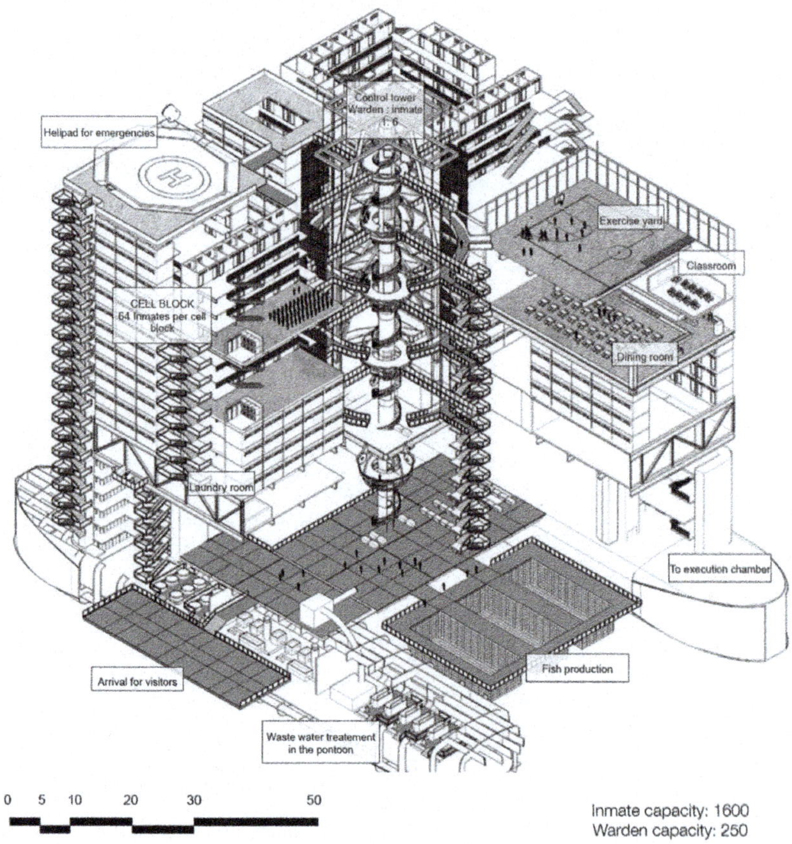

Fig. 20 Spatial components of a semi-submersible prison

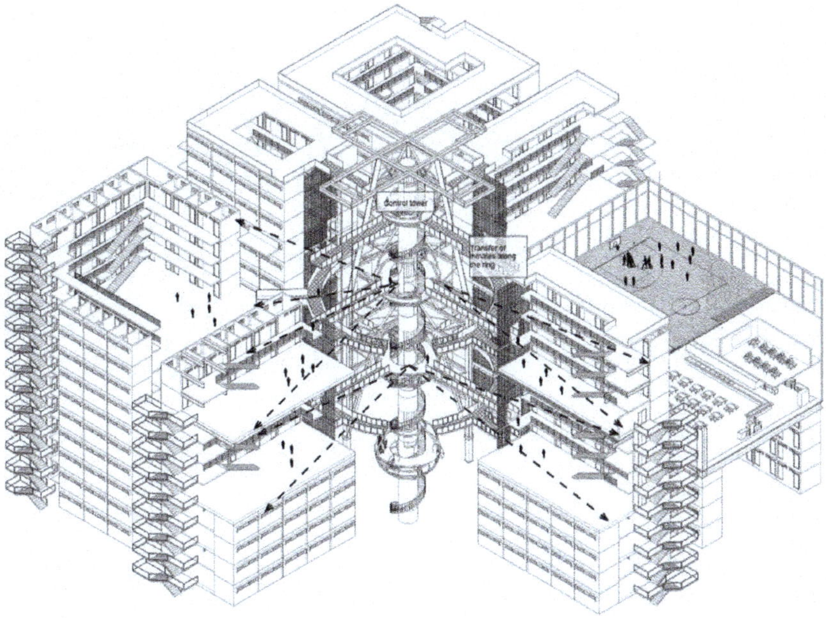

Fig. 21 Relationship between control rooms and cellblocks. *Sightlines* Wardens in the control rooms of an oil rig prison have a view into four floors of cellblocks as compared to two in Singapore Changi prison. This improves the sightlines for wardens thereby reducing the total number of staff needed for daily surveillance. *Security and safety of prison staff* Movement of inmates and prison staff are separate. The transfer of inmates will occur on the ring so that they are under constant watch while the staff have their own dedicated access in the control tower. Access from control room to the ring is via a retractable bridge controlled by the prison staff

Table 5 Operational capacity of a semi-submersible prison

Rig type	Semi-submersible			
Comparison	Changi prison	SS prison	Vegetable farming	843 tonnes
			Fish farming	45 tonnes
Landplot area/Deck area	11,972 m^2	11,660 m^2	Exercise yard	15,700 m^2
No. of cells	768	1600	Dining hall	380 m^2
No. of officers	128	133	Sick bay	325 m^2
No. of prisons to house 48 million inmates in 2050? [world population × average incarceration rate*] *ref: http://www.prisonstudies.org/highest-to-lowest/prison-population-total	62,500 Changi prisons [9.6 billion × 0.005/768]	30,000 SS prisons [9.6 billion × 0.005/1600]	Warden housing	8,200 m^2
			Hanging chamber	100 m^2
			Energy source	wave energy, biofuel, piezo electricity
Footprint in 2050	748,250,000 m^2 (land area that can be saved) [62,500 × 11,972 m^2]	349,800,000 m^2 [30,000 × 11,660 m^2]	Waste reclamation capability	2,987 kg compost per day

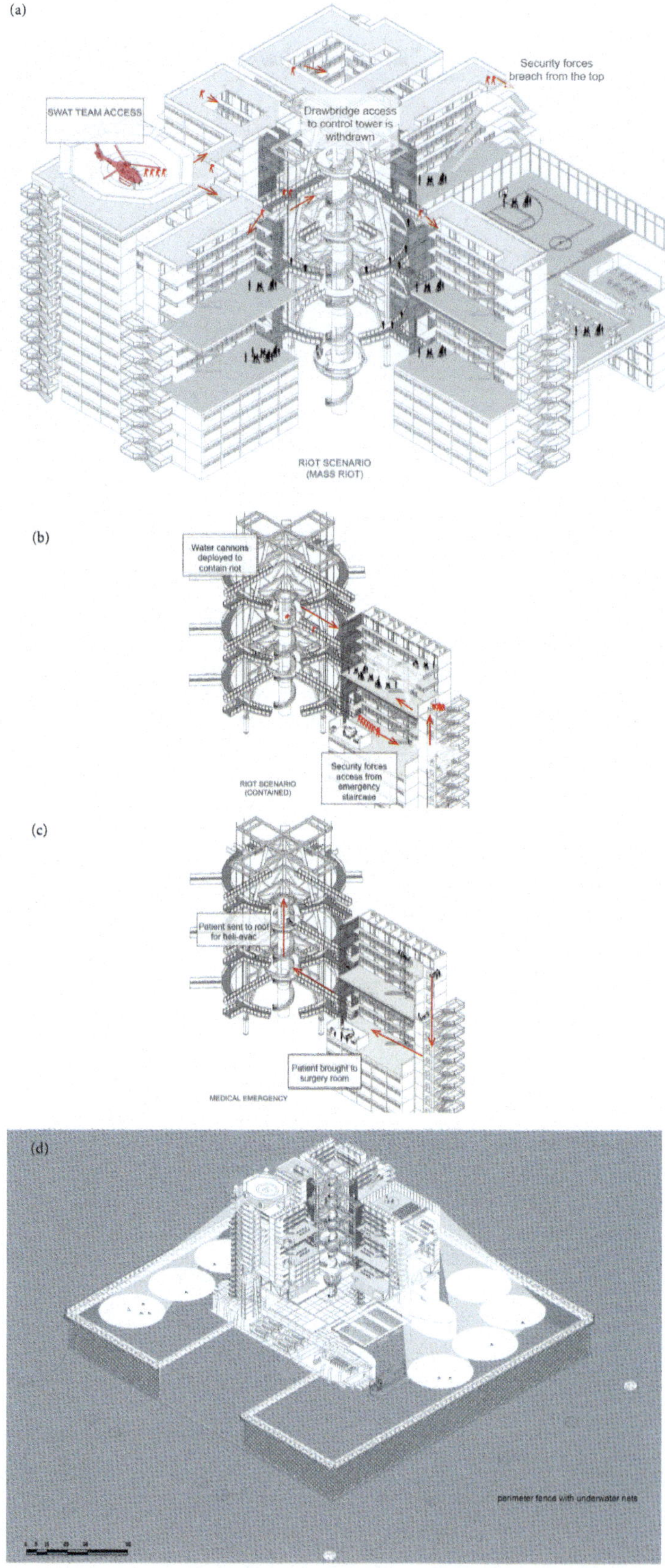

Fig. 22 Emergency scenarios. The layout is designed to effectively lockdown cellblocks to segregate violent outbreaks and provide the fastest routes for insertion of security forces and evacuation. **a** Mass riot scenario: • Control rooms are inaccessible from inmates, • Swat team accesses from the roof via voids of every block, • Water cannons placed at control rooms to control crowds and extinguish fires.
b Riot scenario: • Riot is locked down within a cellblock, • Main access to cellblock closed, • Riot control initiates riot protocol from alternative emergency stair core.
c Medical emergency: • Corridors are stretcher accessible, • Inmate is evacuated via emergency stair core to surgery room, • Inmate is sent to rooftop via main lift core for helicopter evacuation.
d Escape scenario: • Perimeter fence with underwater nets to prevent escape from prison compound. • Spotlights to illuminate perimeter at night

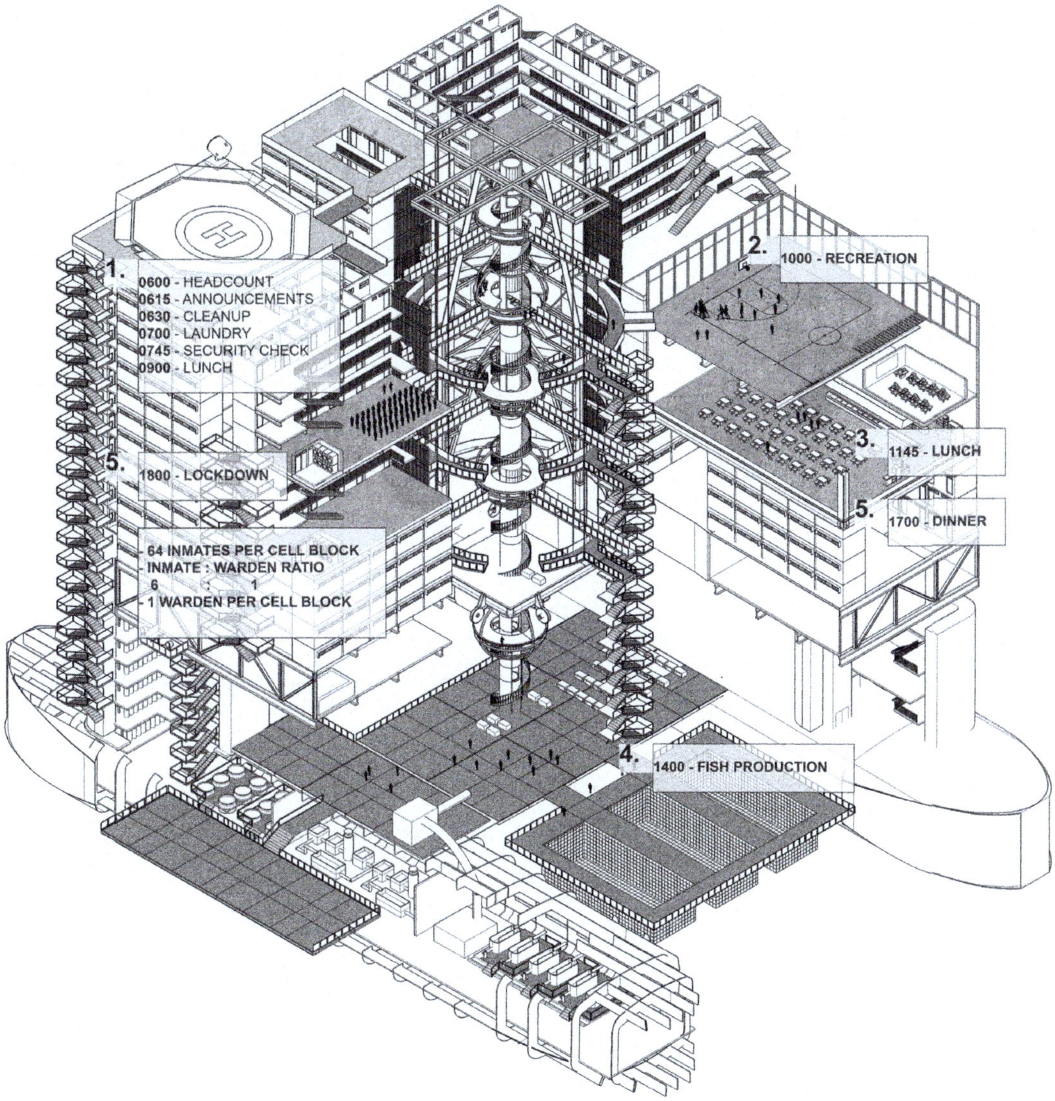

Fig. 23 Day in a life of an inmate

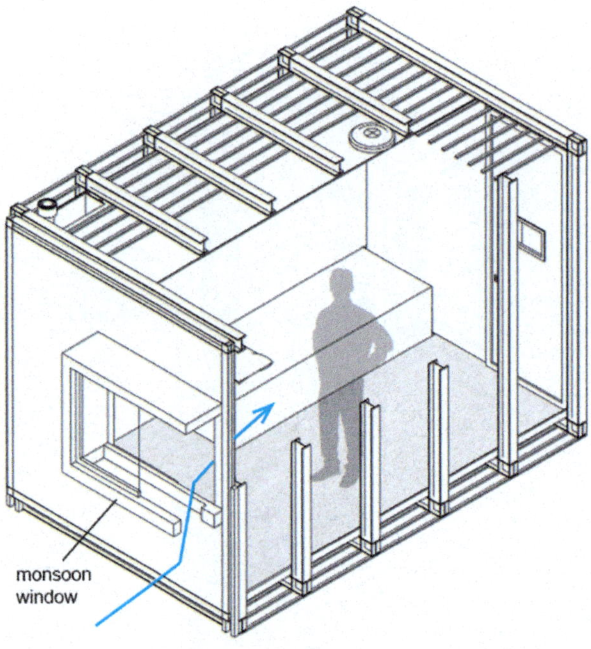

Fig. 24 Design openings for natural ventilated spaces. Monsoon windows are built into each cell to maximize natural ventilation and to minimize energy loads for mechanical cooling. They project out from the external wall with louvers on their underside. These louvers allow airflow without rain penetration during a storm

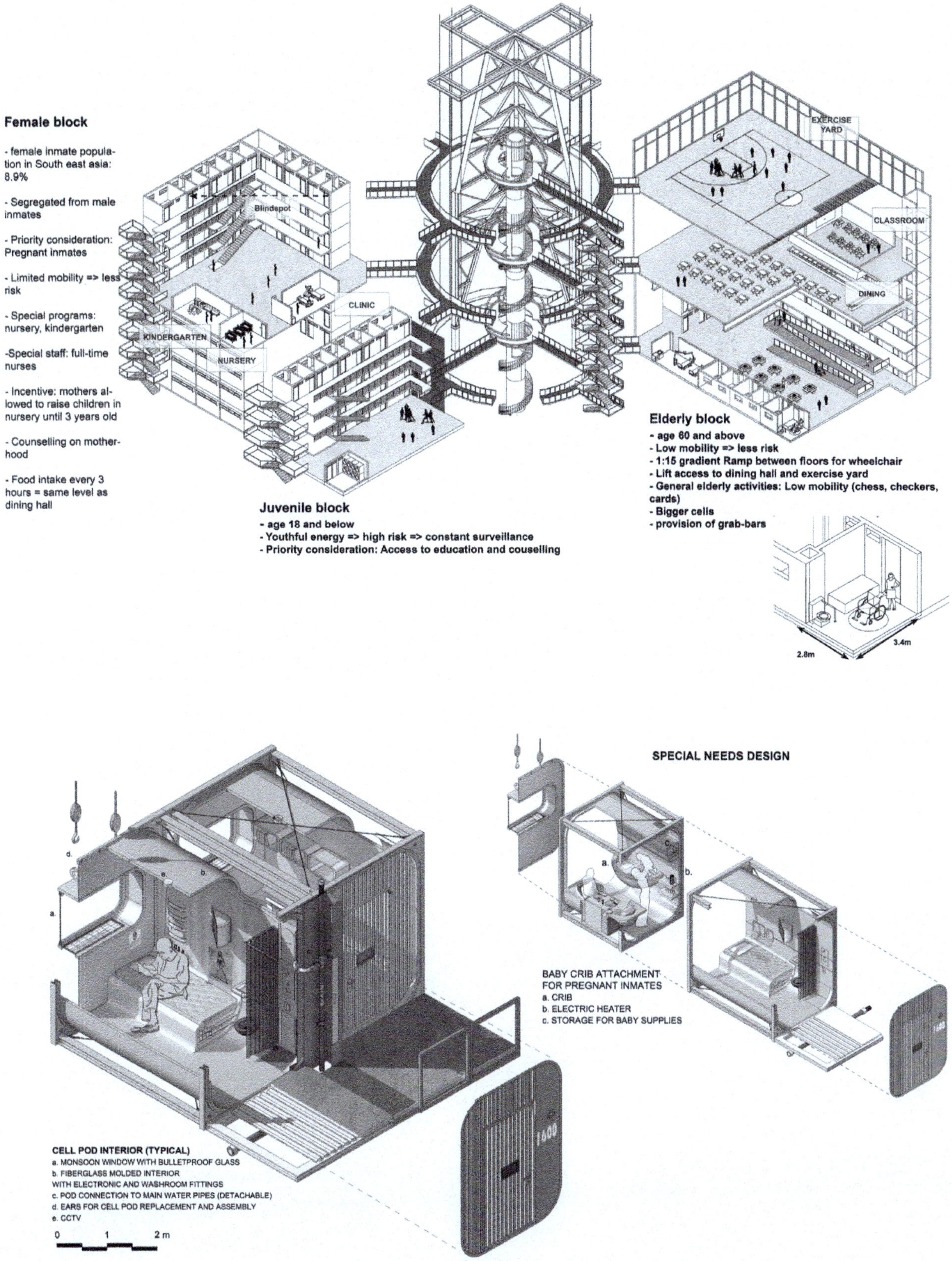

Fig. 25 Special provisions for inmates who are juveniles, expectant mothers and elderly. To anticipate the needs of special inmates, a cell conversion strategy is considered. The space of individual cells can be expanded through external attachments

Fig. 26 A semi-submersible prison with an outer moat of plankton serving as motion detector to prevent escape; and an outermost ring of wave energy harnessers demarcating the boundary of the facility

Fig. 26 (continued)

Fig. 27 Panoptic view of a semi-submersible prison courtyard. Wardens in the control rooms of an oil rig prison have a view into four floors of cellblocks

Fig. 28 The waterbody sheltered beneath the deck and between the pontoons of the semi-submersible can be used for recreation, with surrounding spaces overlooking it in an arena-like configuration. The concentric water body to the outer edge of the pontoons serves as fish-rearing areas

2.4 Buoyancy and Toppling Calculations

The existing structure of a semi-submersible is not designed to accommodate additional dead load from prison cells and live loads. Thus it is necessary to determine the additional pontoon sizes needed to ensure stability. Application of lightweight fiberglass construction is also investigated to reduce the dead load on the semi-submersible (Fig. 29).

The maximum dead and live loads the existing structure can support is 28,173.2 tons, which are approximately 12 storeys of cells. Beyond that, half of the semi-submersible would be submerged underwater which is not ideal. With an additional load of 28,173.2 tons, the metacenter (M) is higher than the center of gravity (CG), which means the semi-submersible would not topple although half of it would be underwater (Fig. 30). Capsizing is likely with strong waves and inclement weather.

In order to prevent sinking and overcome toppling, two strategies are explored. The first is to provide a wider floating base for the vessel. Additional pontoons and legs may be welded to the semi-submersible deck to form a doughnut pontoon formation. The second is the use of lightweight fiberglass for cell construction to reduce imposed loads significantly, compared to the conventional cell made of concrete (Fig. 31).

Construction Sequence A proposed construction sequence of a semi-submersible prison is illustrated in Fig. 34. The cell pods made of lightweight construction such as fiberglass, EPS or MET may be transported to the floating dry dock for direct installation onto the semi-submersible deck. The construction crane may be fitted onto derrick for the replacement of cell pods.

2.5 Energy and Waste Management Systems

Based on studies of existing precedents, a selection of renewable energy systems and waste management techniques are used to create a closed loop system so that the offshore prison is not only self-sustaining but does not discharge toxic effluent to marine environment. The offshore prison can be powered by wave energy harvesting systems

Fig. 29 Lightweight construction of a prison cell

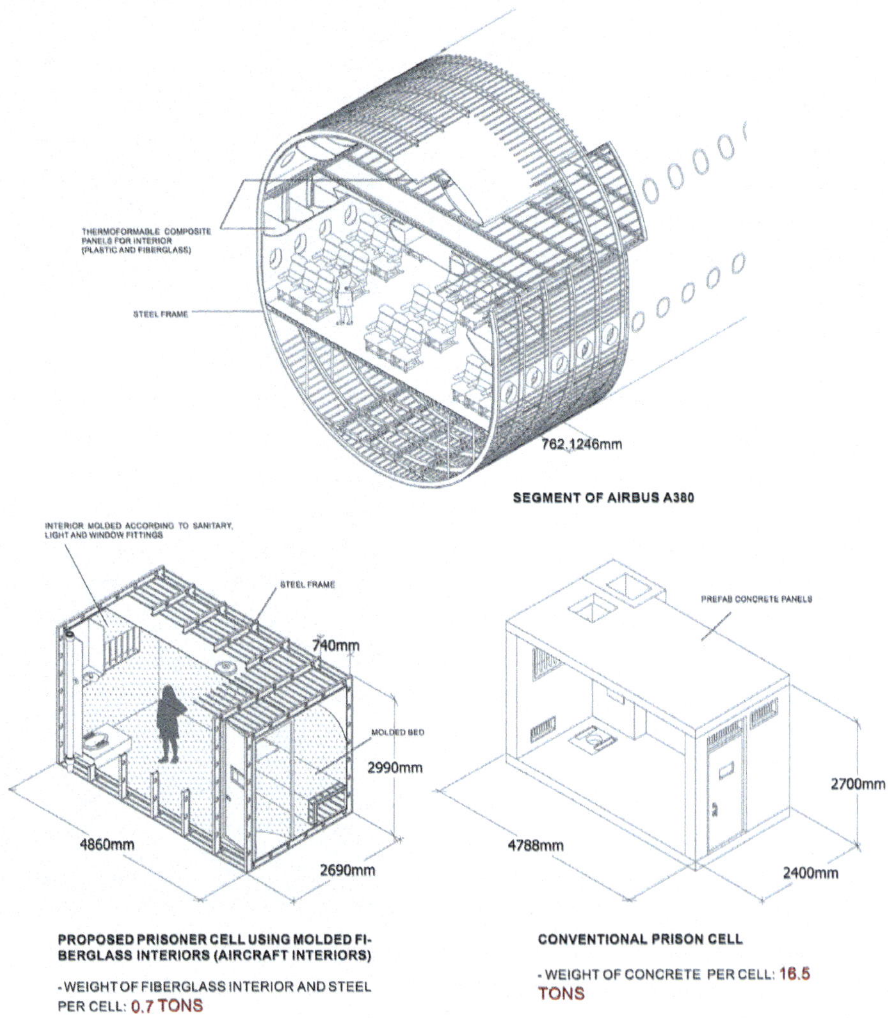

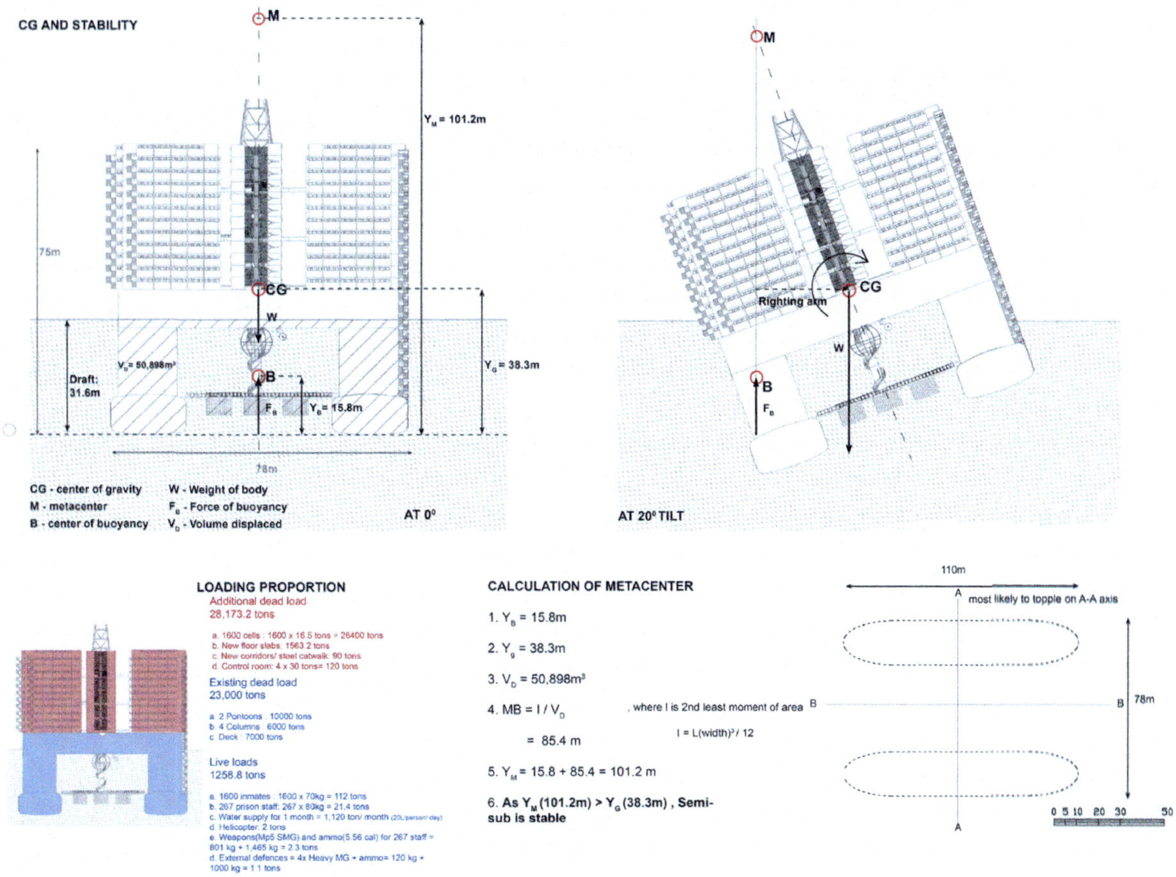

Fig. 30 CG and stability of a semi-submersible

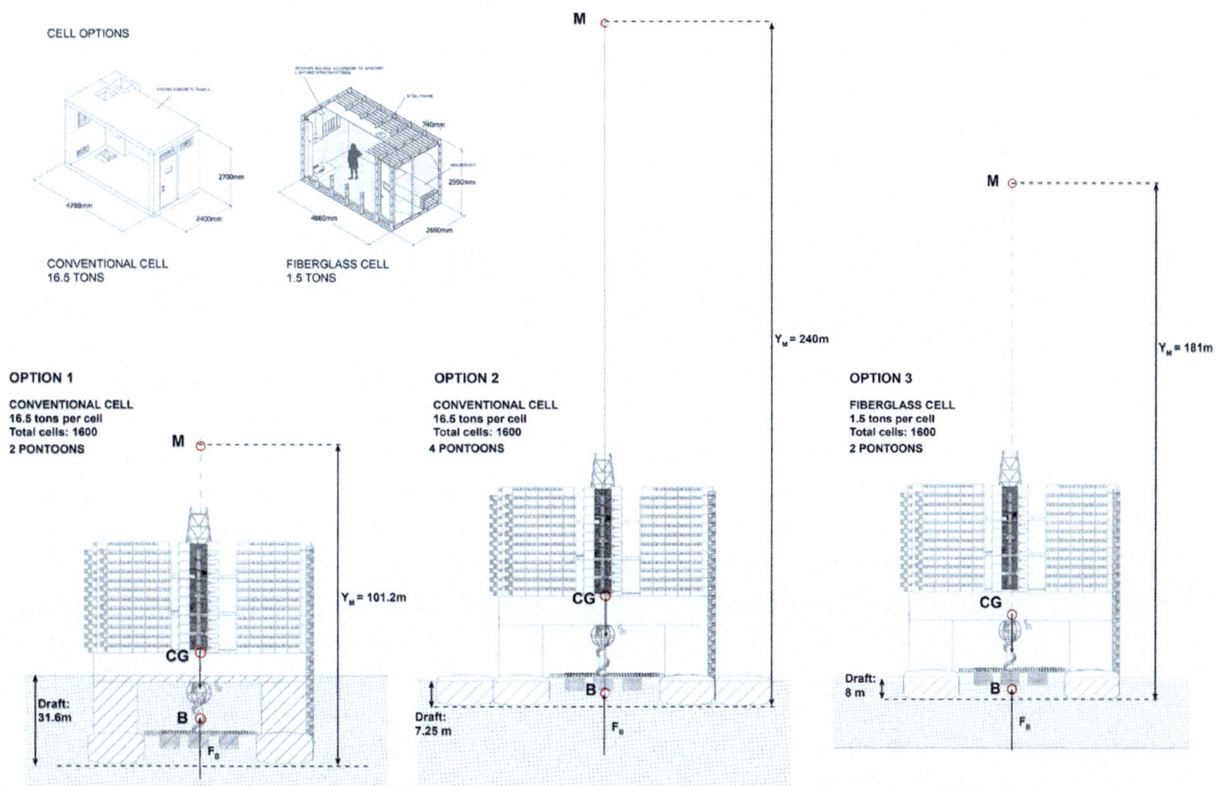

Fig. 31 Strategies to overcome toppling

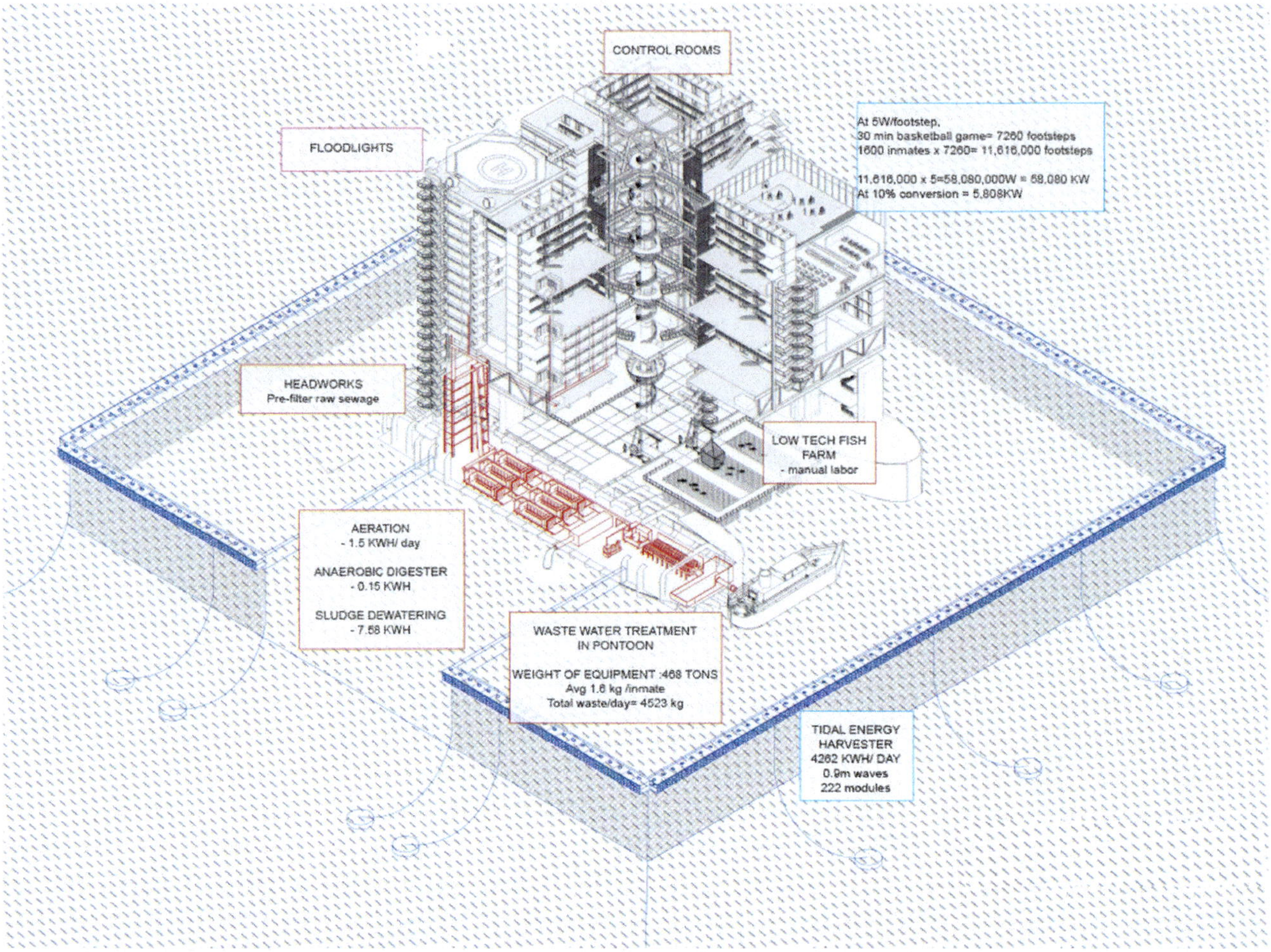

Fig. 32 Components of energy consumption and production

instead of conventional fossil fuels. The application of such systems can be placed along the defense perimeter of the prison (Fig. 33b).

Based on occupancy of 1,600 inmates and 267 wardens, about 2,987 kg of organic waste is produced daily. The waste is processed by a wastewater treatment system—built into the pontoon—into biogas that can be used as fuel for cooking in the kitchen. Piezoelectric energy harvesting and low technology systems (for example, human-powered pulley system to deliver goods) can be considered to reduce energy loads (Fig. 33a).

The overall energy production and consumption is illustrated in Fig. 32 and summarized in Table 6.

Table 6 Overall energy consumption and production

Key prison operations	Quantity	Duration (h)	Daily energy consumption (kWh)	Sources of energy	Daily energy production (kWh)
Wastewater treatment				1. Piezoelectric energy from 1600 inmates exercising at 10% efficiency	5,808 (5 W/footstep)
Aeration		24	1.5		
Anaerobic digester		24	0.15		
Sludge drying		24	7.38		
Lightning					
1600 cells (100 lx)		2	16.4		
Dayrooms (100 lx)		2	2.41		
Dayrooms (10 lx)		10	3		
Surveillance				2. Tidal energy from 222 modules at 20% efficiency	4,262.4 (96 kWh/day) 20% efficiency
CCTV and infra-red	24	24	5.76		
Cooling loads					
Control room	5		43.8		
Emergency					
Emergency lighting	–		7.76		
Floodlights	–		283		
Surgery room	4		44.8		
Anti-riot water cannons	16		–		
	Total energy consumption 416 kWh per day			*Total energy generated 10,074 kWh per day*	

(a)

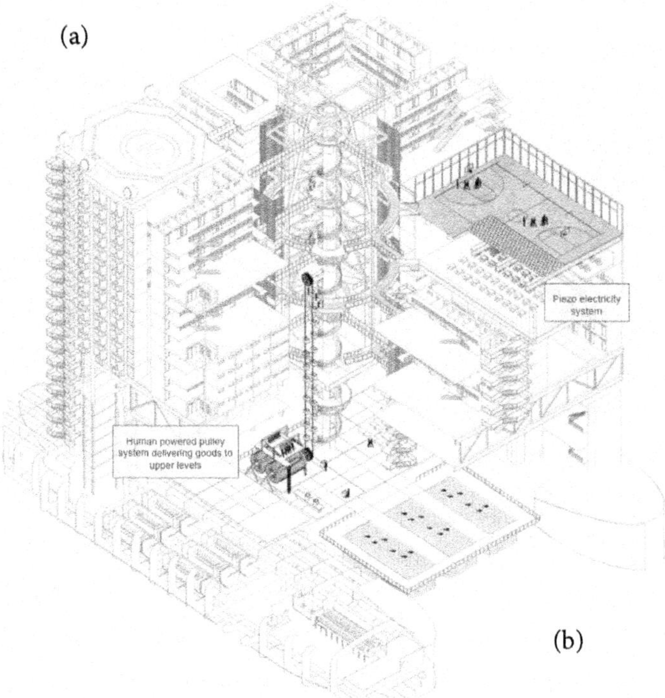

(b)

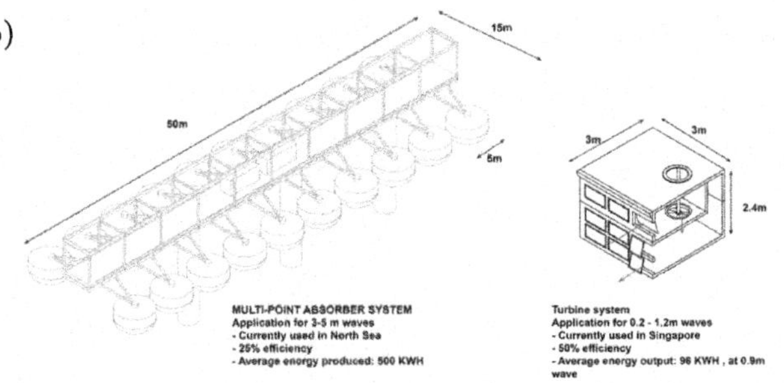

Fig. 33 Harnessing energy on a floating prison. **a** Piezoelectric. The concept is to convert the 'work done' by inmates in daily routines to power their needs. This is possible by integrating piezoelectric converters to harness the potential energy of inmates moving through stairways and corridors; and actuating mechanical pulleys and pumps in daily routine. **b** Wave energy harvesting systems

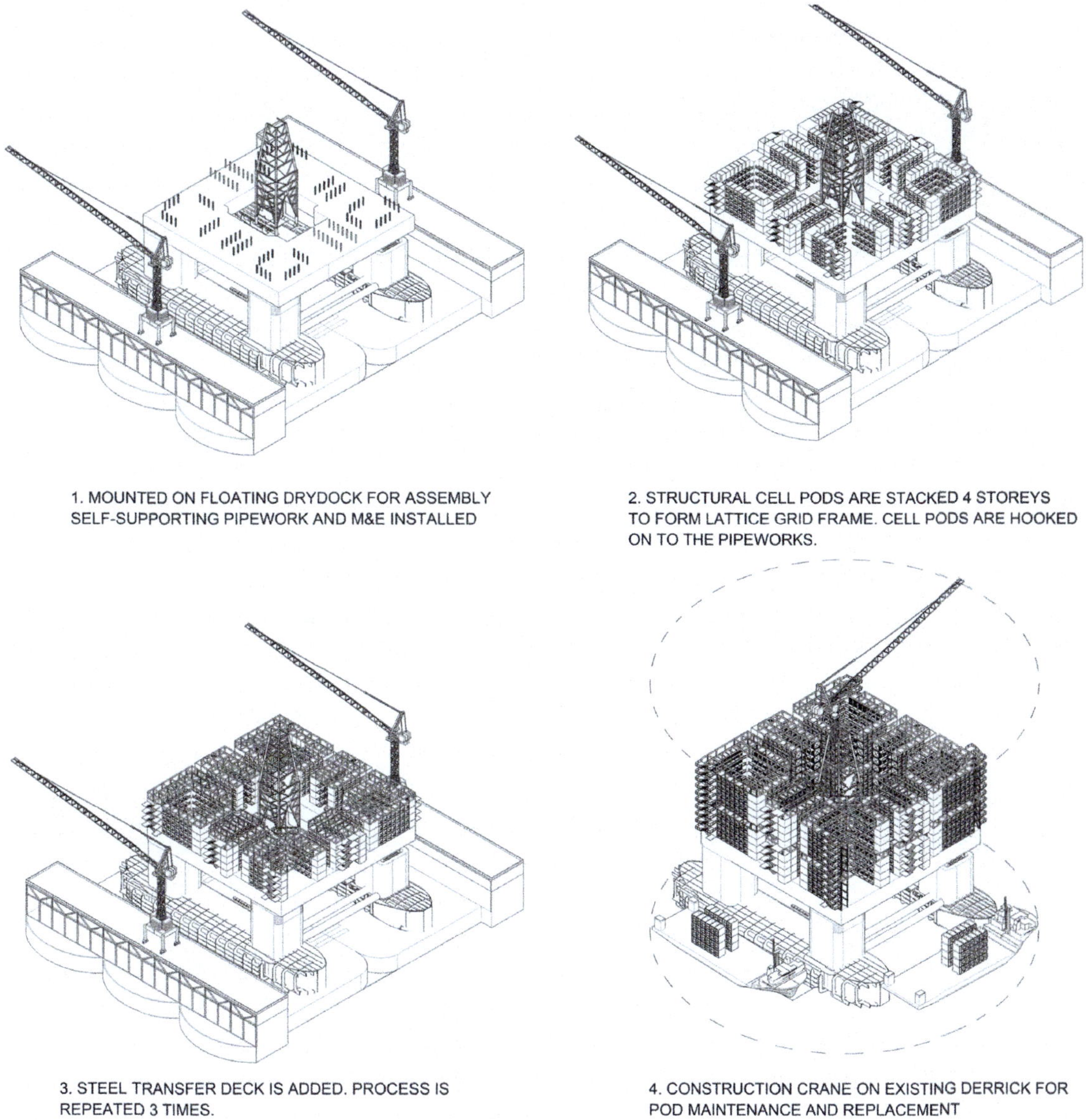

Fig. 34 Proposed construction sequence of a semi-submersible prison. Floating dry docks positioned alongside the rig assemble additional structure, mechanical systems, and cell pods onto its deck

2.6 Jack-up Prison

A similar study is conducted with a jack-up of deck area 4,050 m^2. 2,148 cells can be accommodated on a jack-up leg length of 180 m (105 m with 30 storeys of pods; 50 m depth of water; 30 m for communal pods above water level allowing for fluctuating wave heights). There are two banks of cell pods, which face inwards without a sea view. Although the capacity of the jack-up to accommodate cell pods is larger that the semi-submersible, it has triangulated closed cloisters, which posed challenges to warden surveillance (Figs. 35, 36, and 37).

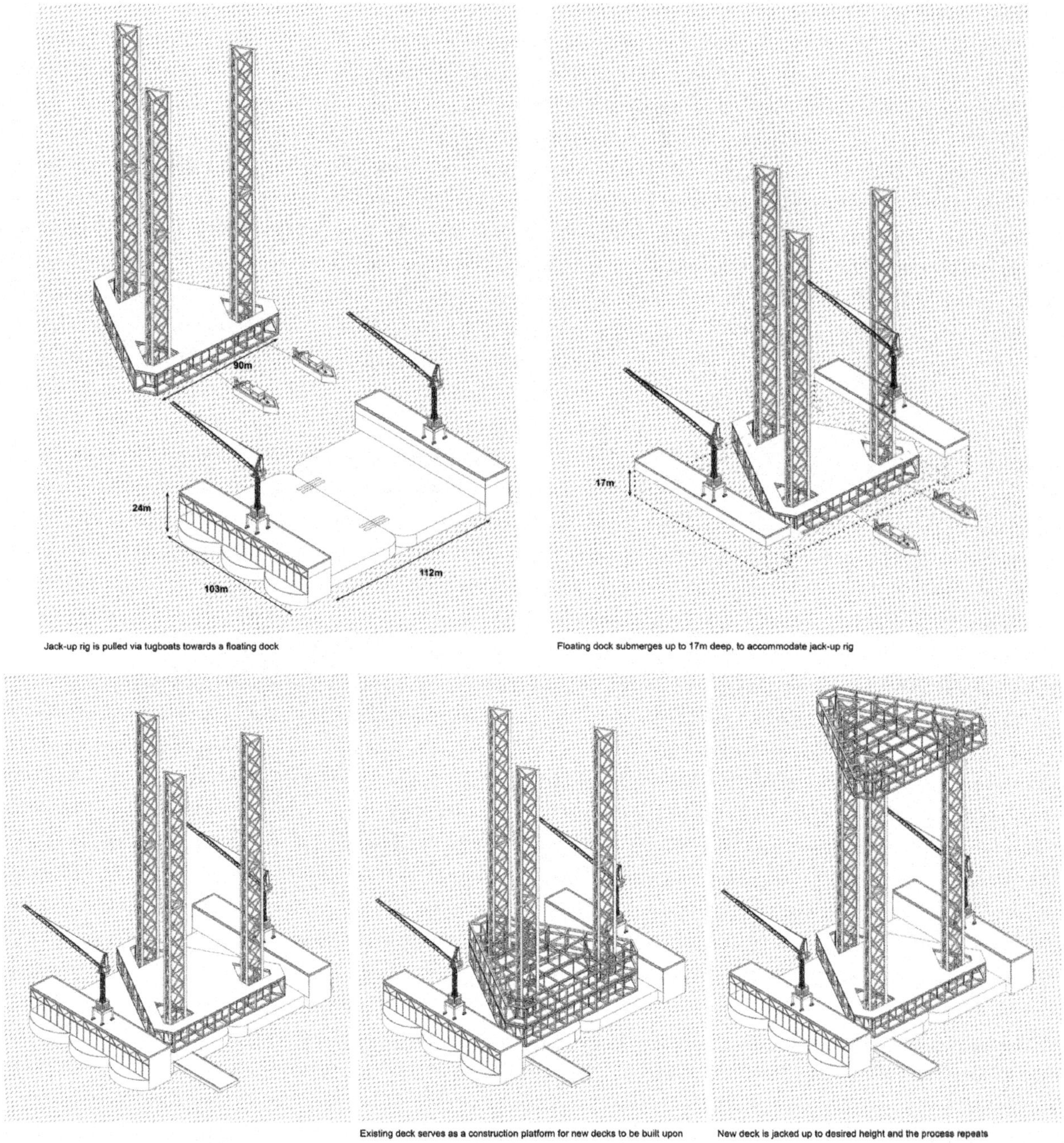

Fig. 35 Proposed construction sequence of a jack-up prison. Floating docks may be used to assemble a number of storeys of cell pods on each hull/deck structure before jacking-up to position using the self-elevating mechanism of the rig. The sequence is repeated for speed of erection

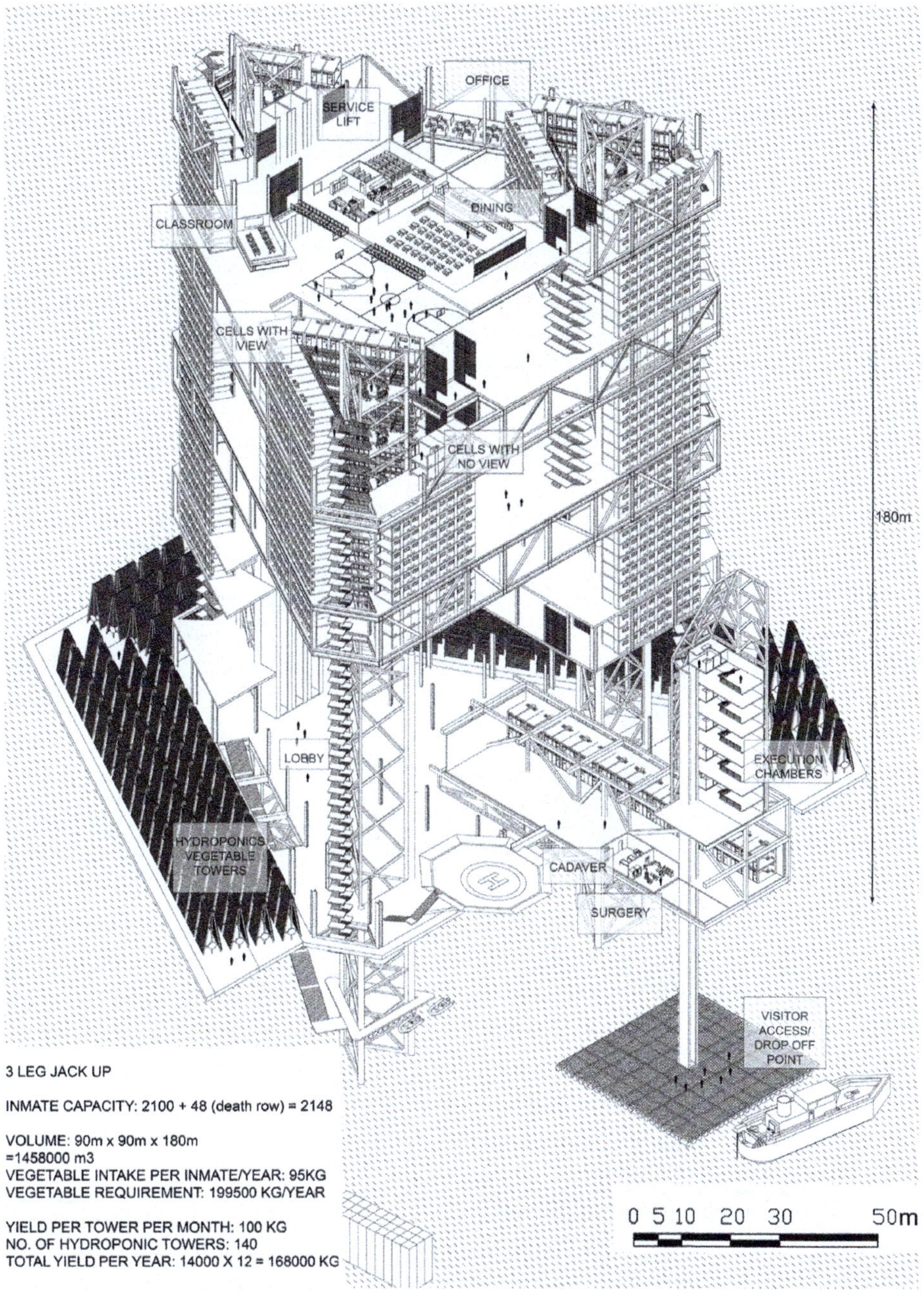

Fig. 36 Comparing jack-up and semi-submersible prisons capacities: the capacity of a jack-up prison alternative is 1.34 times more than the semi-submersible version on a deck area that is also 0.35 times the size of the former. Deck areas for jack-up and semi-submersible prisons are 4,050 m^2 (45 × 90 m) and 11,660 m^2 (110 × 106 m) respectively. Changi prison has a footprint of 11,972 m^2. In comparison, a semi-submersible prison can hold 2.08 times the numbers of cells in Singapore Changi prison for the same footprint

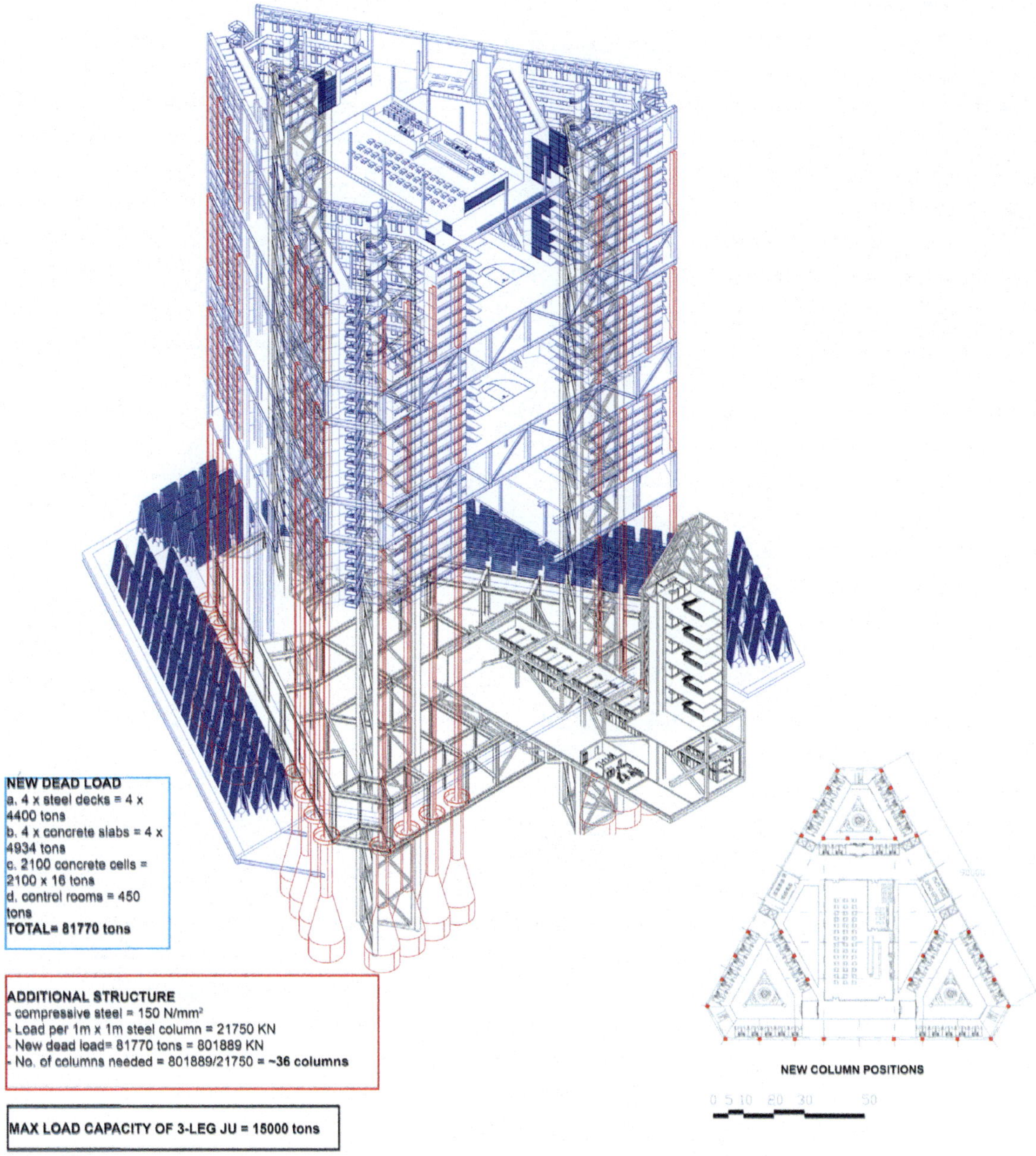

Fig. 37 Additional structure required of a jack-up prison of 2,100 concrete cells. Using additional steel deck and columns with reinforced concrete cells, the additional load is estimated at 81,770 tons or 5.45 times the original load capacity of a three-legged jack-up at 15,000 tons. To reduce this additional dead load, the use of lightweight cells designed for escape-proof would be necessary

3 Semi-submersible Sports Facility

The hosting of World Cup and Olympics events is not a viable proposition for cities that do not have the necessary infrastructure. Host cities would have to invest in infrastructures to provide sufficient hotel rooms and upgrade transportation lines. The construction costs of specialized large scale competition venues to accommodate spectators and athletes over the event period; and the operational costs in hosting both opening and closing ceremonies including security are not easily recovered (Fig. 38).

Large scale sports facilities constructed for hosting the event have limited post-event use and the upkeep of these under utilized facilities over protracted periods of time result in recurrent annual deficits. Opportunity costs of long term debt servicing after hosting the games can burden city budgets and prevent essential public spending. In the example of Greece, hosting the 2004 Summer Olympics in Athens was the contributing factor to her financial crisis [31]. Greece is smaller than most other cities that had hosted the Games prior to 2004; added to this, the Games had to comply with the strict EU regulations. The 2004 Games were held in 32 venues, of which 18 were newly constructed, 12 were renovated and 2 were temporary facilities. Data from the State Budget reported an estimated 5.1 billion euros expenditure in the preparation period between 2002 and 2004.

According to data from the International Olympic Committee (IOC), most of its income can be attributed to broadcast licensing contracts and international sponsorships, of which IOC retains a large percentage. 73% of IOC revenue sources between 2013 and 2016 came from broadcast rights [8]. As the costs of hosting have spiked sharply, revenues cover only a fraction of the hosts' expenditures (Fig. 39). Post-event research reveals that such major events do not in the long term, improve the city economy by creating jobs or drawing tourists. This is because infrastructure building jobs are temporary and usually taken up by workers who are already employed; diminishing the impact on the economy. The security, crowding, and higher prices created by the events also dissuade patronage to the Olympics [14].

Therefore, the idea of converting semi-submersibles into reusable floating sports facilities is to mitigate the risks of incurring long term debts by cities which do not already have existing sports infrastructure but who wish to host such major events. The dimensions of FIFA stadia and IOC track and field events serve as guidelines to match the configuration of semi-submersible deck areas. Off-competition usage of the same deck surface can accommodate a range of activities popular to the public.

Such floating sports facilities occupy footprints smaller than those of land based versions. Leisure parks, retail malls, recreational and dining outlets can be collocated to support the main events and activities and ensure revenue in

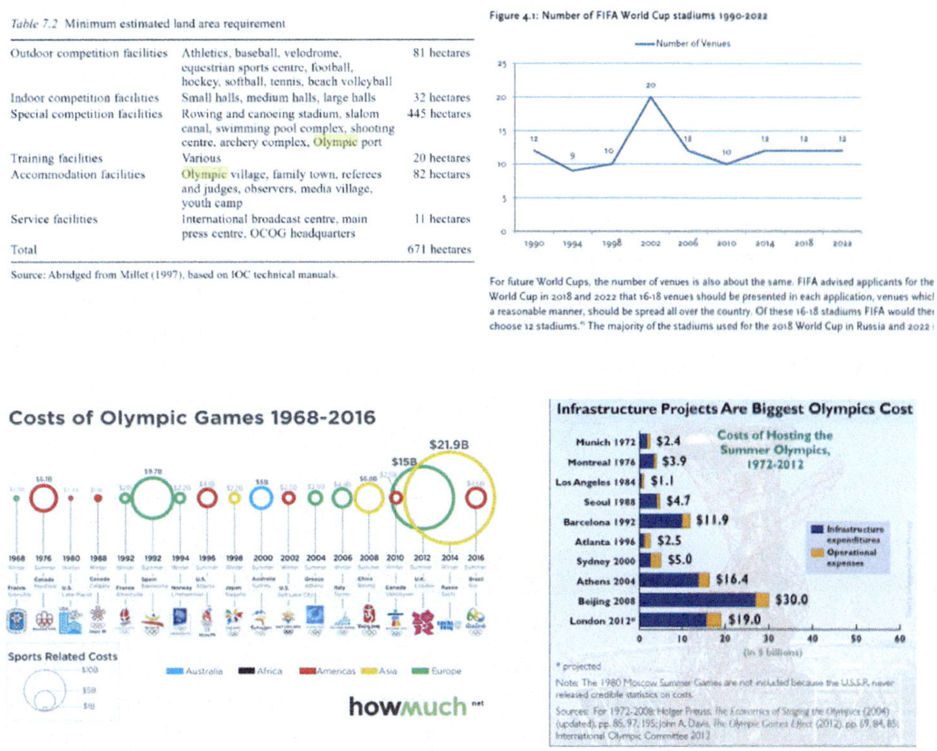
Table 7.2 Minimum estimated land area requirement

Outdoor competition facilities	Athletics, baseball, velodrome, equestrian sports centre, football, hockey, softball, tennis, beach volleyball	81 hectares
Indoor competition facilities	Small halls, medium halls, large halls	32 hectares
Special competition facilities	Rowing and canoeing stadium, slalom canal, swimming pool complex, shooting centre, archery complex, Olympic port	445 hectares
Training facilities	Various	20 hectares
Accommodation facilities	Olympic village, family town, referees and judges, observers, media village, youth camp	82 hectares
Service facilities	International broadcast centre, main press centre, OCOG headquarters	11 hectares
Total		671 hectares

Source: Abridged from Millet (1997), based on IOC technical manuals.

Fig. 38 Expenditure of sporting events

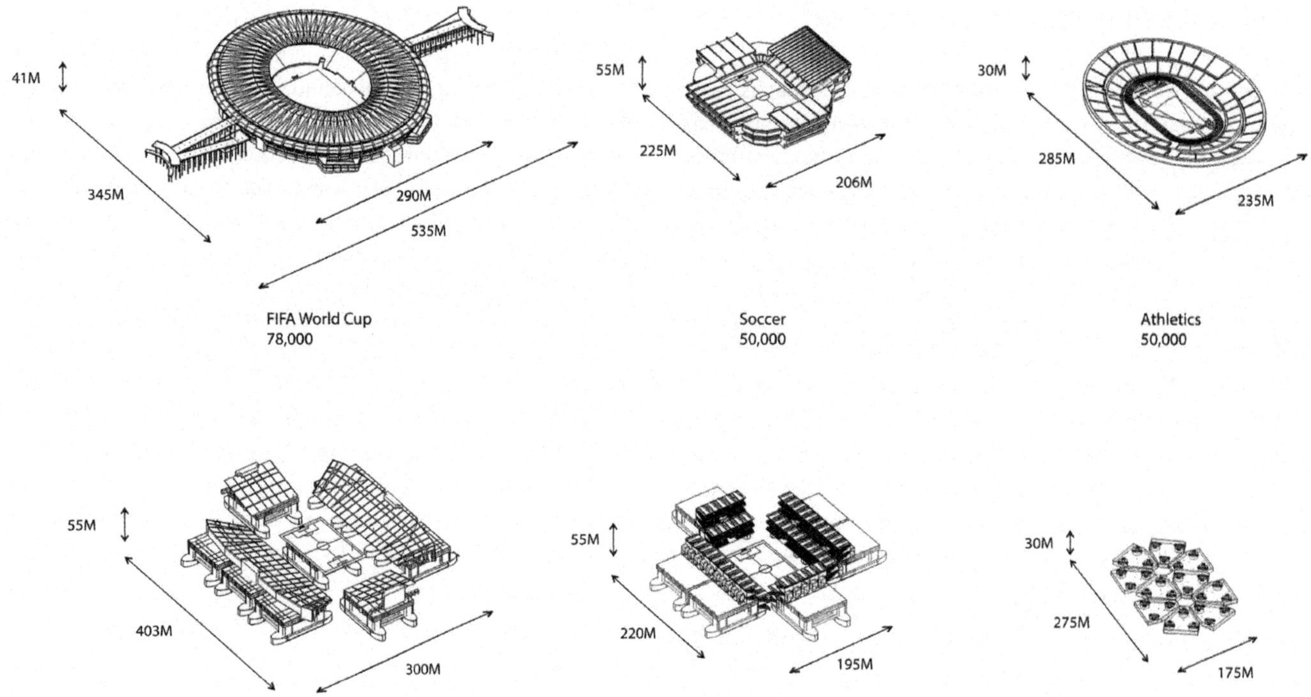

Fig. 39 Comparing plot areas of stadia with semi-submersibles and jack-up arrangements

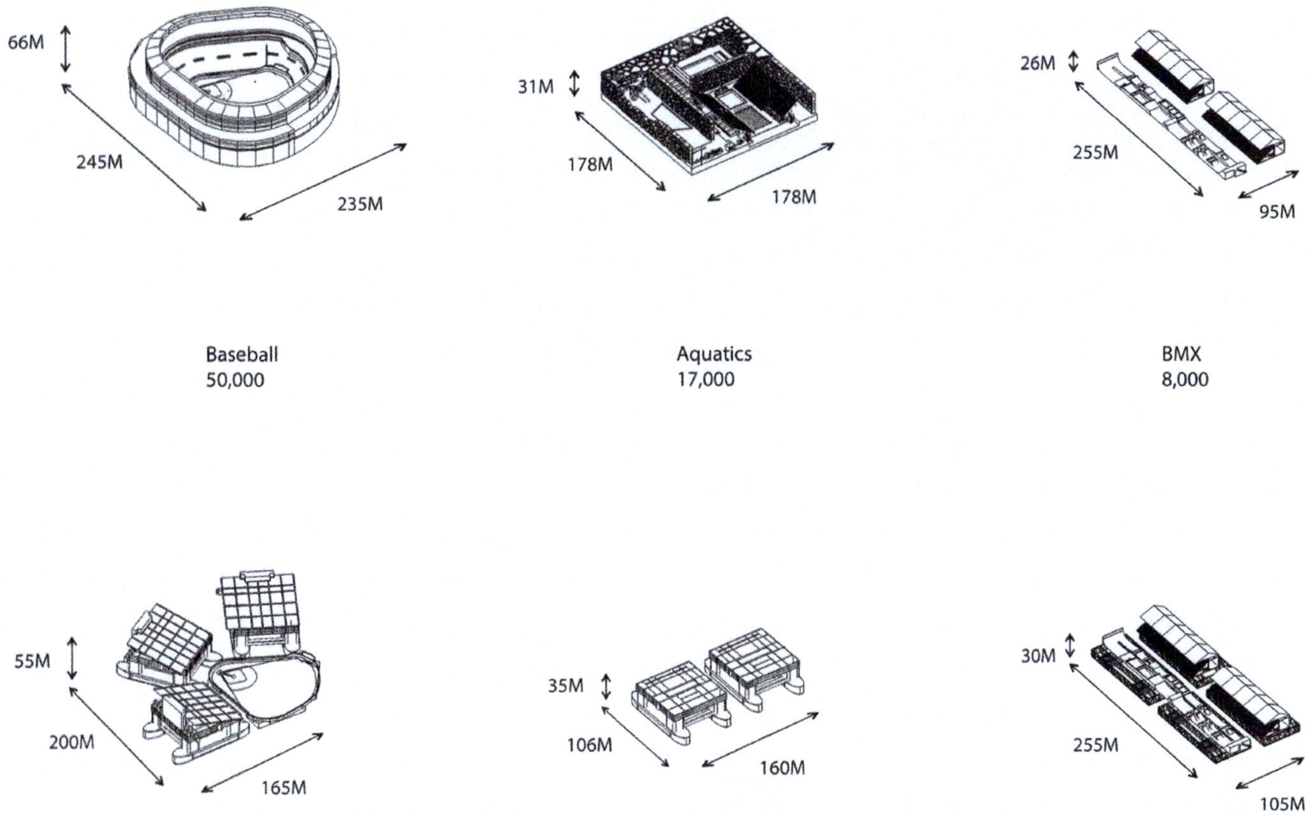

Fig. 39 (continued)

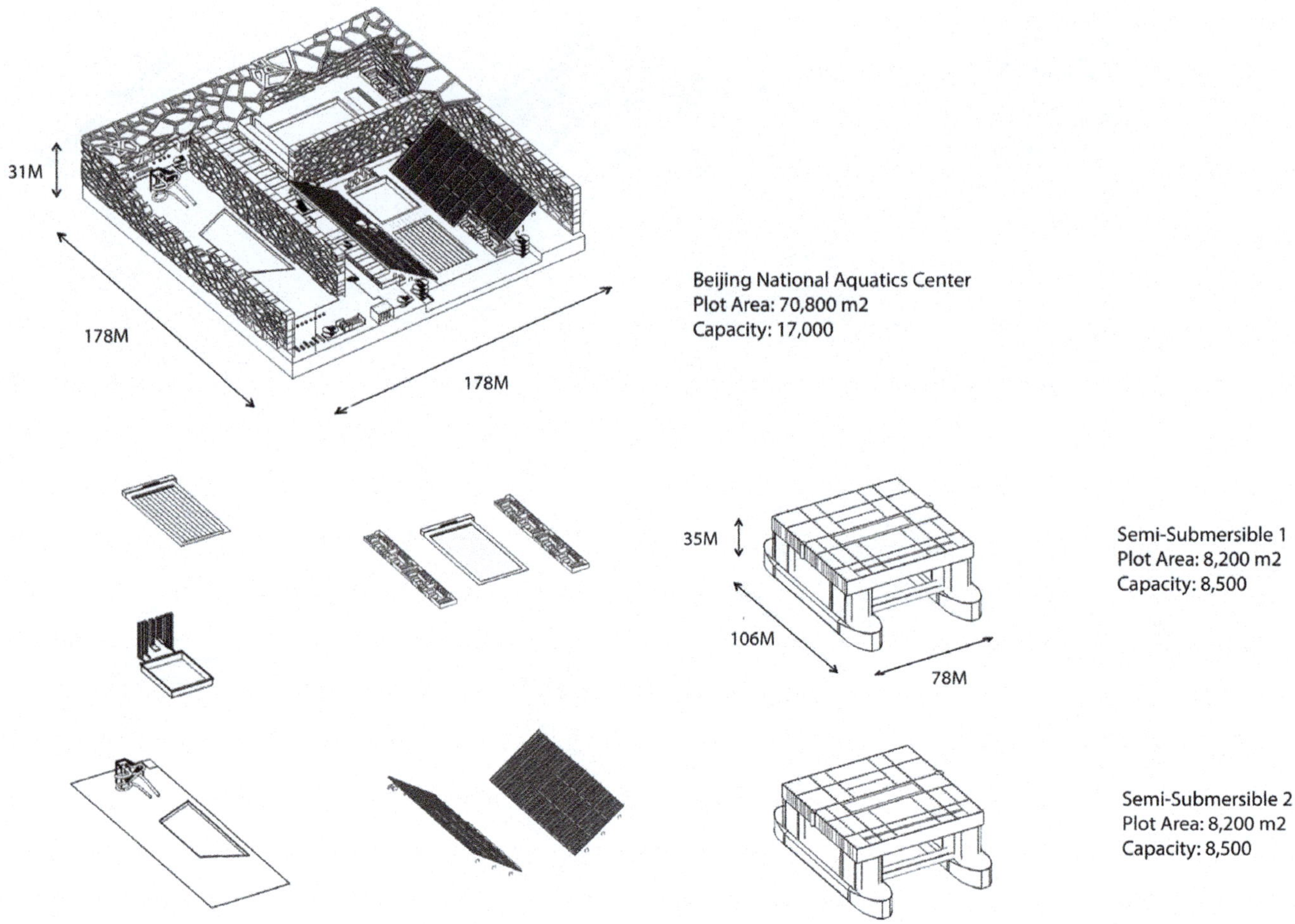

Fig. 40 Comparing plot areas of Beijing Aquatic Center with that of a mid-size semi-submersible

off-competition periods. A semi-submersible sports facility can sail to any offshore or nearshore location and can be leased on flexible time frames until rental costs have been covered to turn a profit.

Figures 40, 41, 42, 43, 44, 45, and 46 illustrates the design capabilities of repurposing semi-submersibles into sports facility.

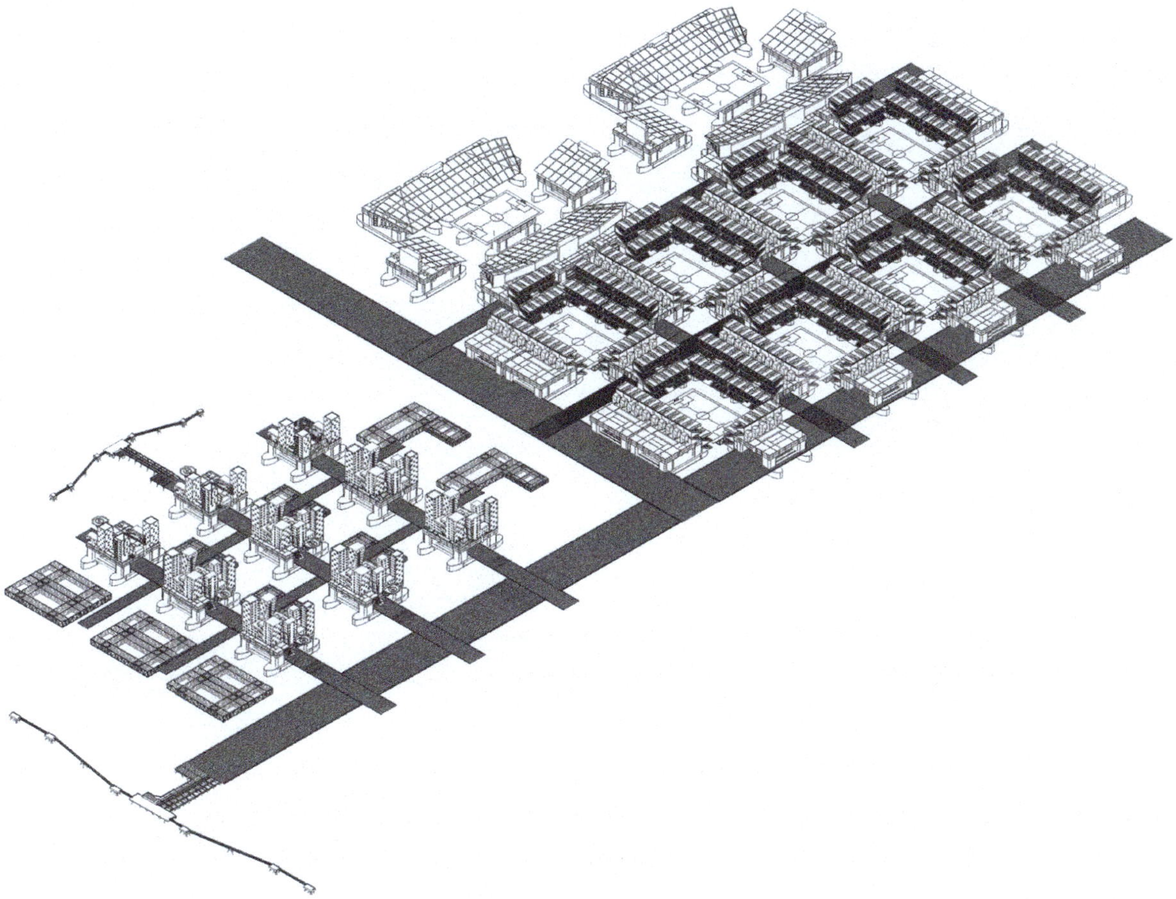

Fig. 41 Semi-submersible arrangement for part of World Cup tournament

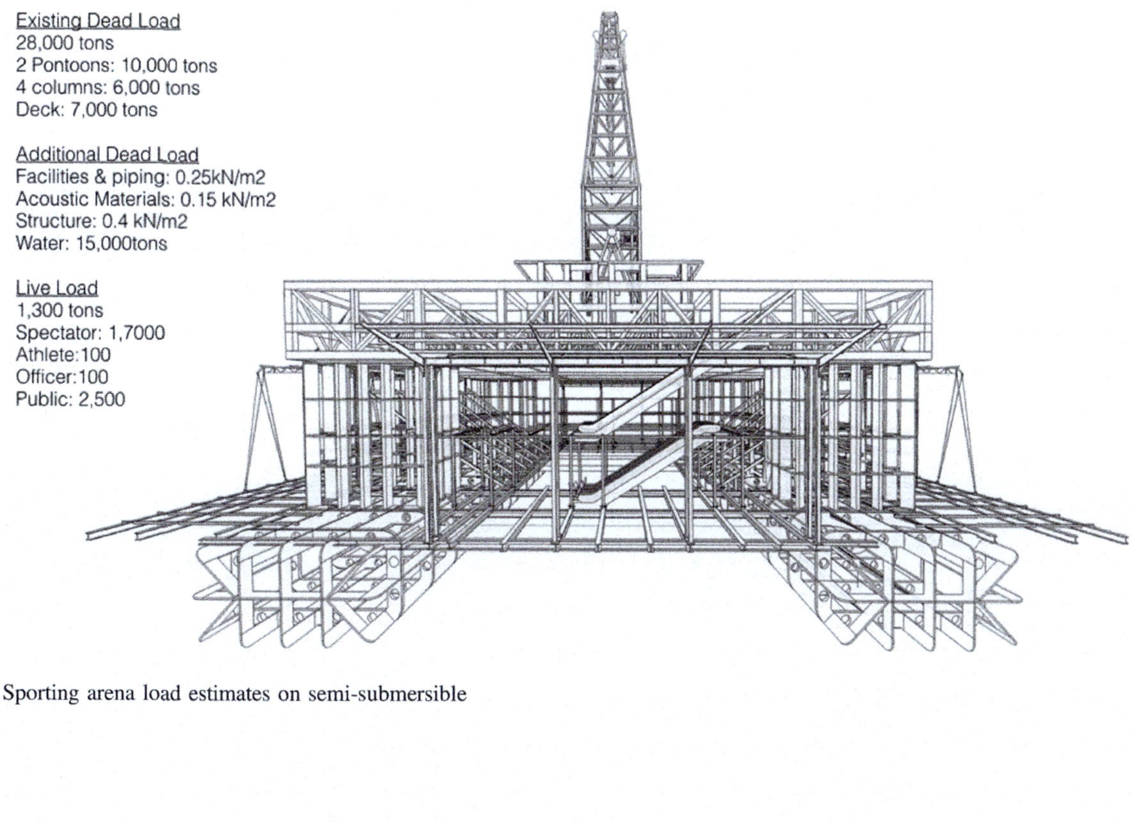

Fig. 42 Sporting arena load estimates on semi-submersible

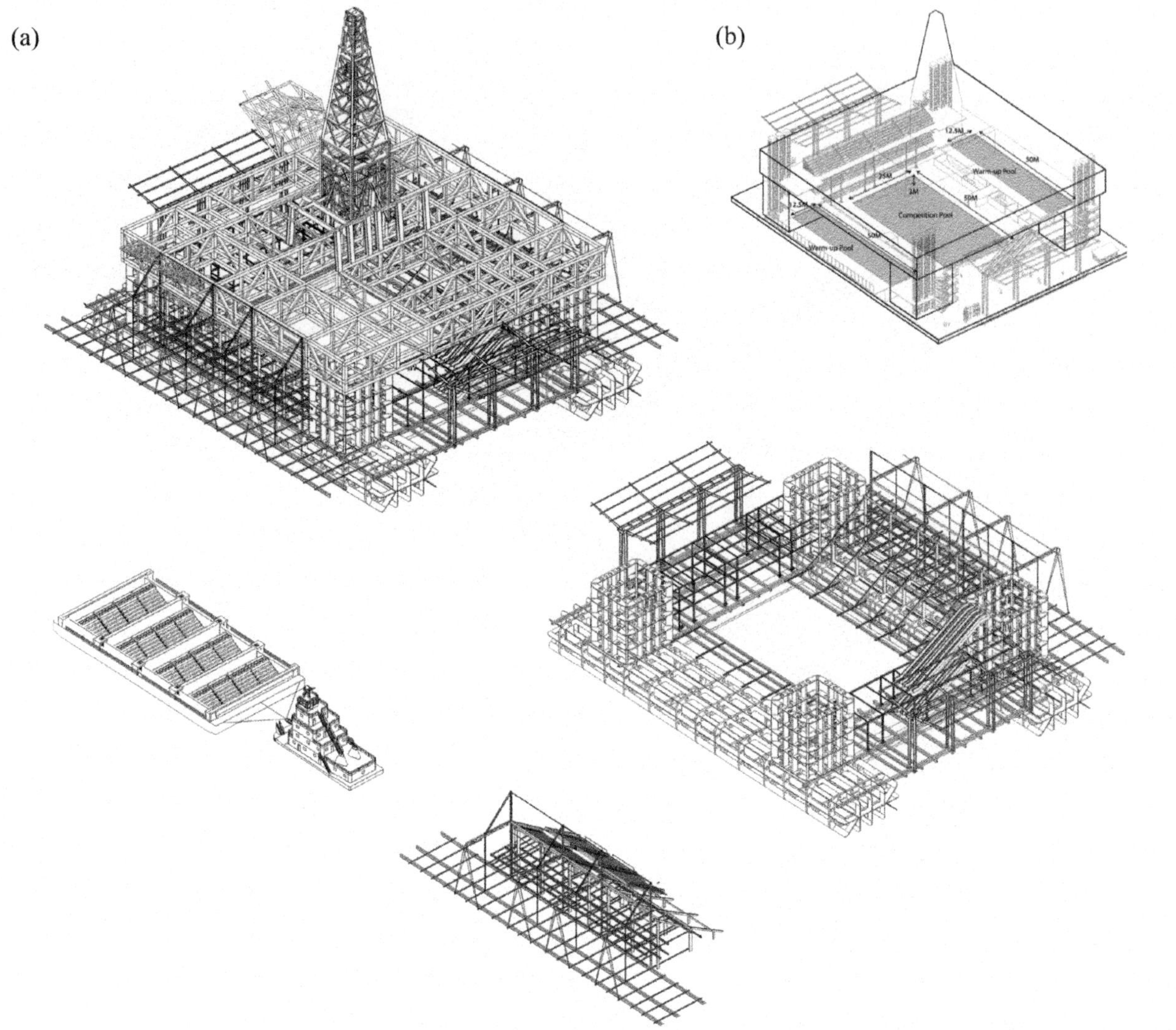

Fig. 43 **a** Retrofitting of Olympic competition pool using floating dry docks and barges, **b** spatial dimensions of competition pool incorporated on a semi-submersible; below hull/deck and using pontoon level

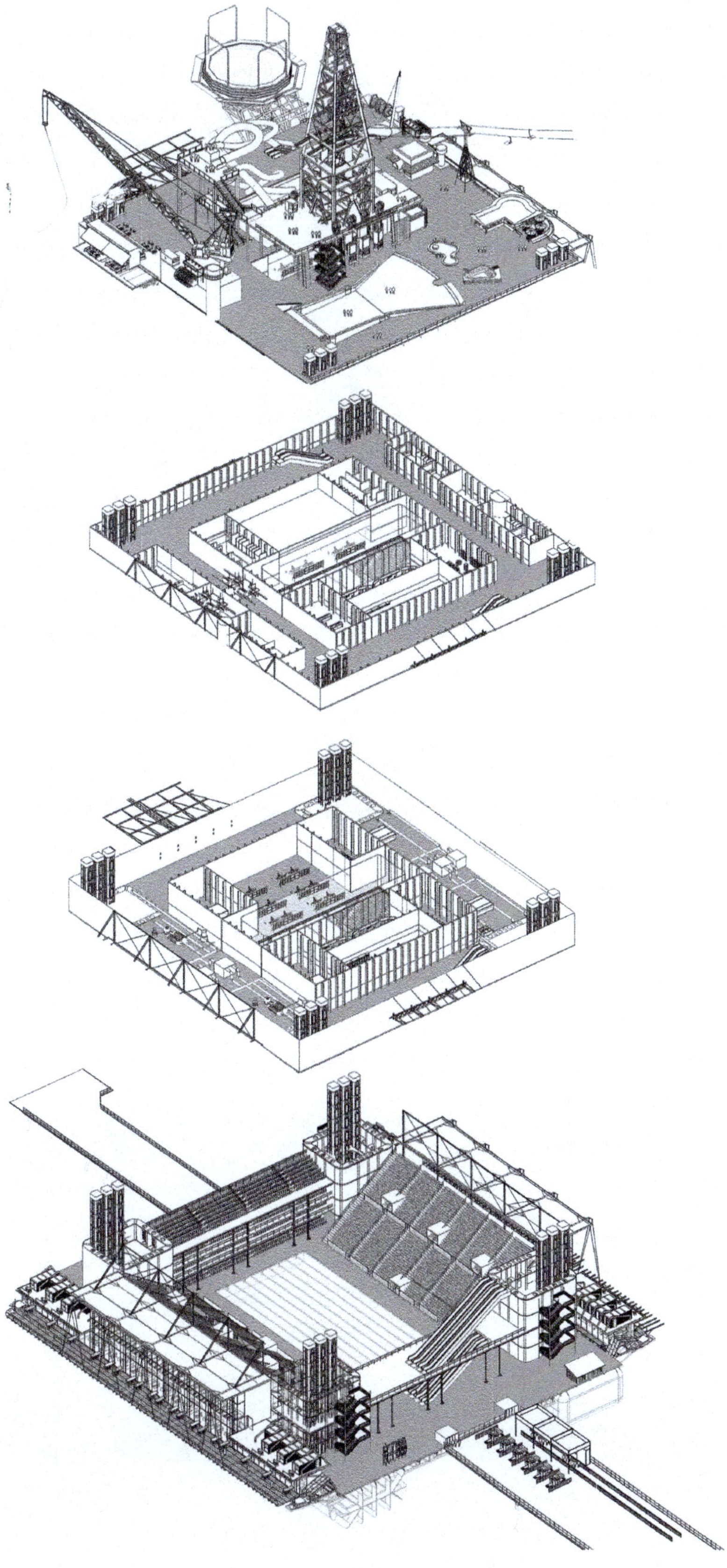

Fig. 44 Study for a recreational facility above a competition pool fitted into a semi-submersible structure

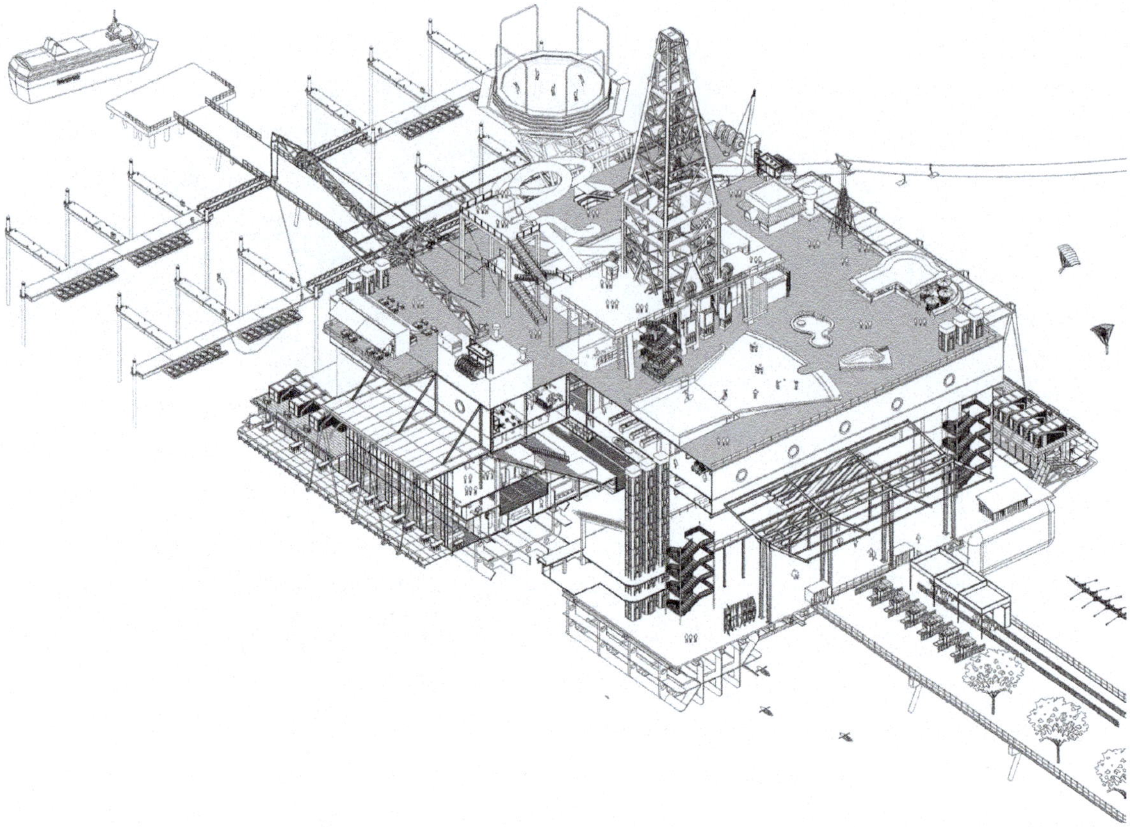

Fig. 45 Detail of waterpark recreational facility above semi-submersible hull/deck

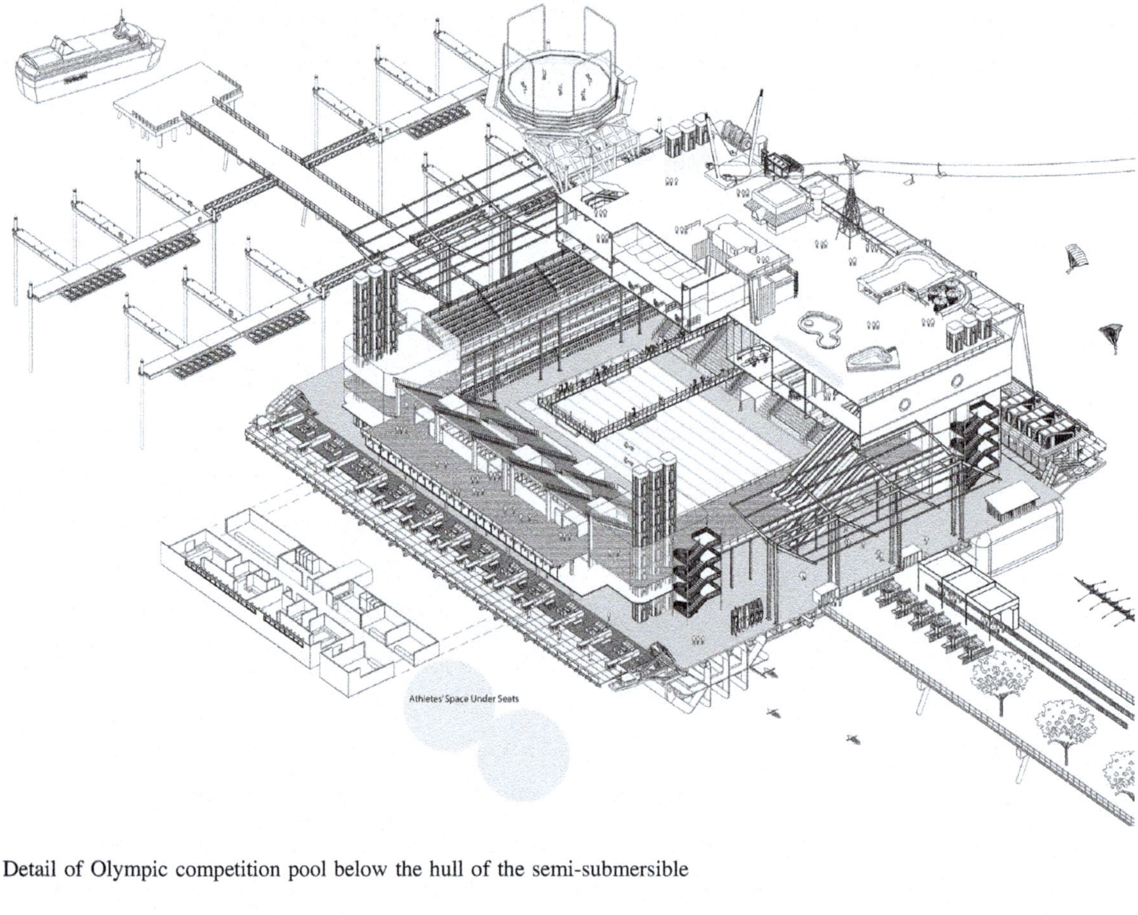

Fig. 46 Detail of Olympic competition pool below the hull of the semi-submersible

4 Semi-submersibles LNG Bunkering Facility

Currently Liquefied Nature Gas (LNG) can be delivered to both LNG supply facilities and the receiving ship. Both onshore and offshore supply includes ship-to-ship (STS), and truck-to-ship (TTS) operations. Shore-to-ship supply of LNG into port storage bunkering systems is enabled by LNG carriers. In Yun's [30] conceptual design of an offshore LNG bunkering terminal, merchant ships are refueled by a four segment supply chain (Fig. 47). Long distance LNG vessel carriers supply to port bunker terminals followed by in-port distribution using shuttle vessels. Receiving ships then draw LNG from the bunker terminals or they can draw directly from LNG carriers.

In Singapore, the workflow of vessel to berth in the port, cargo loading/reloading is separate from the refueling process, which is not conducted at the port. Instead, small bunkering vessels shuttle between the oil storage facilities located on Jurong Island, Singapore and the receiving vessel. The process is time-consuming and labor-intensive. The port berths are congested while the sea space near the port serves as anchorages.

This proposal is for an offshore LNG bunkering facility using six semi-submersibles or six jack-ups, each about 6,000 m^2. Each offshore facility is arranged in a radial hexagonal plan form to provide twelve berths for LNG receiving ships. The bunkering facility is equipped with emergency shut down systems with non-pressure and pressurized LNG tanks made of structural concrete tube sections upcycled from the hulls of barges or pontoons.

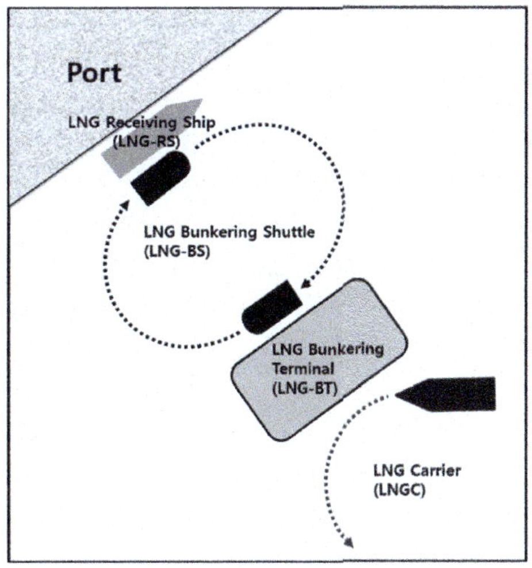

Fig. 47 LNG supply chain proposed by Yun et al. [30]

With 85% of the containers that arrive at Singapore being transshipment cargos, collocating a container transshipment hub at this offshore bunkering facility can increase TEU capacity without increasing port yard area. To obviate the process of bunkering after unloading cargo, the use of mechanized container freight systems in the airspace above both types of rigs allows cargo ships to break bulk concurrently while bunkering. This would reduce the turnover time for port berthing and refueling. Nearshore transshipment

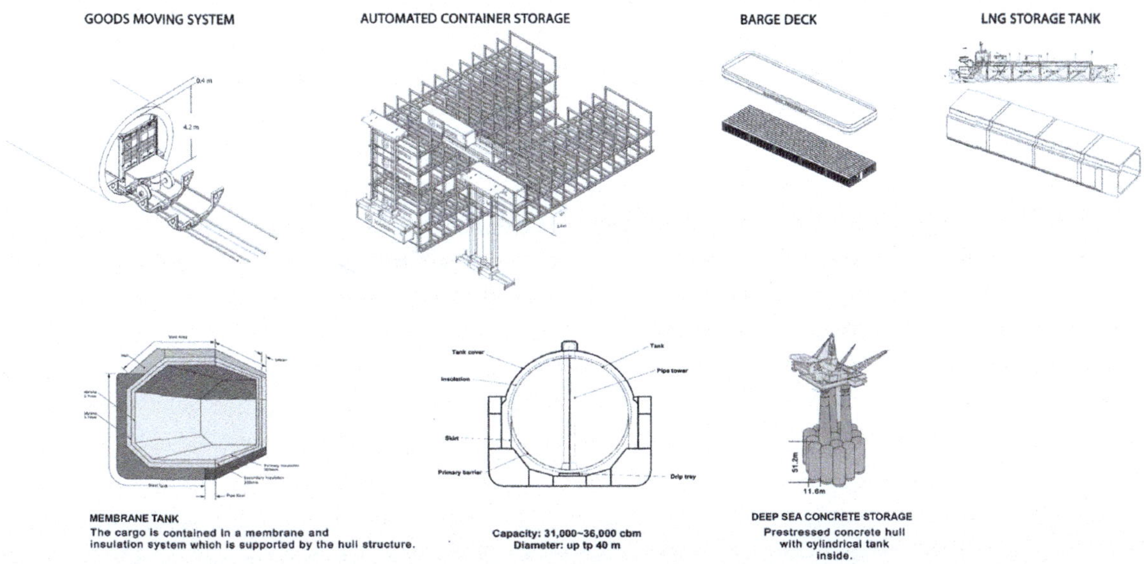

Fig. 48 The transshipment and LNG bunkering facility incorporate the components of a goods mover system connected to a mainland network. An automated container storage system and prestressed concrete LNG storage tanks integrate with barge decks to form berthing areas for large merchant vessels

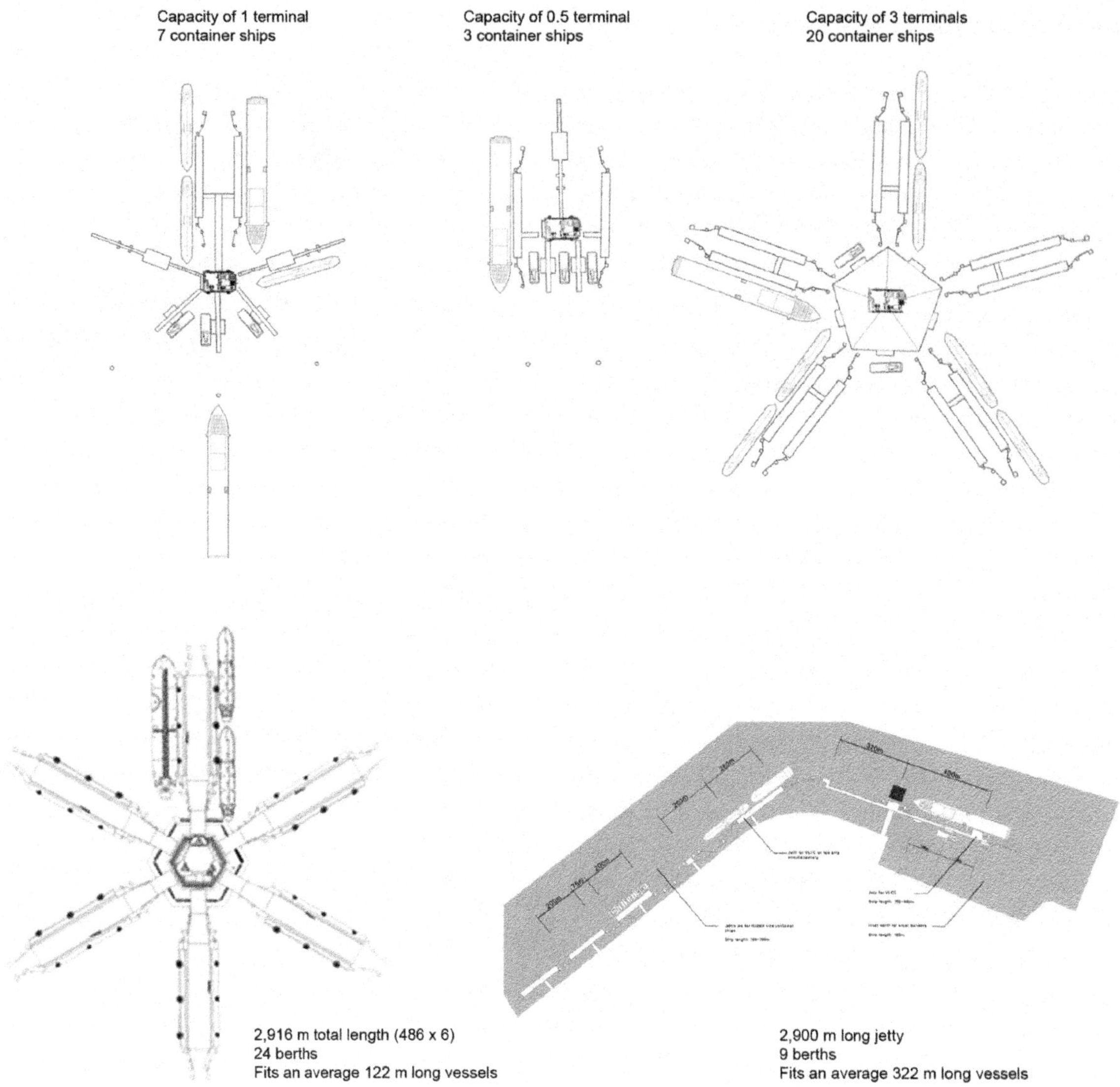

Fig. 49 Determining jetty berth lengths from container ship sizes. Jetties allow berthing of 200–400 m long container ships

hubs can further increase TEU capacity by using a goods delivery system network below ground level to connect to inland break bulk centers or Singapore distriparks. The transshipment operations will need to comply with the strictest of LNG bunkering safety measures.

Figures 48, 49, 50, 51 and 52 illustrate preliminary explorations on the design concept of semi-submersibles LNG bunkering facility.

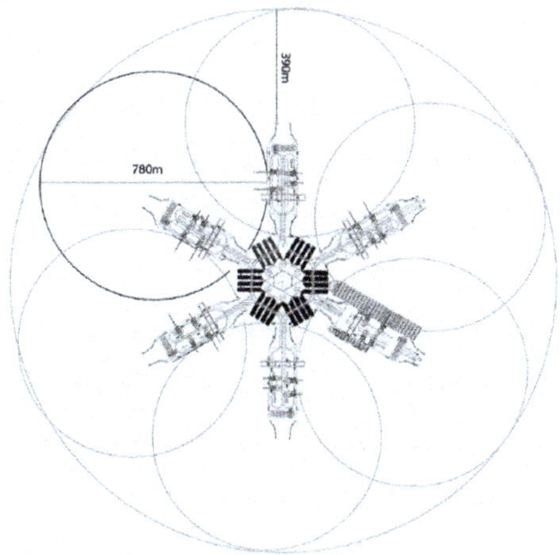

Fig. 50 Maneuvering dimensions of large vessel with tugboats used to establish radial berthing lengths

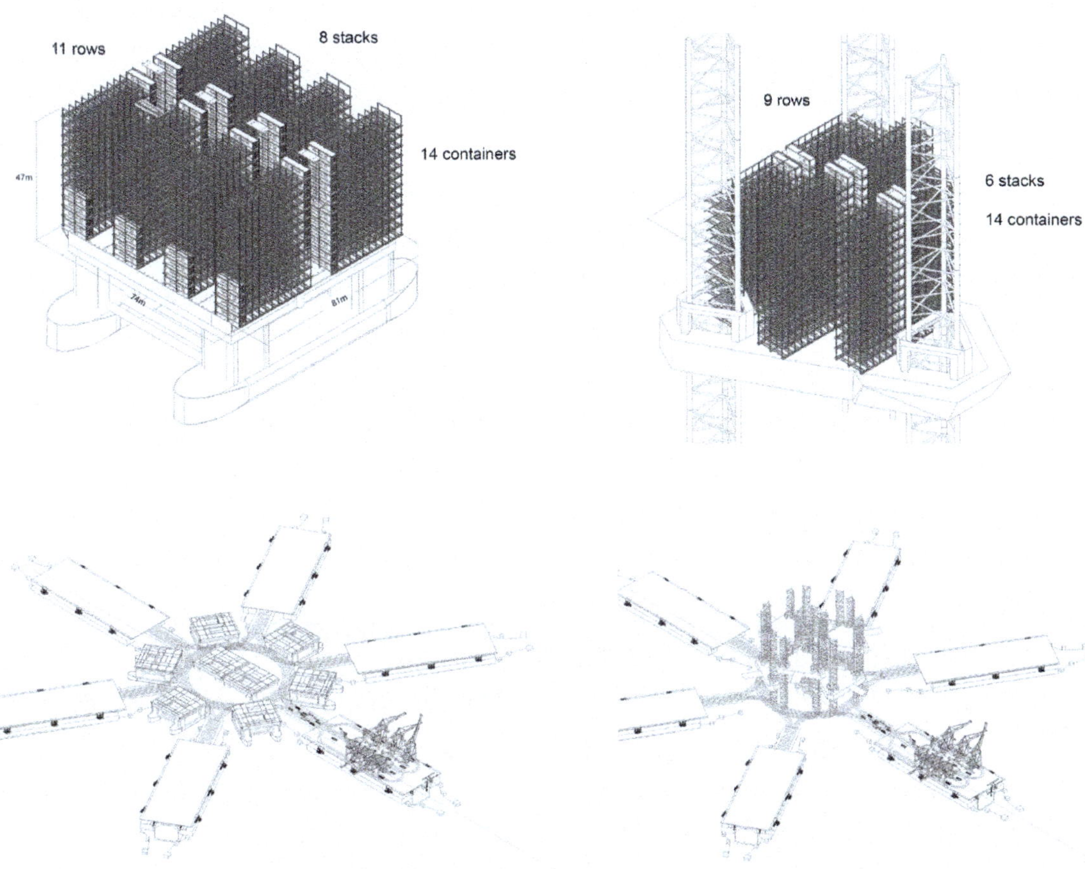

(a) Semi-submersible
Estimated storage capacity = 1,232 40 ft. containers or 2,464 TEUs (20 ft. equivalent)
Area of deck = 5,994 sqm
Platforms required = 7.2

(b) Jack-up
Estimated storage capacity = 756 40 ft containers or 1,512 TEUs (20 ft equivalent)
Area of deck = 6,067 sqm
Platforms required = 11.7

Fig. 51 Comparing container capacity between semi-submersible and jack-up rigs. Storage capacity was based on Singapore's annual container throughput in 2016 [13]. Container throughput = 30.9 million TEUs, Vessel arrival = 17,700, Average vessel capacity = 1,745 TEUs, 85% transshipment rate = 1,483 TEUs per month; or 17,796 TEUs annually

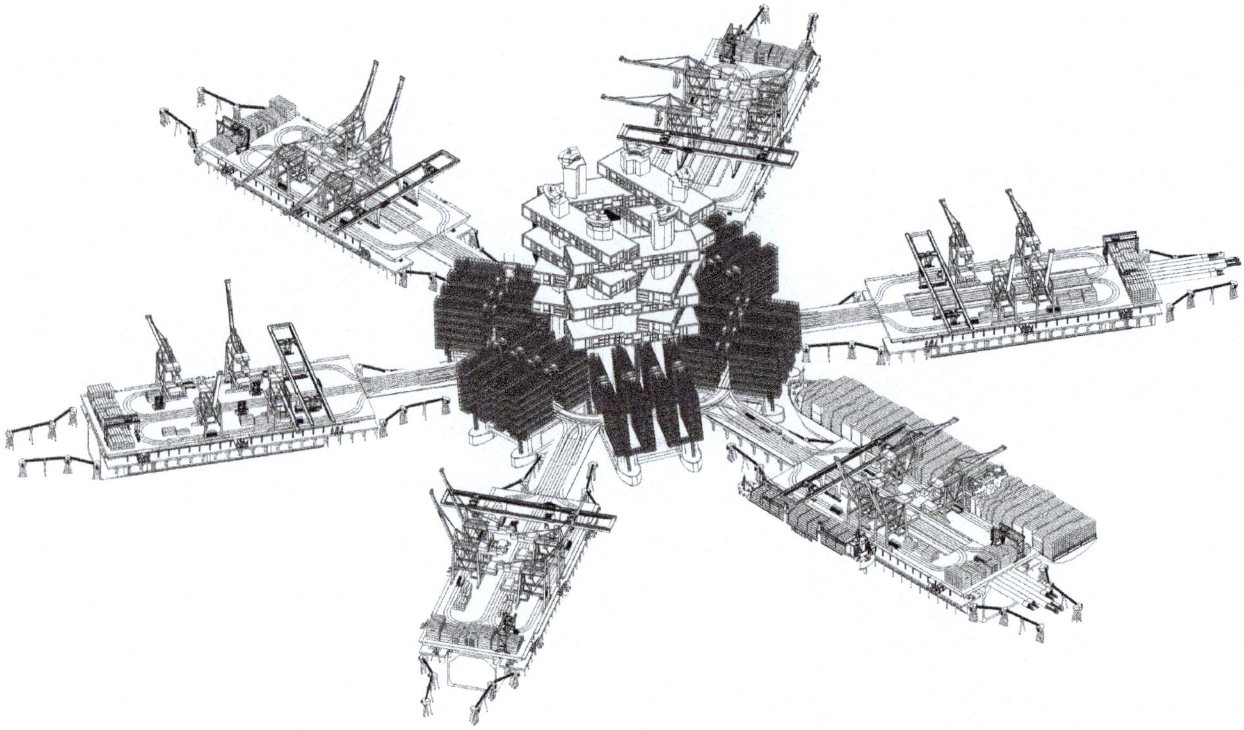

Fig. 52 A transshipment hub assembled with six semi-submersibles each with a deck of 5,994 m^2 supporting an automated container storage system

5 Jack-ups Waterpark

5.1 Resource-Intensive Parks

Waterparks are a twentieth century industry dominated by Walt Disney Parks and Resorts, Universal Studio Recreation Group and Six Flags. There are 600 large scale waterparks worldwide and have been able to sustain their businesses in Asia as a popular form of family recreation and entertainment. In Singapore, the two prominent waterparks are Adventure Cove on Sentosa Island and Wild Wild Wet in Pasir Ris, the eastern part of the city state dedicated for sea sports and housing.

Waterparks are notorious for operating expenses due to the extensive consumption of energy and water resources. More significantly, they are laid out in large plots of land where water rides are integrated with natural and artificial landscape. Coastal locations have desalination plants to process seawater. The bulk of energy consumption is in powering water pumps, filters, sensor controls and large areas of air-conditioned space. Water levels and pressures for the range of water rides are maintained by a system of water tanks and water reticulation networks. The system compensates for water losses through evaporation, pool water dilution, the filtering of back wash, overflow and splash out.

The capex of a theme park on a repurposed oil rig would have been reduced by the absence of land costs, and primary structure construction costs: as water slides can be composed to be supported off the jack-up masts in between their air-space or on the deck areas of the semi-submersibles (Fig. 53). Water rides can be composed in ways, which reduce water losses by positioning and layout described in Sect. 5.5 Design Development (Fig. 54).

5.2 Waterpark Studies

Land Sunway Lagoon and Wild Wadi waterpark are studied and compared for scale, configuration and water management processes.

In spite of a small plot, Wild Wadi waterpark has a much greater number of visitors per square meter (Fig. 55). A compact waterpark project can be successful even with lesser attractions per sq. m. This is useful in projecting a nearshore waterpark that attempts to avoid the sprawl associated with floating platforms typologies. Floating structures must keep away from maritime traffic zones designated in Singapore waters.

Layout Water consumption is attributed to four activities: splash out, evaporation, deck wash down, and backwash loss. Splash out is defined as loss of water from human interaction with the water system. The loss of water through

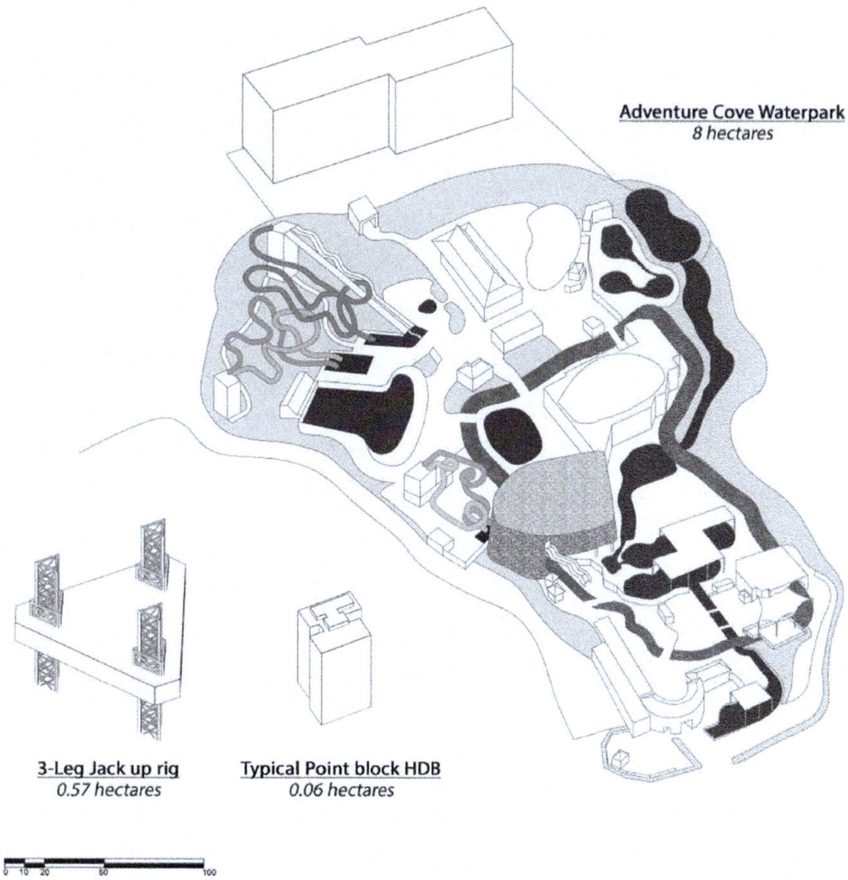

Fig. 53 Visual comparison of physical footprints of a jack-up rig, a housing block and a land plot occupied by a conventional waterpark. The housing block provides the scale of the land plot to accommodate a housing estate. The jack-up rig has air space between its masts to consolidate the sprawl of the waterpark on land

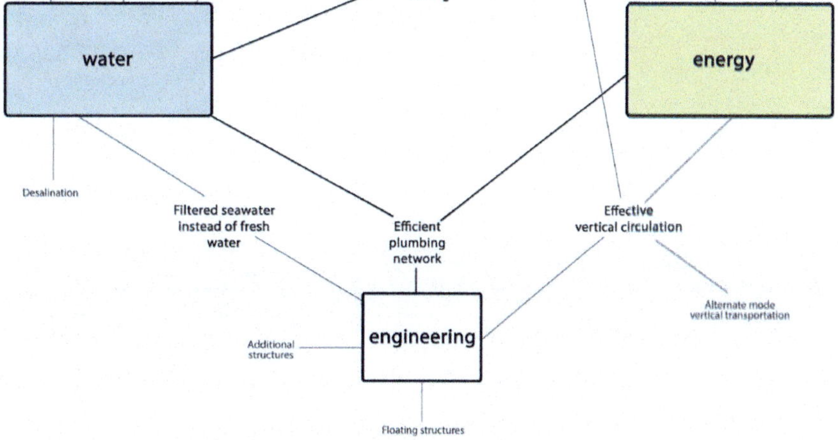

Fig. 54 Design considerations for a vertical waterpark to reduce resource consumption

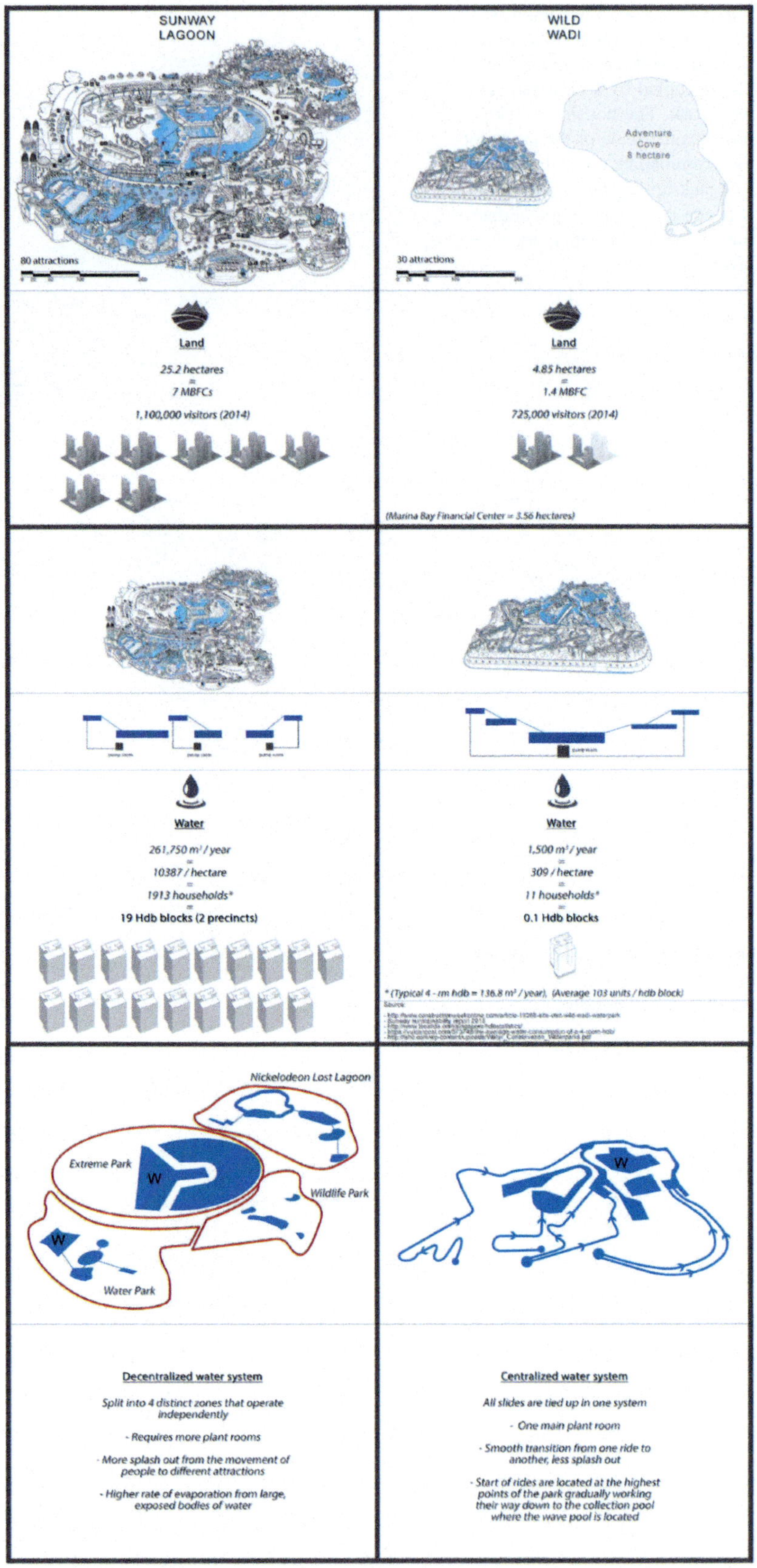

Fig. 55 Comparisons in land area, water consumption and water systems between Sunway Lagoon and Wild Wadi waterparks

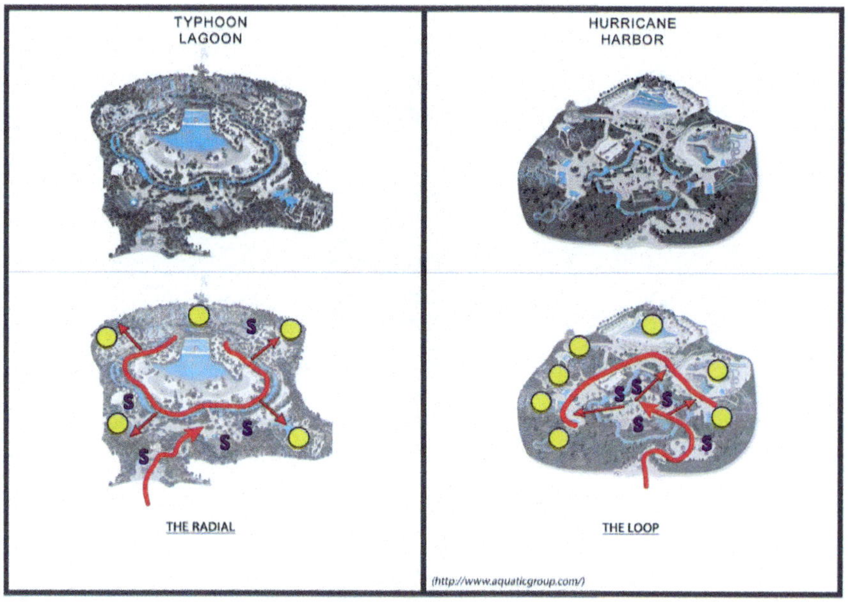

Fig. 56 Pedestrian circulation of Typhoon Lagoon and Hurricane Harbor rides

evaporation is greatest at tropical outdoor waterparks. Deck wash is a maintenance operation for cleaning the solid surface areas surrounding the aquatic landscape. Backwash maintenance involves the cleaning of the particle filters in a water reticulation system and is the major component of waterparks' water consumption. In the overall waterpark water system, the maintenance and topping up operation accounts for 2–3% of total water use on a daily basis. In other words, a waterpark is re-using approximately 97–98% of its water system [7].

There is a large discrepancy in the water consumption between Sunway Lagoon and Wild Wadi waterparks. As visitor numbers are not significantly different, their layout and the types of filtration systems used affect the water usage of these parks.

Sunway Lagoon has a decentralized water circulation network due to its large waterpark area made up of multiple zones. This leads to high water usage due to backwashing of replicated sets of pumps, greater evaporation loss in more waterbodies and possible leakages in a extensive reticulation network.

In contrast, Wild Wadi Waterpark integrates a modern filtration plant in its centralized system for water efficiency. Attractions are arranged consecutively to provide smooth transition between one another, and to reduce splash onto pavements and the subsequent loss of water through evaporation. Water splash and spillage is collected at a main collection pool which is the wave pool located at the lowest point of the park.

This forms the basis of a design strategy which attempts to arrange a vertical stacking and staggering of waterbodies to capture splash outs from each other and to optimize reticulation in shorter vertical or inclined runs within the confines of the oil rig. Stacked/staggered waterbodies minimize energy required for reticulation and reduce the frequency in which water is pumped into waterbodies to replace the water loss from splash out.

Circulation Pedestrian circulation can be organized into two general categories: radial and loop [21]. Radial is a 'Hub and Spoke' concept which has a central core and paths that extends outwards to major attractions, promoting cross circulation while alleviating overcrowding at attraction locations, for example, Six Flags Hurricane Harbor in New Jersey. In loop, all major elements are organized on a single path that circulates around one or two core attractions, driving guests to feature areas along the path of the full loop, for example, Disney's Typhoon Lagoon in Florida (Fig. 56).

In avoiding large groups of people accumulating at the ends of rides, the attractions are staggered to separate them from starting queues and to avoid crowding at zones of possible collision between ride goers and passers-by. Safety issues in the evacuation of injured customers and minimizing injury in mass stampede situations are also important. Therefore, the entry point and pedestrian flow paths are carefully laid out to enable ease of orientation and for safety considerations.

Amenities such as F&B and retail opportunities are strategically positioned near changing room and ride end-points as well as at mid-circulation paths for convenience in accessibility.

Types of attractions Table 7 is a compilation of waterpark attractions. Specifications such as the rider capacity and required water flow are used to determine the location within a vertical context. The layout seeks to provide a choice of major features forming destination rides as optional sequences in water rides. Water intensive features can be

Table 7 Water slide profiles

	BOWL	FUNNEL	SERPENTINE RIDES		RACERS	SPEED SLIDES		PLAY

	BEHEMOTH BOWL	TORNADO	MAMMOTH	PIPELINE	KRAKEN RACER	SUPERLOOP	MULTIBUMP	RIDEHOUSE
Min. rider height (m)	1.2	1.2	1.1	1.1	1.1	1.2	1.2	
No. of riders	5/4	6/4/2	6/5/4	4/3/2/1	1	1	1	
Vehicle conveyor	yes	yes	yes					
Hourly ride capacity	900/720	1080/720/360	1440/1200/960	1200/900/600/300	900	180	180	4068
Pool dimensions (m)	6.1 × 12.2	6.1 × 12.2	6.1 × 15.2	7.6 × 12.8				
Pool depth (m)	0.9	0.9	0.9	0.3				
Required water flow (m^3/h)	2044	1254	1590–2271	570–680	159 per lane	228	228	1544
Attendants required								6
Proposed location		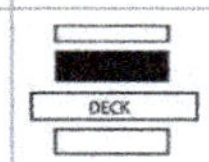						

located at the lower levels of the oil rig elevation to reduce the energy needed for pumps.

5.3 Study on Adventure Cove

Design issues unique to the Singapore context are identified in Universal Studio Singapore's Adventure Cove waterpark (Fig. 57).

Water The primary water supply is potable grade and supplied by government's Public Utilities Board. Surface run-off rainwater and drainage water is captured in eco lagoons and underground water storage tanks for treatment in order to reduce the reliance on potable water. Treated recycled water is used for irrigation and fire-fighting. In 2014, the facility collected 731,280 cubic meters of rainwater, equivalent to 293 Olympic-sized swimming pools.

Water consumption increased by 6.7% due to high evaporation from prolonged drought, and frequent plant irrigation to prevent the landscape from drying out. As such, an online monitoring system was installed to reduce the water consumption and also, to reuse them in attractions [6].

Energy Liquefied petroleum gas is used as a main source of energy, while diesel is used to operate backup power generators. Grid electricity formed the bulk of energy consumption in the facility. In 2015, solar panels were installed on rooftops into generate 679,911 kWh of electrical energy [6].

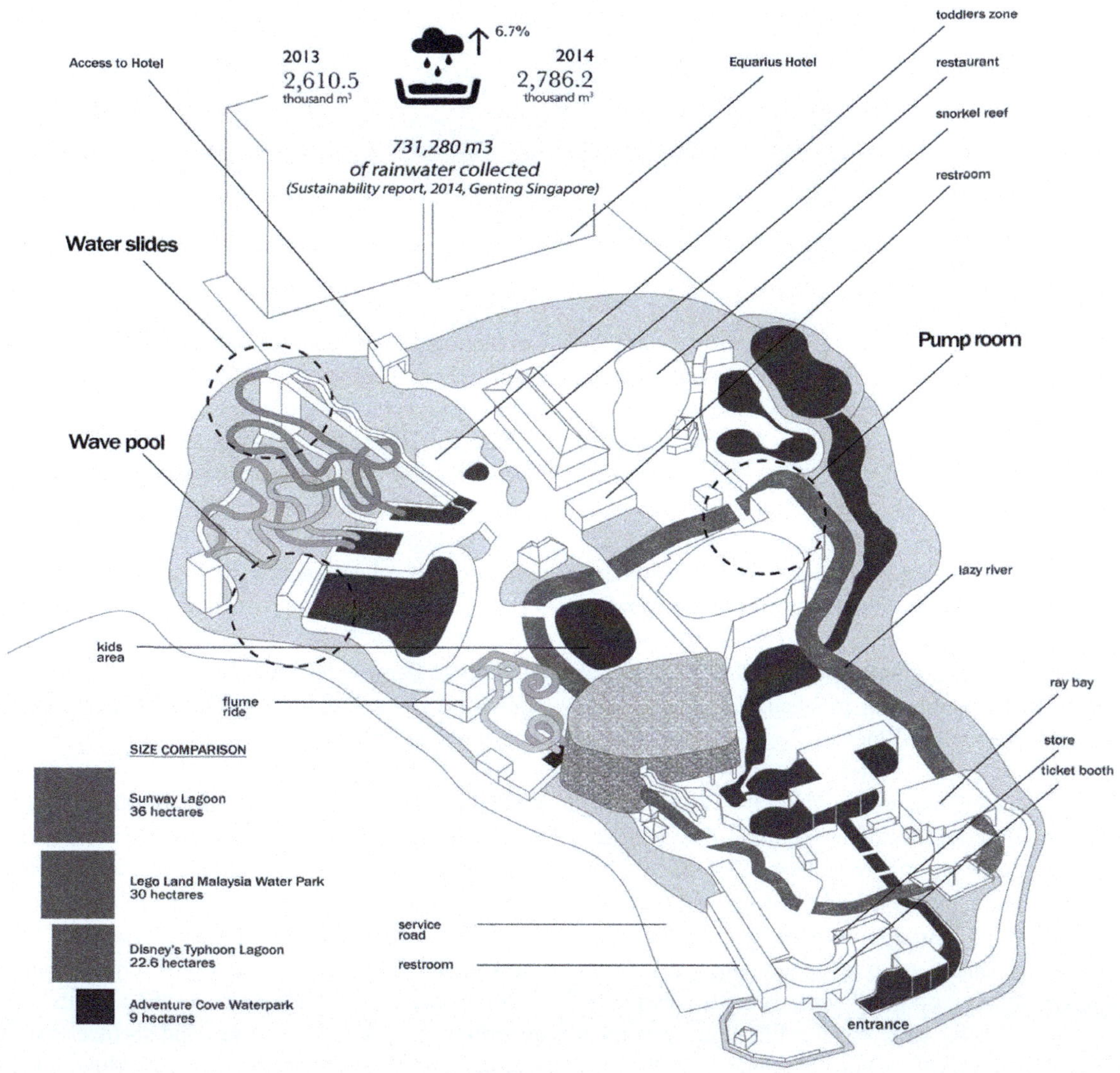

Fig. 57 Layout of Adventure Cove, Singapore. An aerial map shows there is no room for expanding the waterpark

Fig. 57 (continued)

5.4 Resource Network

Water A waterpark uses three grades of water supply quality in the pool of commercial waterparks: chlorinated freshwater, saltwater (3000–5000 ppm) and filtered seawater (35,000 ppm).

In order to achieve a higher degree of self-sufficiency, filtered seawater would be ideal for a nearshore oil rig waterpark. However, the high salinity content is not suitable for all users. Saltwater pools are chlorinated but the chlorine level is typically lower than that of a freshwater pool. The advantage over typical chlorinated pools is the reduction of chloramines, which poses health problems for people with asthma and other respiratory illnesses.

As such, a hybrid supply system is proposed which utilizes saltwater pools and changes to filtered seawater during periods of low rainfall. Rainwater harvesting is then a design determinant in augmenting this reticulation system.

In waterparks, pump rooms filter pool water and transport water up the elevated starting positions of various rides, while reserve water tanks for top up and monitoring systems to maintain the quality of pool water are also essential. A typical pump room for backwashing filtration systems is illustrated in Fig. 58. Backwashing amounts to a third of water consumption that can be recycled for general maintenance and landscape irrigation.

The difference in water usage between Wild Wadi and Sunway Lagoon can partly be attributed to the type of filtration system. Figure 59 is a comparison between a conventional high-rate sand filter and a regenerative media filter system. The latter is shown to require less energy and water while occupying a small footprint.

Energy The use of renewable energy sources focus on PV panels and Ocean Thermal Energy Conversion (OTEC).

Solar panels produced in local fabrication plants may improve their economic viability for solar power generation compared to other sources such as wind energy. Current research attempts to improve intermittency issues and low efficiency levels from 14 to 25%. The research proposal attempts to harness 59% of the energy needed to power the rides through optimizing solar collection in horizontal surfaces onboard the rig.

The roof surface is used for PV panels as well as for harvesting rainwater. The design of the roof attempts to maximize surface area for solar collection and rainwater catchment while allowing sufficient openings for daylight into spaces below the rig.

OTEC is a process that produces electricity by using the temperature difference between deep cold ocean water and warm tropical surface waters. OTEC plants pump large quantities of deep cold seawater and surface seawater to run a power cycle and produce electricity. Although still largely untapped, OTEC is one of the world's largest renewable energy resources that is available to many countries.

A 10 MW OTEC plant in deepwater can provide surplus energy and freshwater to the waterpark. The use of solar

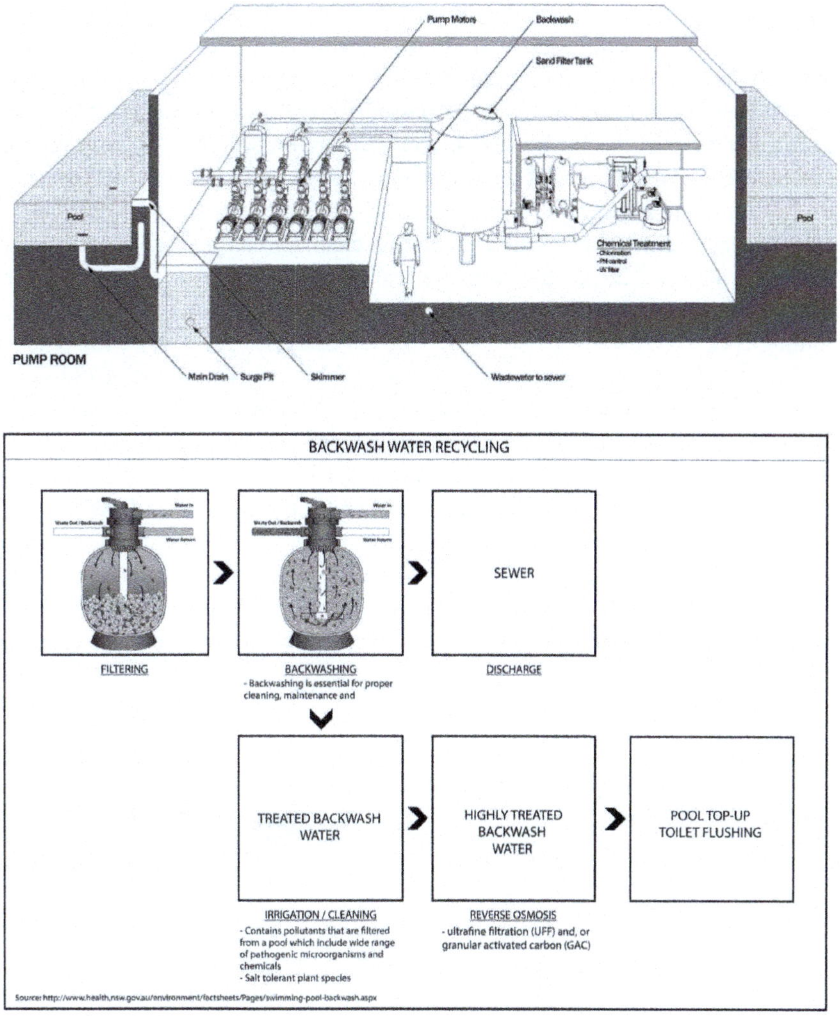

Fig. 58 A typical pump room for backwashing filtration systems

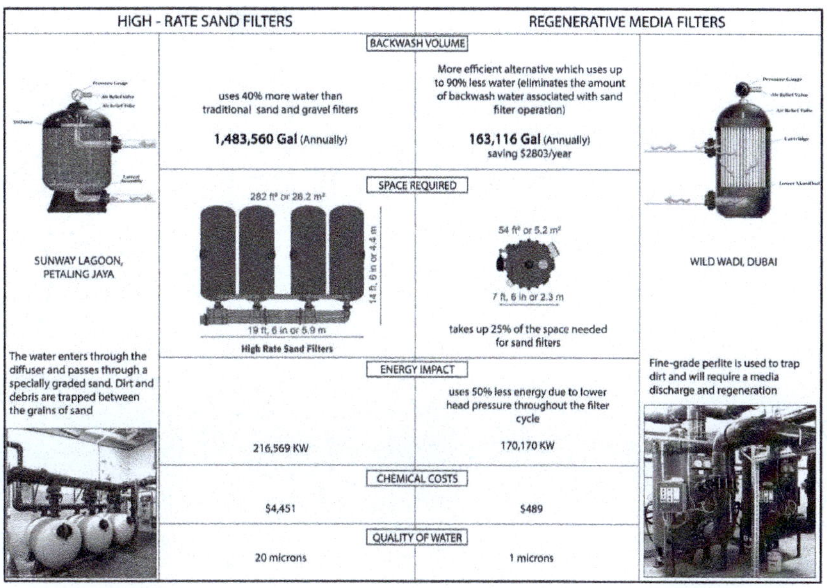

Fig. 59 Comparing filtration systems of Sunway Lagoon and Wild Wadi

distillation process in generating energy for the waterpark produces freshwater for pools/washrooms/restaurants usages as well as for cooling of pump room mechanisms and spaces. Discharged cold seawater is passed through a heat exchanger to chill the freshwater in a closed circuit, which can be used to cool spaces. The surface seawater evaporated to drive the turbine in an OTEC plant sheds its load of sea salt. The salt retrieved from the evaporation chamber can be sold as a product in its raw form but can also be further refined to extract metals and trace elements termed 'ocean mining' [12].

The OTEC facility on a jack-up has to be in close proximity to ocean depths of >800 m to reach cold seawater. Another option is to locate the OTEC on a floater platform and feed into a nearshore jack-up waterpark. The seawater pumps dominate the energy consumption of an OTEC cycle. These pumps and other auxiliary equipment consume an estimated 20% of the total electricity produced. For safety issues, an OTEC plant may be separated from a high occupancy facility, but is not an issue in settlement planning.

The three by-products of the OTEC plant is fed to the waterpark as follows:

(1) Treated seawater of a minimum 100 m^3 can be used to top up a swimming pool that has a volume of 3000 m^3. Expected water loss from the pool is estimated at 90 m^3 per day.
(2) 6000 m^3 potable water provided per day with an estimated consumption rate of 120 m^3 for washrooms and 22 m^3 for restaurant. Excess water would be piped away for other usage such as irrigation and maintenance.
(3) Cold seawater can be used for air conditioning and heat exchange systems to cool the pumps.

An offshore OTEC plant is capable of generating 100 MW, but for the oil rig waterpark, a 5 MW plant is sufficient to generate 40 million kWh per year [19]. The energy consumption is estimated at 1 MW for attractions, 200 kWh for amenities/restaurant and 300 kWh for lifts, lighting and other electronics. This leaves 3.5 MW in excess (Fig. 60).

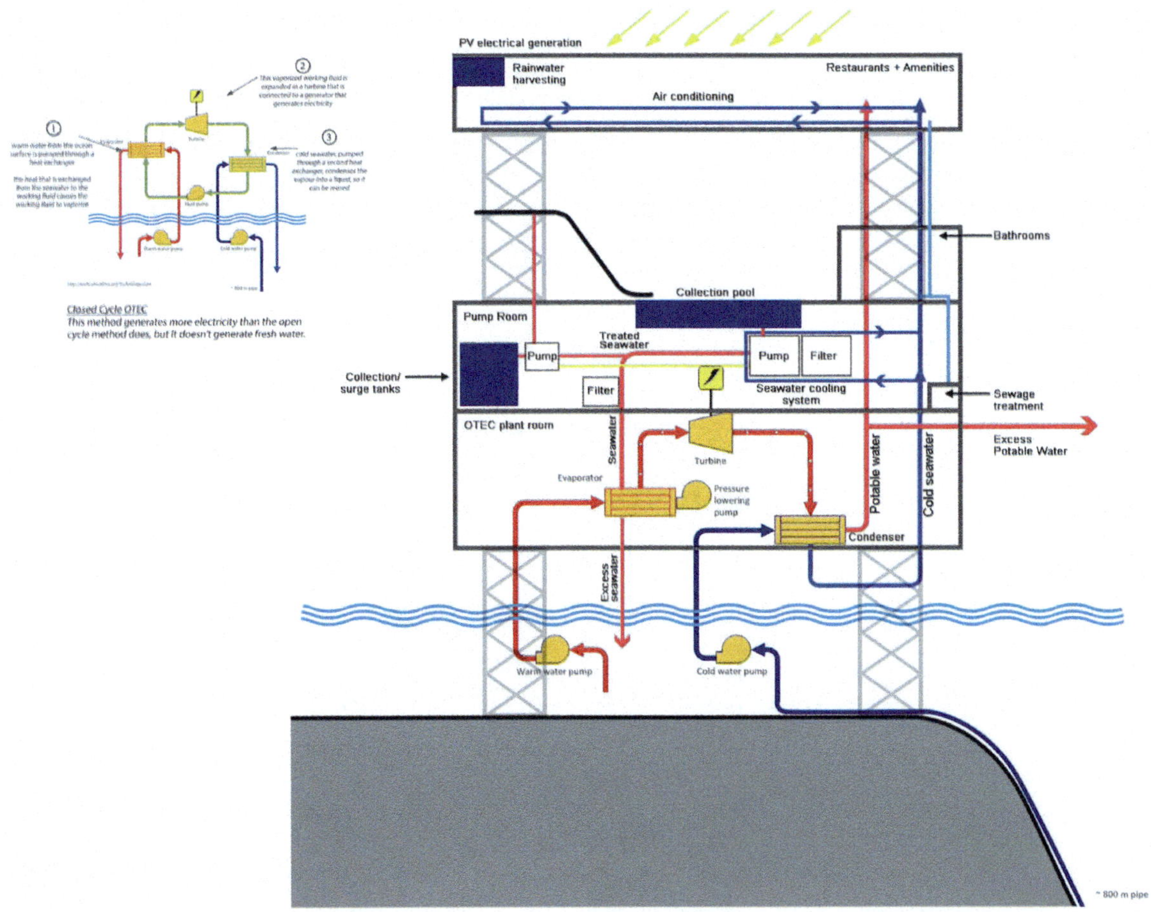

Fig. 60 If the OTEC plant is integrated with the jack-up waterpark, it needs to be on a seabed typography as above; on a continental shelf where nearby water depths is in excess of 800 m

5.5 Design Development

Design layout takes on the following considerations: (1) ride lengths are maximized by the airspace above oil rig decks areas; (2) staged pumping is used to reduce energy consumption to pump water to tall positions; (3) tall positions are minimized where water is to be pumped up, instead optimize reticulation with shorter vertical or inclined runs; (4) water loss through splash out may be reduced by overlapping splash out positions so that lower levels capture splash out from upper levels; (5) self-shading of pool areas in the tropical context to reduce solar gains; (6) where rainfall conditions permit, rainwater harvesting for energy; (7) deploying surface areas for solar panels to augment energy consumption; (8) minimize additional structures to free up space for rides to fly over large voids; and (9) selecting the appropriate pumping system for multiple zones that can be connected vertically.

Early explorations of repurposing jack-ups and semi-submersibles are illustrated in Fig. 61. Three basic configurations are then studied with extension of deck area created by combining two jack-ups.

Skewer (Fig. 62a) This design calls for modules attached to the masts of the jack-up rig linked by skybridges, which takes advantage of the airspace in composing waterslides. The geometry allows for the location of a wave pool on the deck of the rig. Infinity pools are cantilevered off the sides to increase surface area for visitor accommodation. Main vertical circulations are nestled within the masts while the entry to the waterpark is along a floating bridge connected to the mainland. The disadvantage of this scheme lies in the lack of expansive space for activities and gatherings. Vertical transportation would be crucial such as an inclined tramway.

Trio (Fig. 62b) An optional design with a radial array of three rigs with decks raised to different heights. This creates different zones with which to accommodate different rides. A hotel is located at the center of the waterpark together with supporting amenities and vertical circulation in the form of escalators. Varying the heights and zones of water rides, in order to enhance patron experience, differentiates spaces. The huge deck area requiring multiple supports is a disadvantage.

Slender (Fig. 62c) In developing solutions on a single jack-up rig, deck form up zones and inclined slides are intertwined within the airspace in this scheme. This results in a slender, towering design that has amenities and spaces interspersed with water rides. A central atrium space draws daylight into the deeper recesses of the plan while slides and spaces are weaved around the masts. The drawback of a stacked waterpark is the energy needed to pump up large volumes of water and in powering the vertical transportation for users to constantly access the elevated positions of water rides.

Stumpy (Fig. 62d) This version is an improvement from the 'slender' scheme. By densifying the lower spaces, it brings activities closer to the water, enabling various sea activities to take place not possible with traditional land-based versions. The periphery of the rig is extended which requires additional structures to enlarge the roof area needed for rainwater and solar energy collection. Additional features can integrate a water energy device in a floating breakwater lining the waterpark. This breakwater provides protection from swells as well as generating wave and tidal energy.

Paired combination 1 (Fig. 62e) Further explorations considered repurposing two jack-up rigs. The advantage lies in the larger surface area for not only human circulation, but also greater catchment area for solar and rainwater. Paired combination 1 is a variant of the 'stumpy' scheme by placing them apart, to create an entry point for transport vessels and ferries. A new massing block at the intersection between the rigs houses a funicular system for energy-efficient, high-capacity vertical transportation.

Paired combination 2 (Fig. 62f) A more condensed version than 'paired combination 1' features a sweeping atria and retractable roof for all-weather protection. Gangways radiate from the waterpark, culminating in large folding rain harvesters and an amorphous solar skin. This innovative binary scheme merges the benefit of a relatively small footprint while maximizing surface areas for attractions, amenities and energy harnessing equipment.

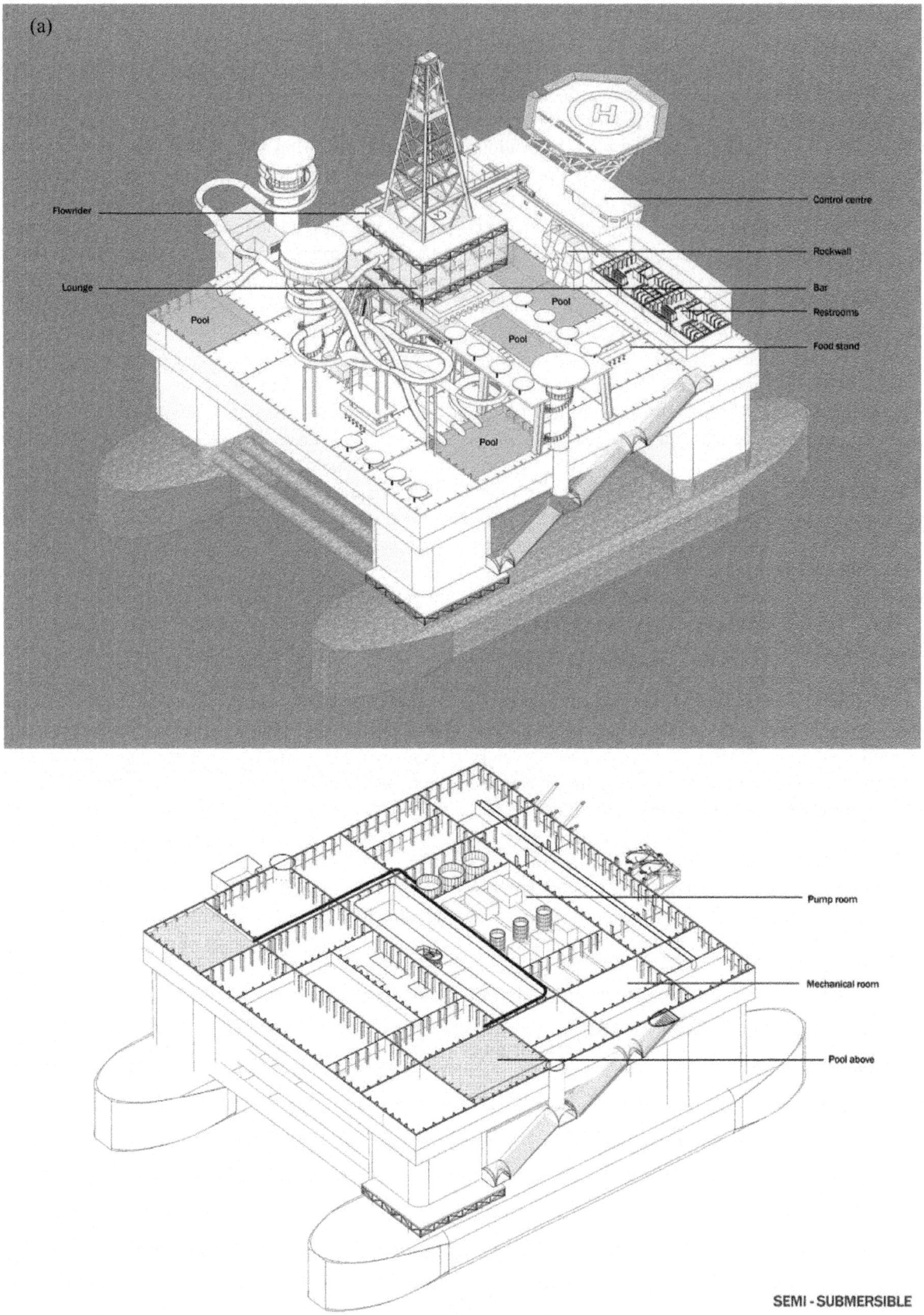

Fig. 61 Early waterpark design explorations on **a** semi-submersible **b** three-legged jack-up and **c** four-legged jack-up

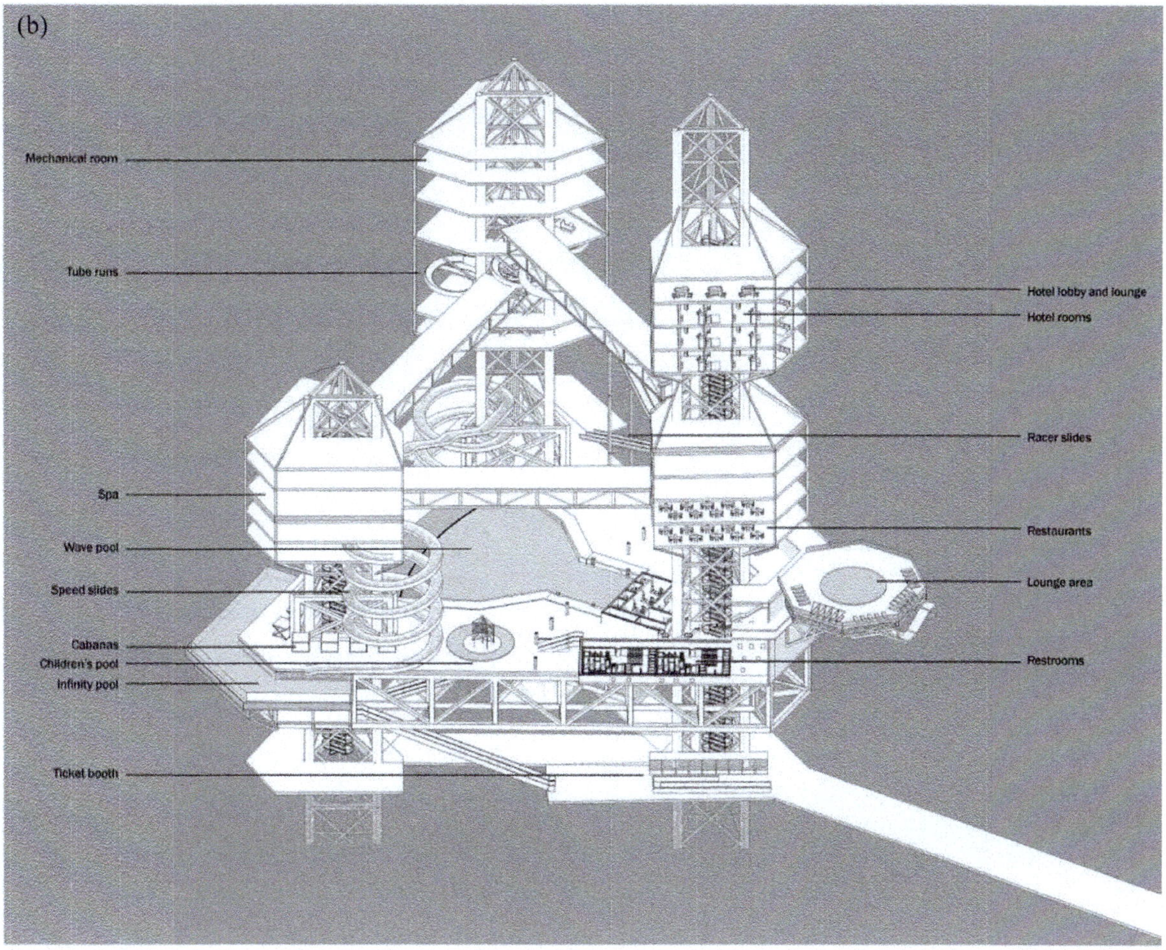

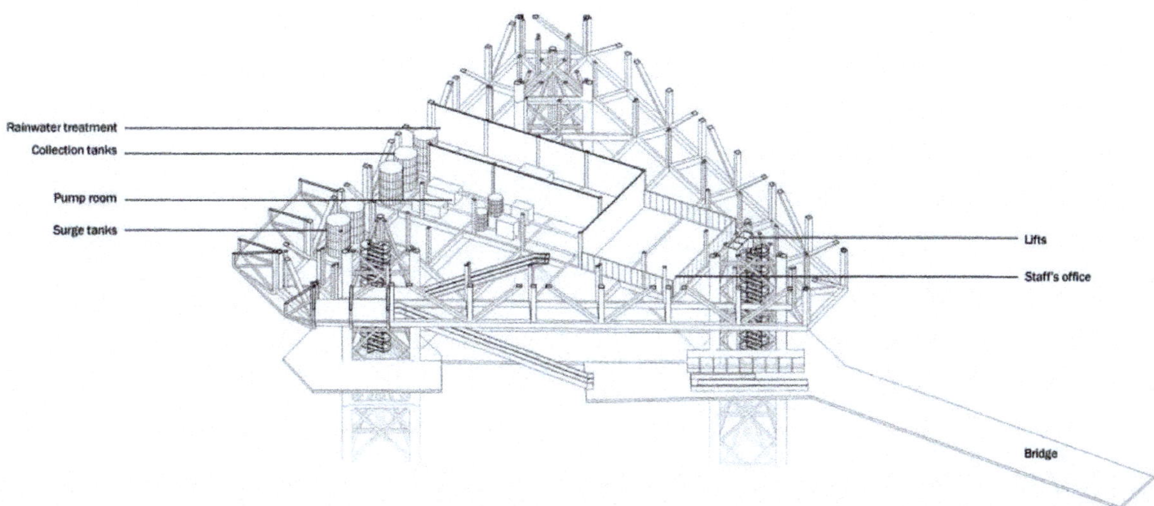

Fig. 61 (continued)

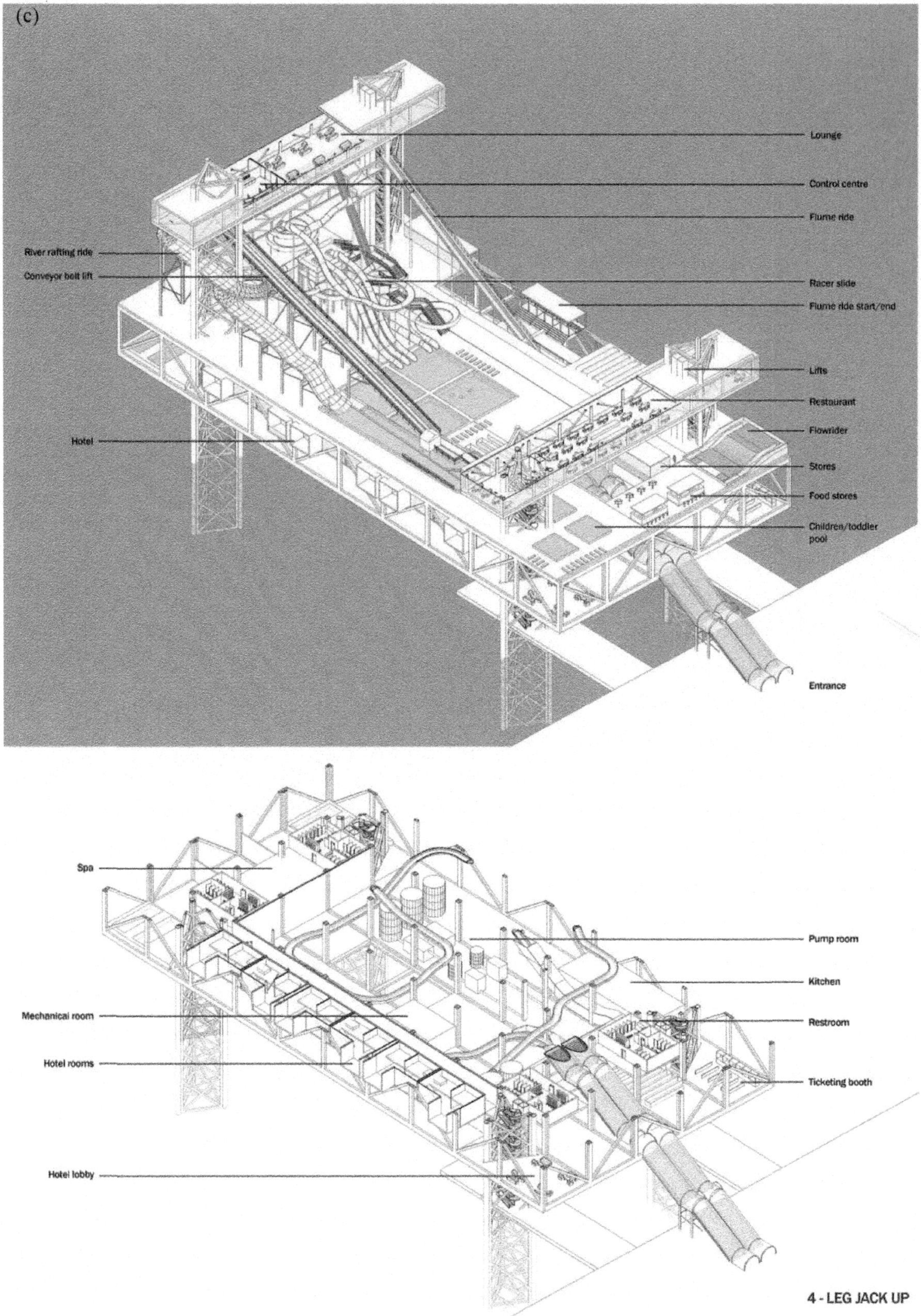

Fig. 61 (continued)

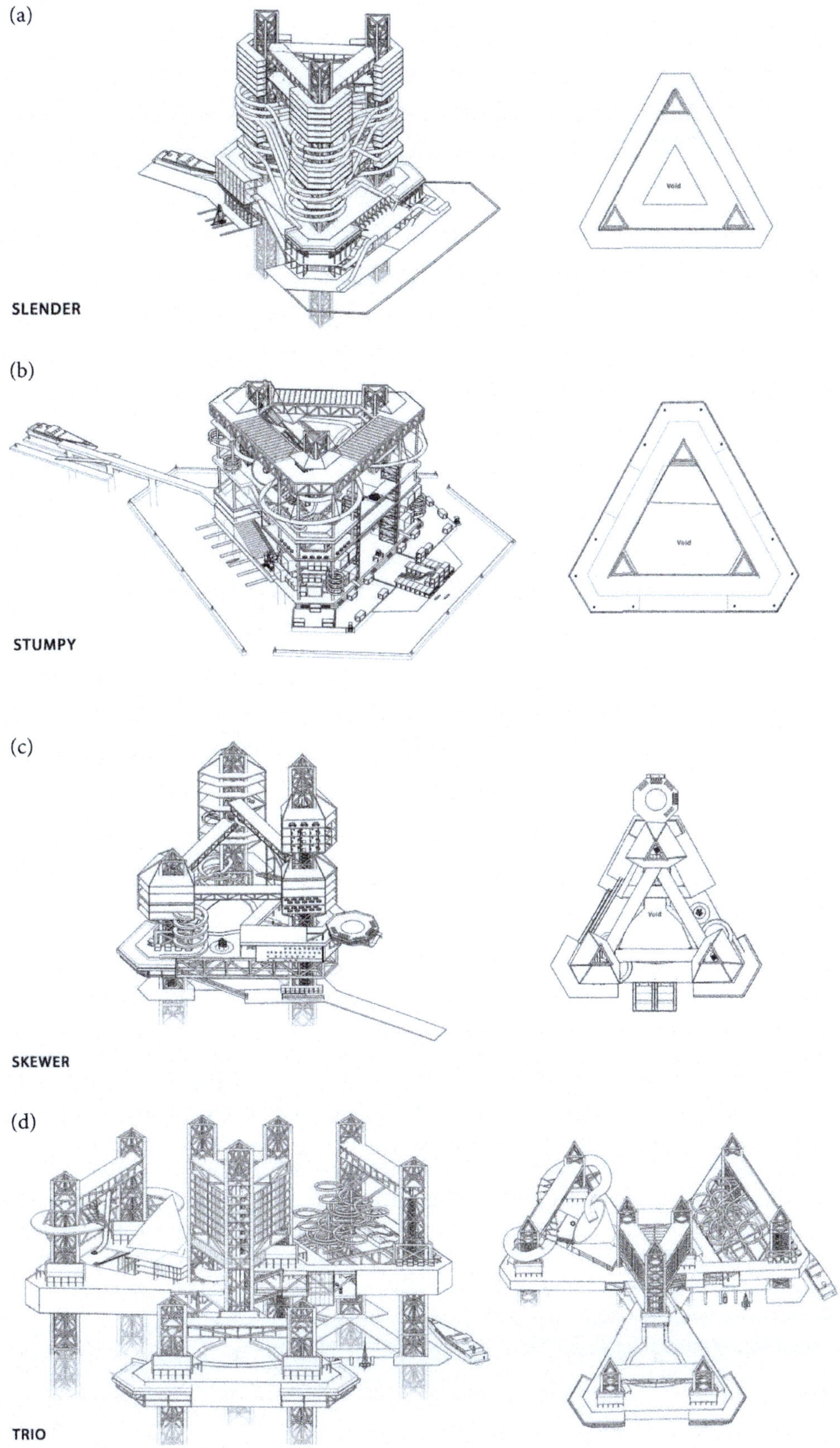

Fig. 62 Configurational study of jack-up waterparks

Fig. 62 (continued)

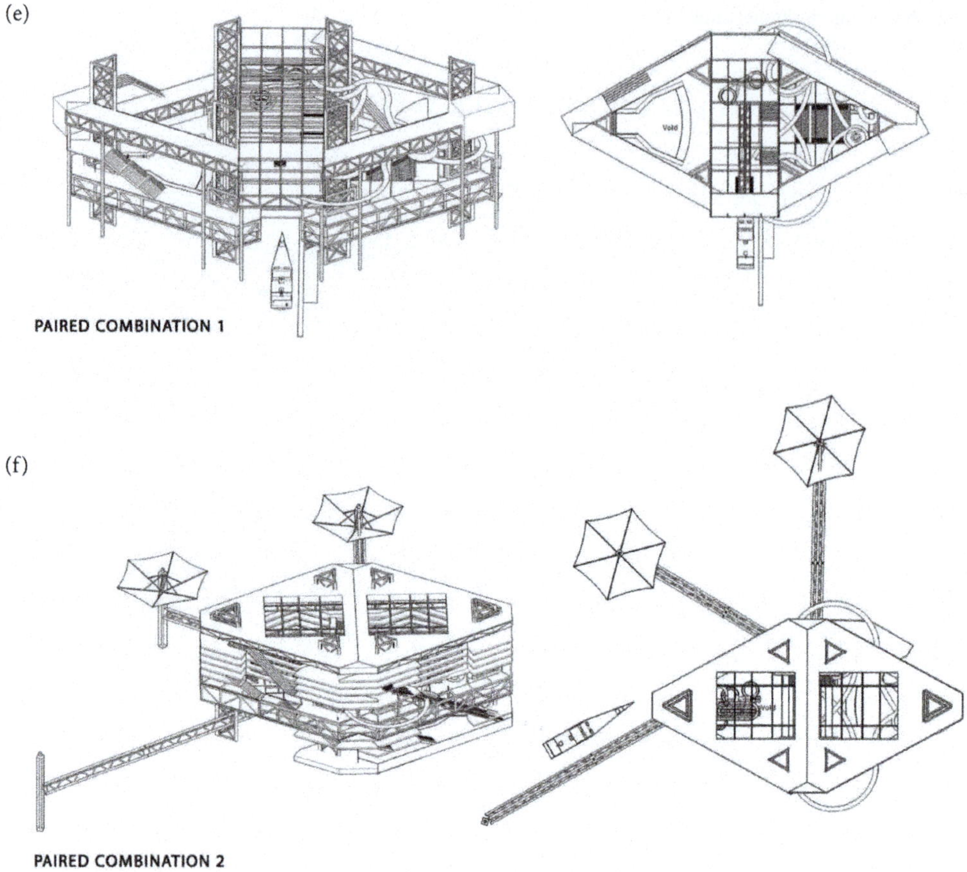

5.6 Energy Consumption and Production

Pump Systems Fig. 63 examines the correlation between the placement of rides, height of structure and pumping energy required. A staggered plumbing system is more effective than a single pump booster system. The location of high water recirculating rides should be strategically located in close proximity to the main pump rooms. Central mains, water tanks and pumps are to be located remotely from high occupancy zones. System layout is to avoid complex plumbing routes with several bends and kinks resulting in energy loss and complicating maintenance issues.

Rainwater and solar yield Rainwater harvesting and solar energy collection at the roof area (Fig. 64) can offset the consumption of water and energy. Table 8 calculations are based on a tropical context of Singapore with average monthly rainfall and solar irradiance. It is noted that these yields are intermittent and largely dependent on climatic conditions. A more reliable source of grid energy is required to power the waterpark. During periods of low rainfall, filtered seawater can be used to augment water supply. Figure 65 illustrates an example of a 'stumpy' jack-up configuration that integrates mechanical systems and water attractions.

Final Design Energy and space efficiency are achieved in an ideal vertical stacking configuration with the generous deck areas nearly equivalent to conventional parks. Spaces can be created from the geometry and structure of the rig to reduce the need for extensive retrofitting. Thus, a novel waterpark typology attempts to improve on the efficiency of inland water parks in terms of land plot usage, water and energy production and recycling (Figs. 66, 67 and 68). The rethinking of resource intensive integrated theme parks as a resource-producing entity is possible with the repurposing of jack-up oil rigs used in combined configurations. The operational capacity of a jack-up waterpark is summarized in Table 9.

Fig. 63 Shaded areas denote rainwater and solar harvesting

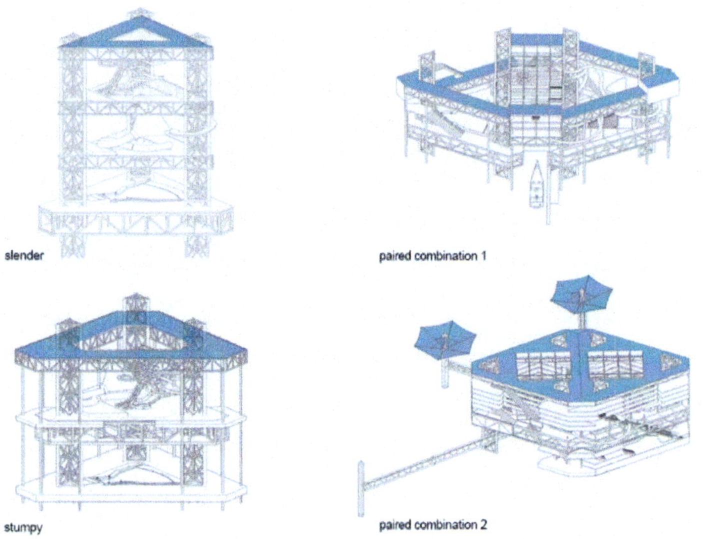

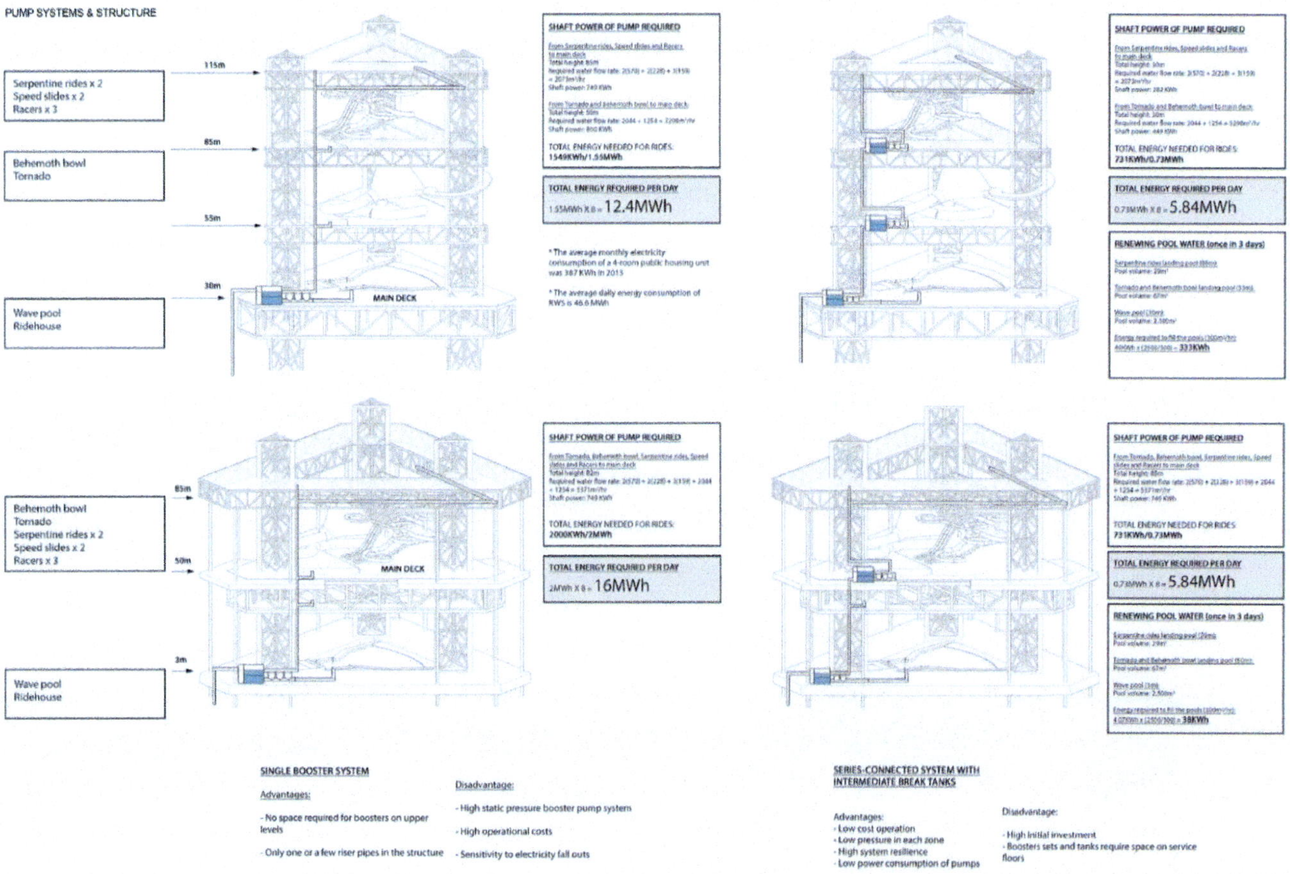

Fig. 64 Pump systems and structure

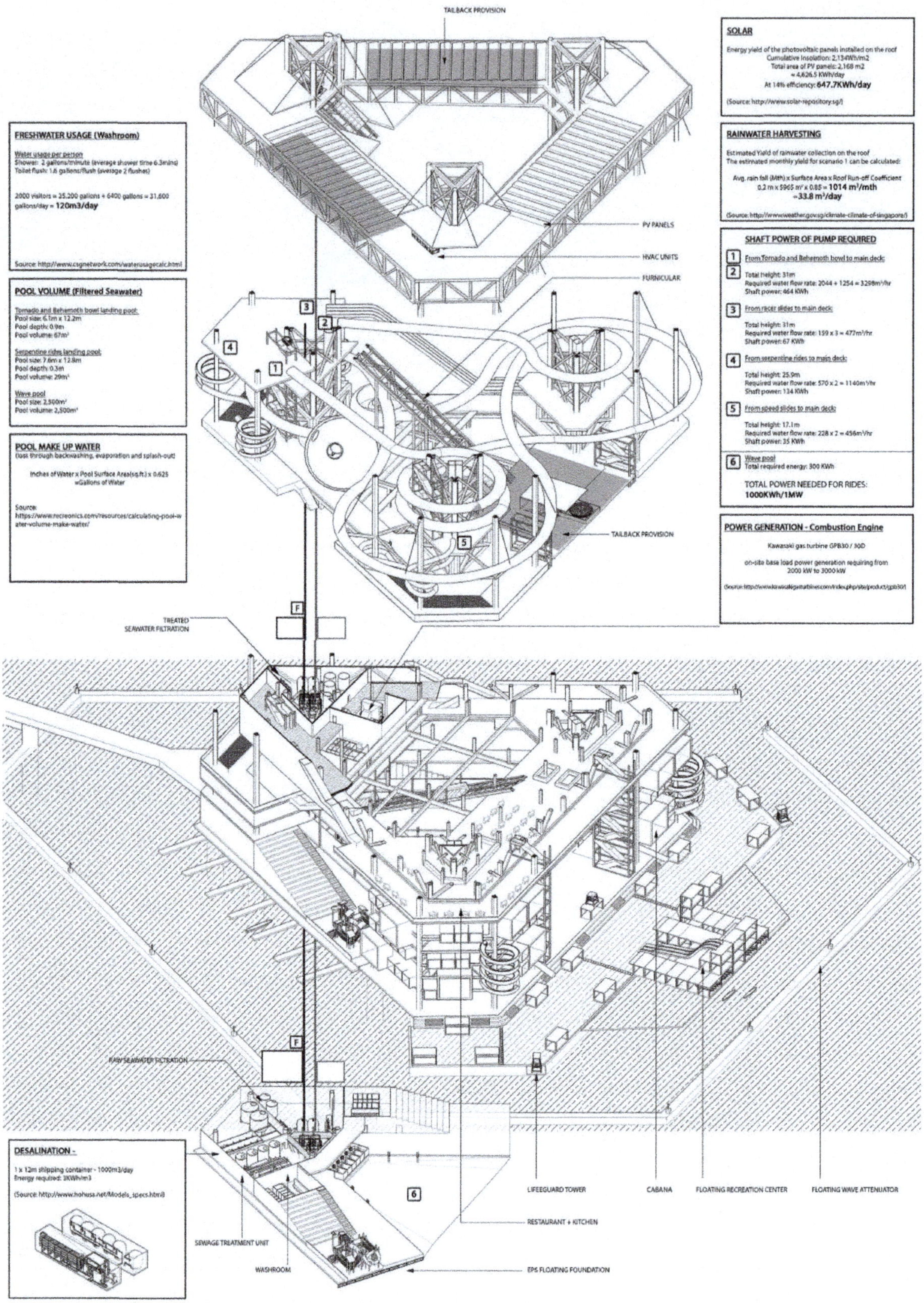

Fig. 65 Operation and spatial arrangement on a 'stumpy' jack-up configuration

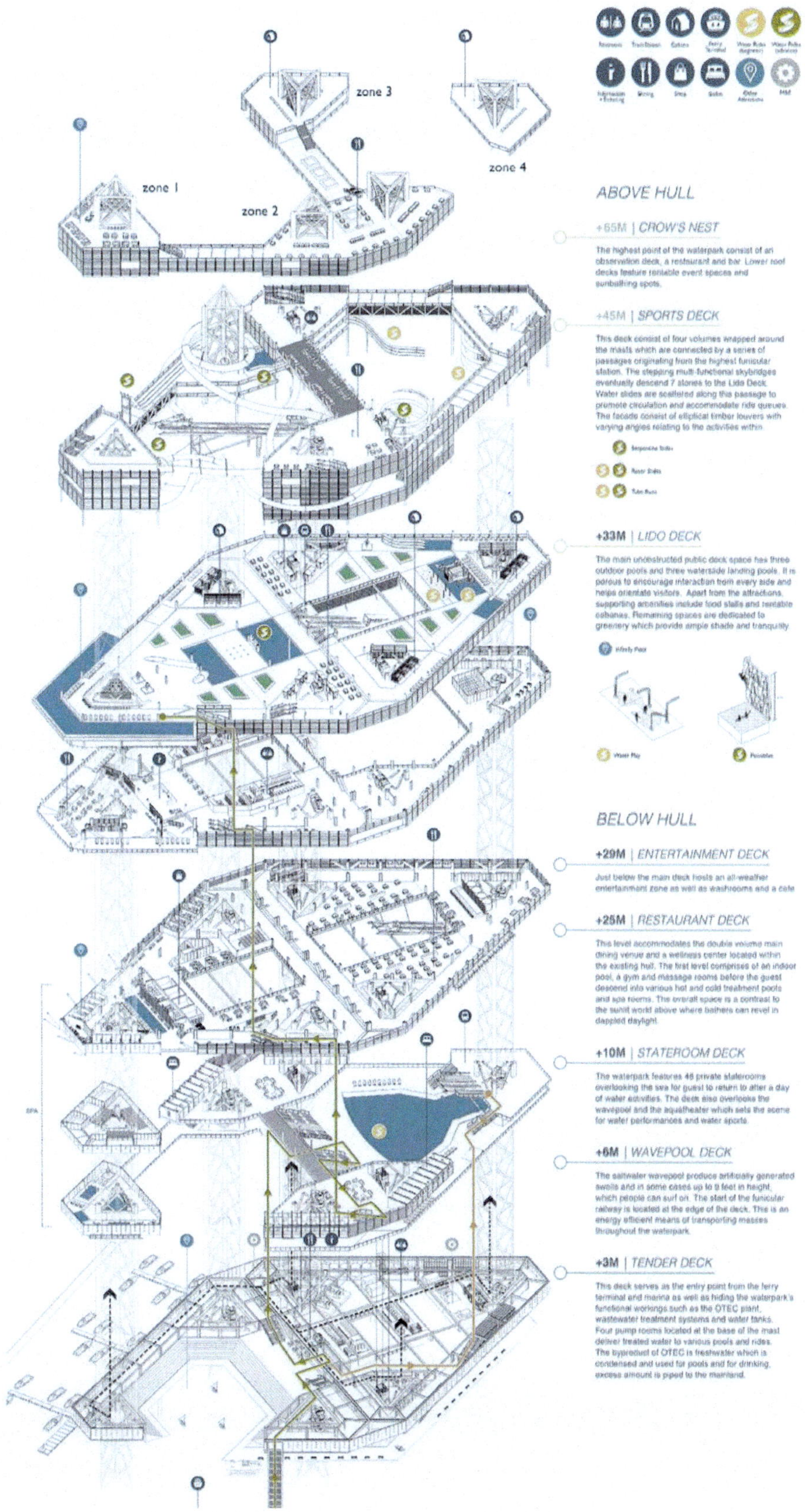

Fig. 66 The exploded axonometric plans indicate the location of each attraction and the main user circulation. The final design is divided into two segments with the corresponding decks above and below the existing hull. It aims to improve the legibility of the plan and user wayfinding

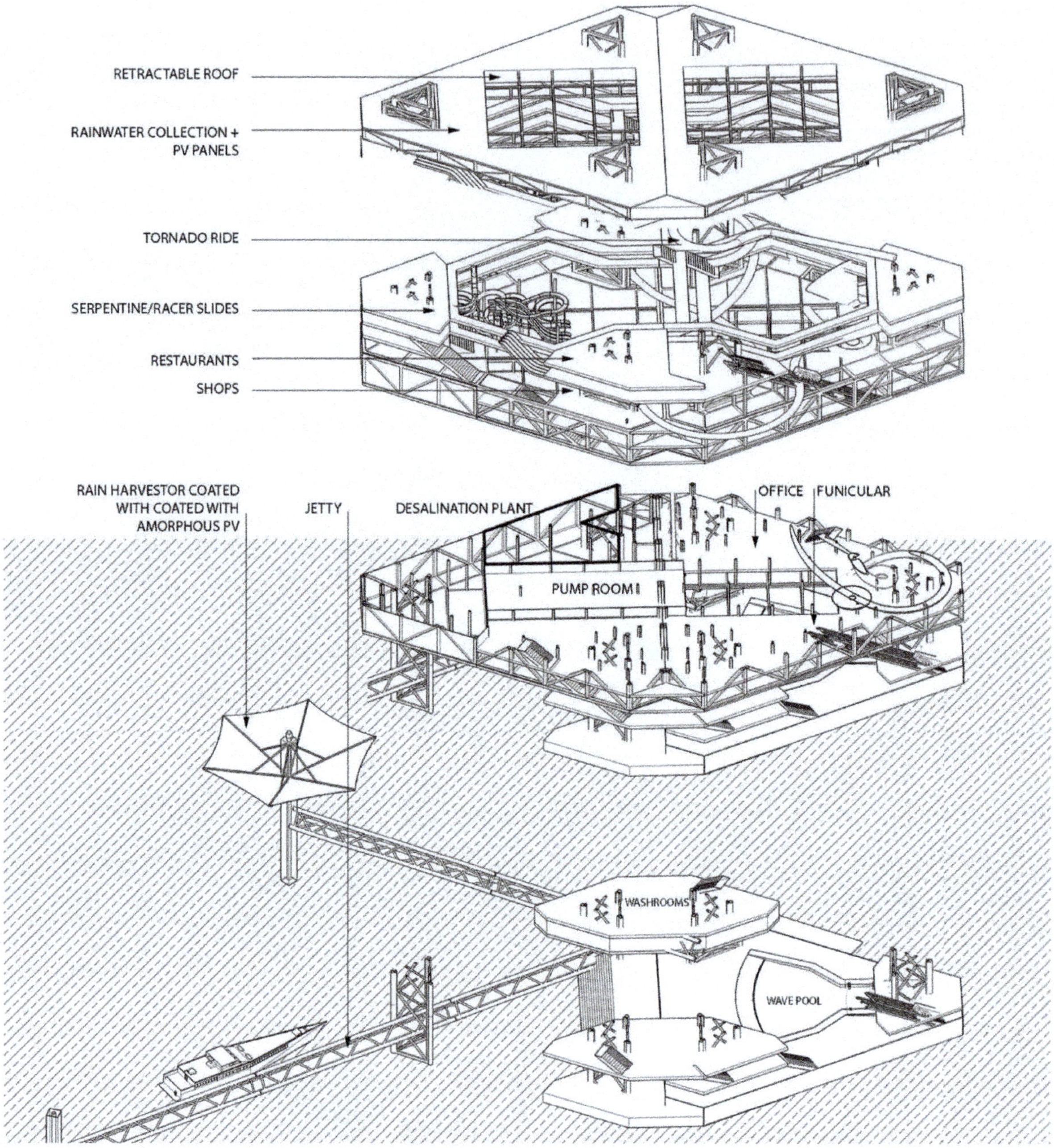

Fig. 67 The mechanical deck that houses the pump rooms and all M&E services are confined to the lowest level of the waterpark

Fig. 68 Perspectives of a jack-up waterpark

Table 8 Rainwater harvesting and solar yield of different permutations

	Roof area (m^2)	Rainwater harvesting (cubic m per month) Average rainfall (0.2 m) × Surface area × Roof Run-off coefficient (0.85)	Supplement water use in washroom (120 m^3 per day)	Solar energy (kWh per day) Cumulative insolation at 2.134 Wh per m^2	Supplement energy required for booster pumps (5,840 kWh per day) (%)
Slender	2,454	417	11.6%	734	12.5
Stumpy	5,965	1,014	28.1	1,779	30
Pair combination 1	8,068	1,372	38%	2,410	41
Pair combination 2	11,630	1,977	55%	3,475	59

Table 9 Operational capacity of a jack-up waterpark

Rig type	3-legged jackup	*Energy source OTEC (10 MW)*	
Length of water slide	480 m	Pros	Cons
No. of restaurant	1	• Discharged cold seawater is passed through a heat exchanger to chill the freshwater in a closed circuit, which is subsequently used to cool spaces • The surface seawater evaporated to drive the turbine in an OTEC plant, sheds its load of sea salt. Termed as *ocean mining*, the salt retrieved from the evaporation chamber can be sold as a product in its raw form or further refined to extract metals and trace elements [ref: www.makai.com/ocean-thermal-energy-conversion/]	• The energy consumption of an OTEC cycle is dominated by the seawater pumps. These pumps and other auxiliary equipment consume 20% of the total electricity produced
No. of cafes	3		
No. of bars	2		
No. of short-stay cabins	40		
Land saved	>4 hA		
Volume of surplus freshwater	200 m^3/day		
Freshwater usage in washroom	120 m^3 per 2000 visitors		
Pool volume (filtered seawater)			
Tornado and Behemoth bowl landing pools	67 m^3 [6.1 m × 12.2 m × 0.9 m depth]		
Serpentine rides landing pool	29 m^3 [7.6 m × 12.8 m × 0.3 m]		
Wave pool	2,500 m^3		

6 Jack-up Luxury Resort

The resort hotel industry taps on the wealthy class of travellers seeking novel vacation experiences in exotic locations. This leads to the proliferation of luxurious hotels built on natural uninhabited coastal land, a worldwide niche market continuing today. Coastal tourist industries attract influxes of tourists to a relatively small area. This additional load to local infrastructure would be problematic if provisions are not made to upgrade limited wastewater treatment capacities to control coastal water quality.

Coastal ecosystems are susceptible to the environmental impact of such resort island developments. These resorts often cover large land or sea areas, which affect the living environment of coral reefs and marine ecosystems. For e.g. mangrove forests and seagrass meadows are removed to create open beaches, piers and jetties, while boardwalks are built directly on coral reefs for viewing or snorkeling.

Thus an exploration on the possibility of converting an oil rig into a luxury hotel, a standalone structure with minimum footprint on water (Fig. 69) and which can accommodate resort activities away from sensitive marine ecosystems would be an alternative means to coastal developments. The design objectives focus on the creation of a hotel experience with the structural shell of oil rigs, for resort and family types of clientele.

In repurposing, the original oil rig structure is retained as much as possible for all the hotel spaces. Permutation studies of different guestrooms configurations are illustrated in Fig. 70. The 5000 m^2 rectangular decks of four-legged jack-up rigs such as the Seafox 5, are ideal for atrium typologies. The atria has potential for a spectacular interior space with overlooking corridors and all rooms having uninterrupted views of the sea. Decks may be positioned at the base, mid-height or top of masted supports in any combination. Banks of guest rooms are positioned for sea views and spaced between atria incorporating open-air deck facilities. With the four-legged jack-up rig, the upper end capacities reach 210 rooms with a leg length of 106 m in 34.5 m depth of water. Capacity is lowest at 166 rooms when rooms are cantilevered off masted supports as a collection of four short towers.

Triangulated configurations with three-legged jack-up rigs where elevators are located within masted supports, can accommodate 369–564 guestrooms. This is comparable to guestrooms capacity in current resort hotels, which range from 100 to 500 rooms [4]. The range is also applicable for vacation ownership schemes. The back-of-the-house and public areas for resort hotels account for 14% and 16% of total built-up area respectively; with guest rooms occupying 70% of the area [26]. Public areas include F&B outlets, landscape, gym and pool.

The efficiency of the design conversion depends on the shape of the plan per floor and buoyancy limits; to stacking the number of floors. Shape efficiency is measured in terms of the number of rooms that can be planned on one floor.

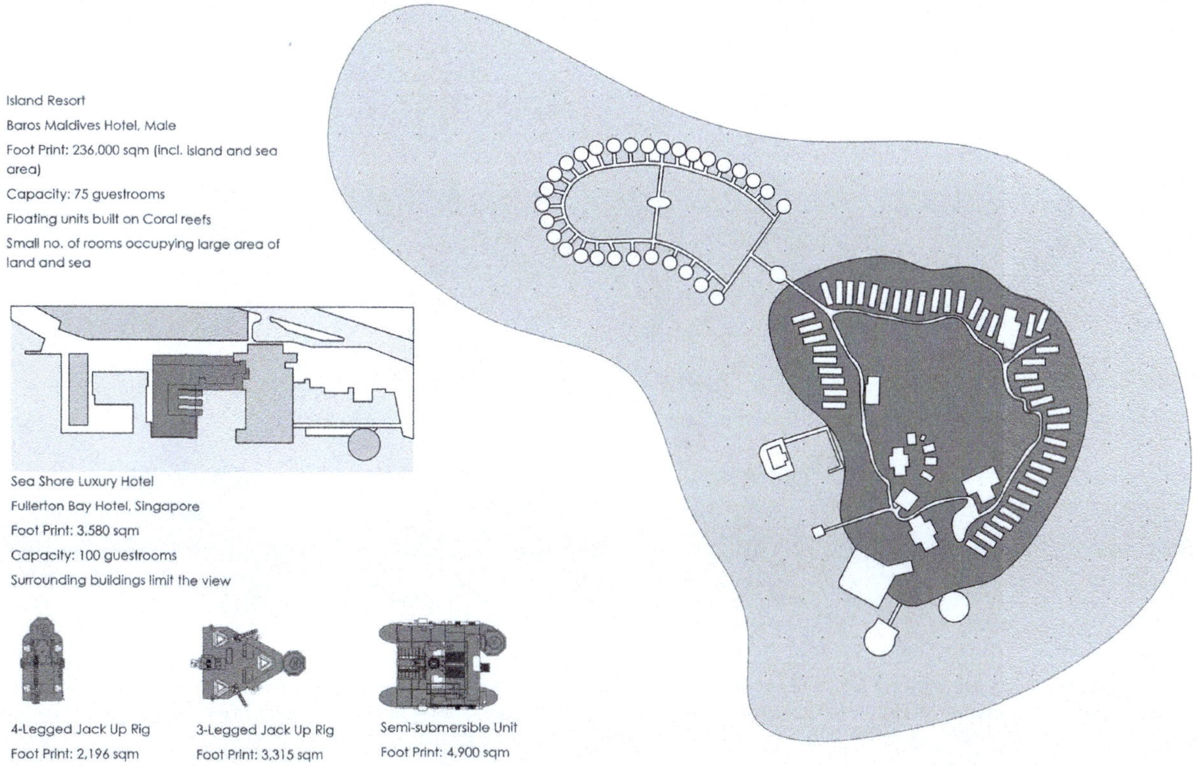

Fig. 69 Comparison of resort island footprints with those of oil rigs

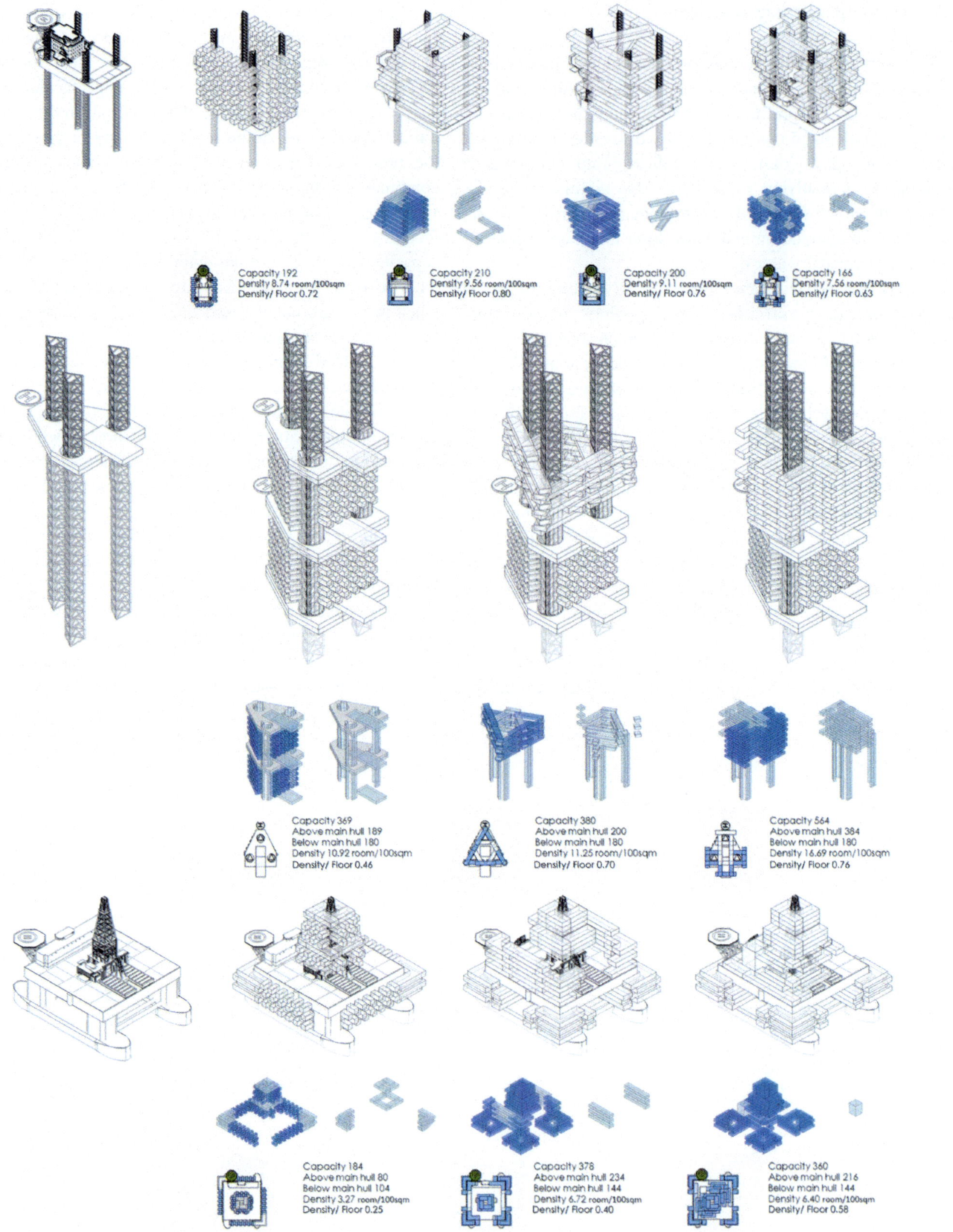

Fig. 70 Permutation studies using oil rig structures to accommodate guest rooms within the range recommended for resort hotels. Both jack-up and semi-submersibles are used to study room capacity. For types of rigs used, see Table 4 in chapter "Intensifying Food and Housing at Sea"

Because deck areas differed between rig types, a rate was used for comparing shape efficiency, in this example, the number of hotel rooms per 100 m^2 of oil rig deck area.

Figures 71, 72, 73, 74 and 75 illustrate the component studies that informed the design development of the four-legged jack-up luxury resort.

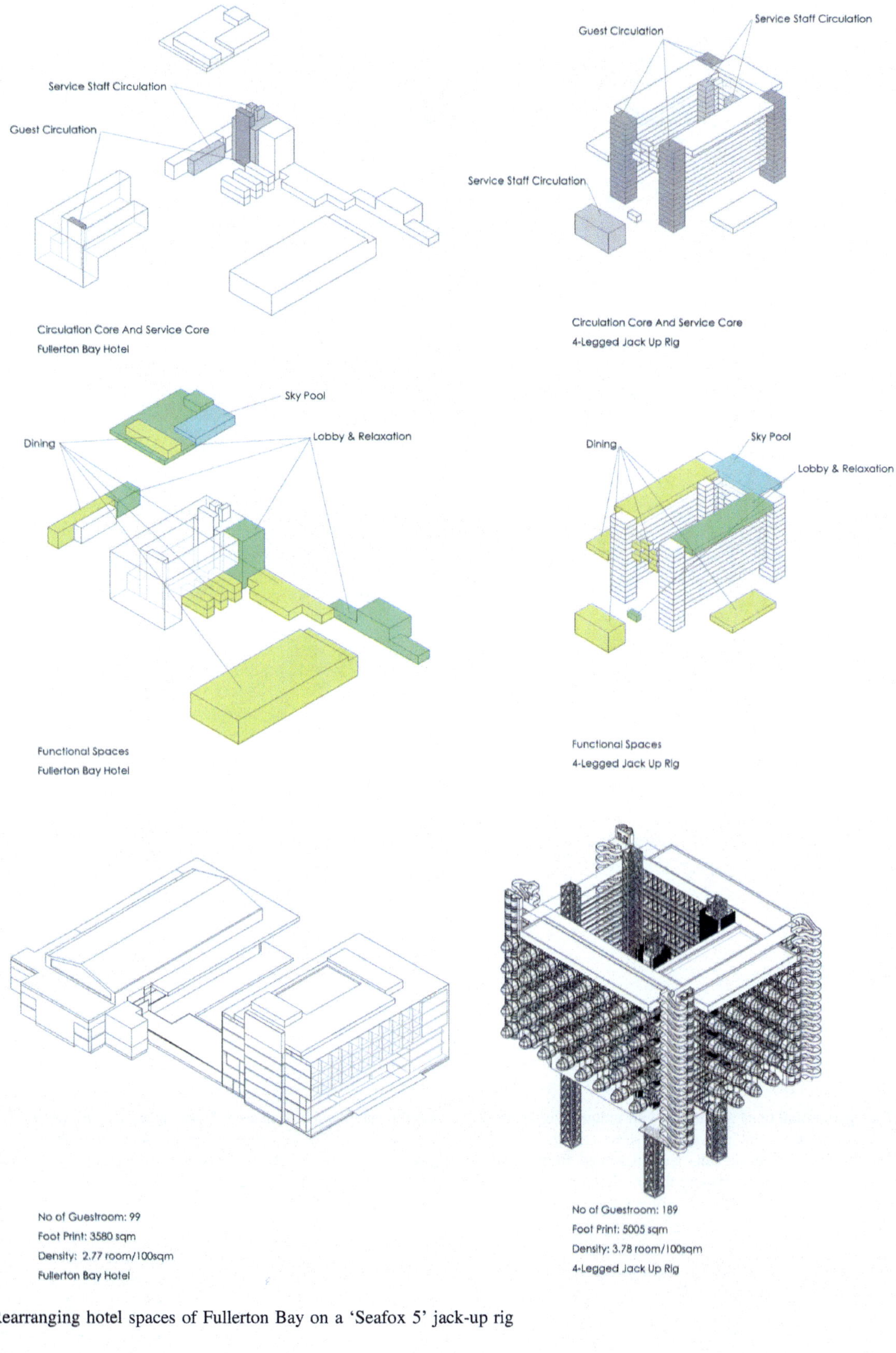

Fig. 71 Rearranging hotel spaces of Fullerton Bay on a 'Seafox 5' jack-up rig

(a)

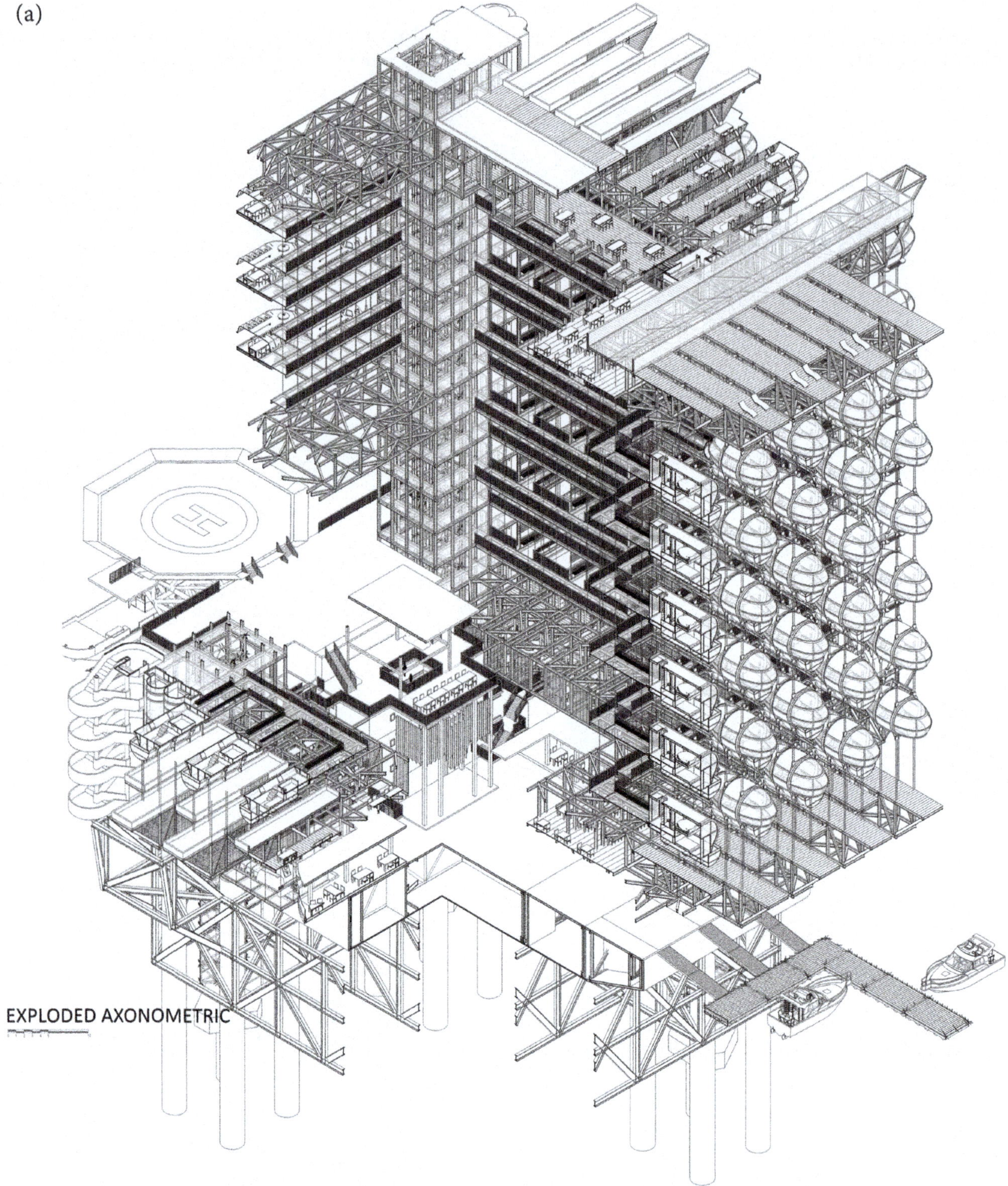

Fig. 72 **a** The spatial quality of an atrium with unique proportions, is framed by guest rooms with panoramic views of the sea. All grill-floored walkways and guestrooms are suspended from outrigger trusses to minimize visual obstruction by the oil rig structures. **b** Hotel rooms are made from recycled capsules of cantilevered observation wheels such as the Singapore Flyer or the London Eye. The additional structures to hold the capsules depend on their proportion and number in fitting into overall rig dimensions. Capsules provide good enclosure for private accommodation, keeping consistent with a nautical nature of the vessel as hotel

(b)

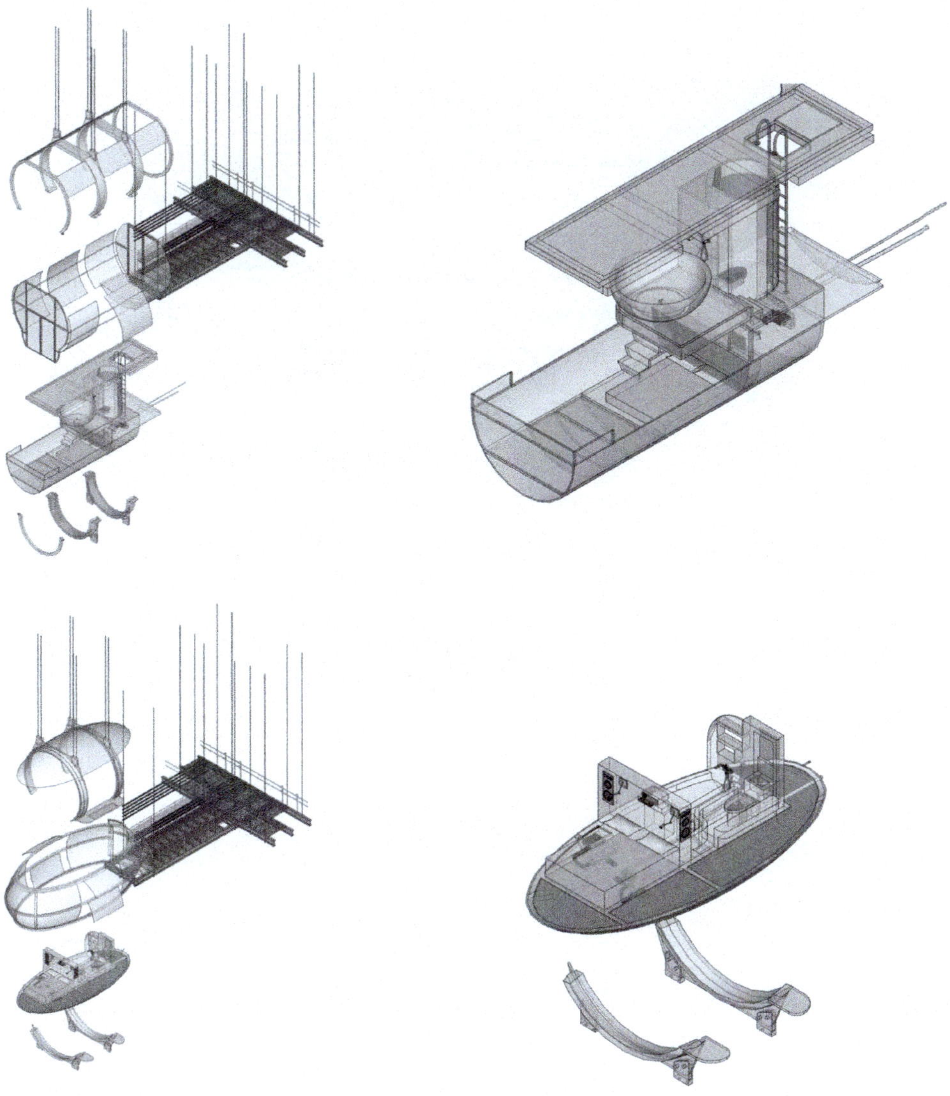

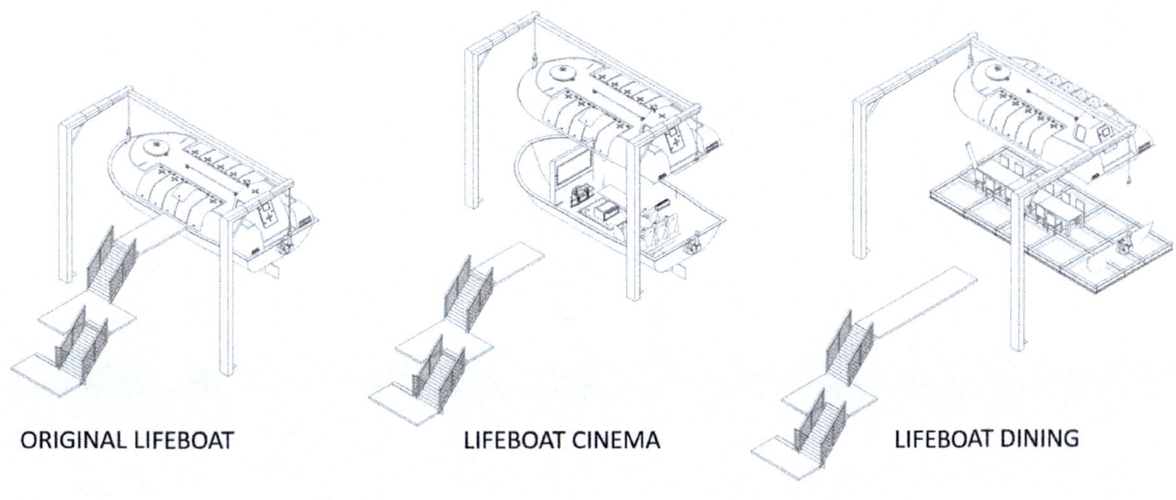

Fig. 73 Repurposing a lifeboat into a small cinema and dining facility

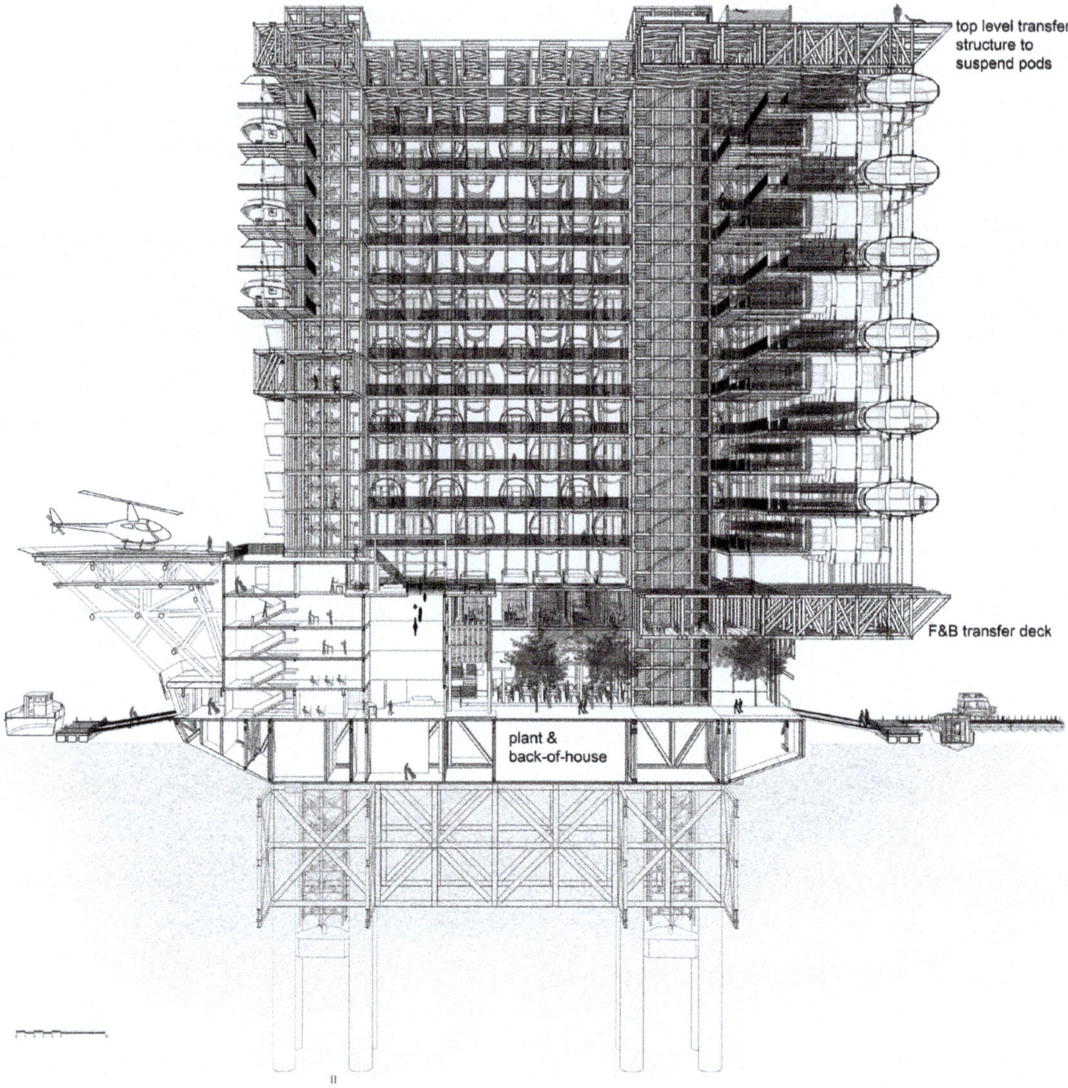

Fig. 74 The jack-up luxury resort can be accessed by helicopter or water vessel. Sea level entry is via a reception lobby next to a courtyard. An additional transfer deck accommodates dining facilities. The main hull/deck is for back of the house functions and mechanical plant

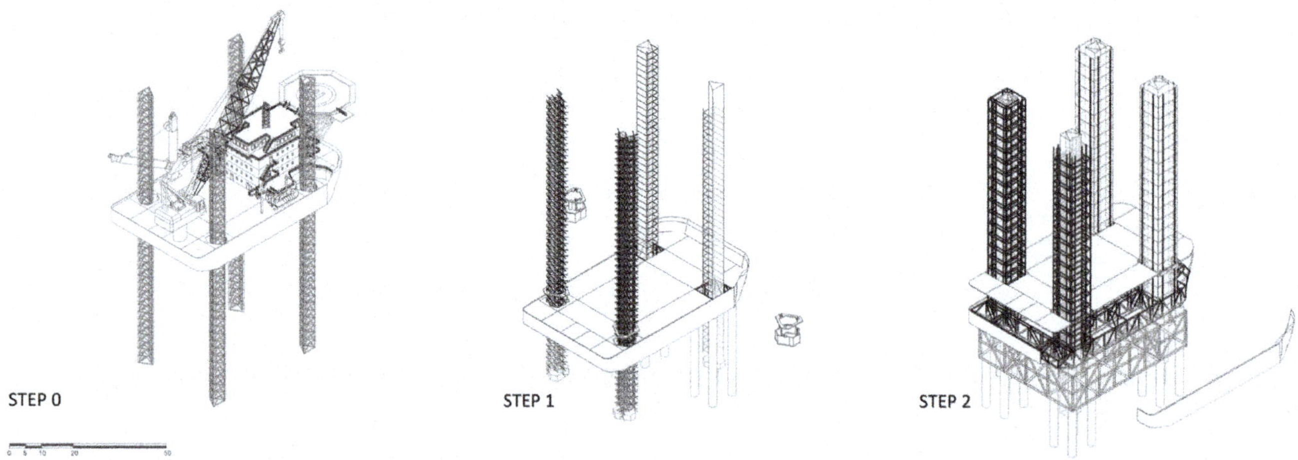

Fig. 75 Construction sequence

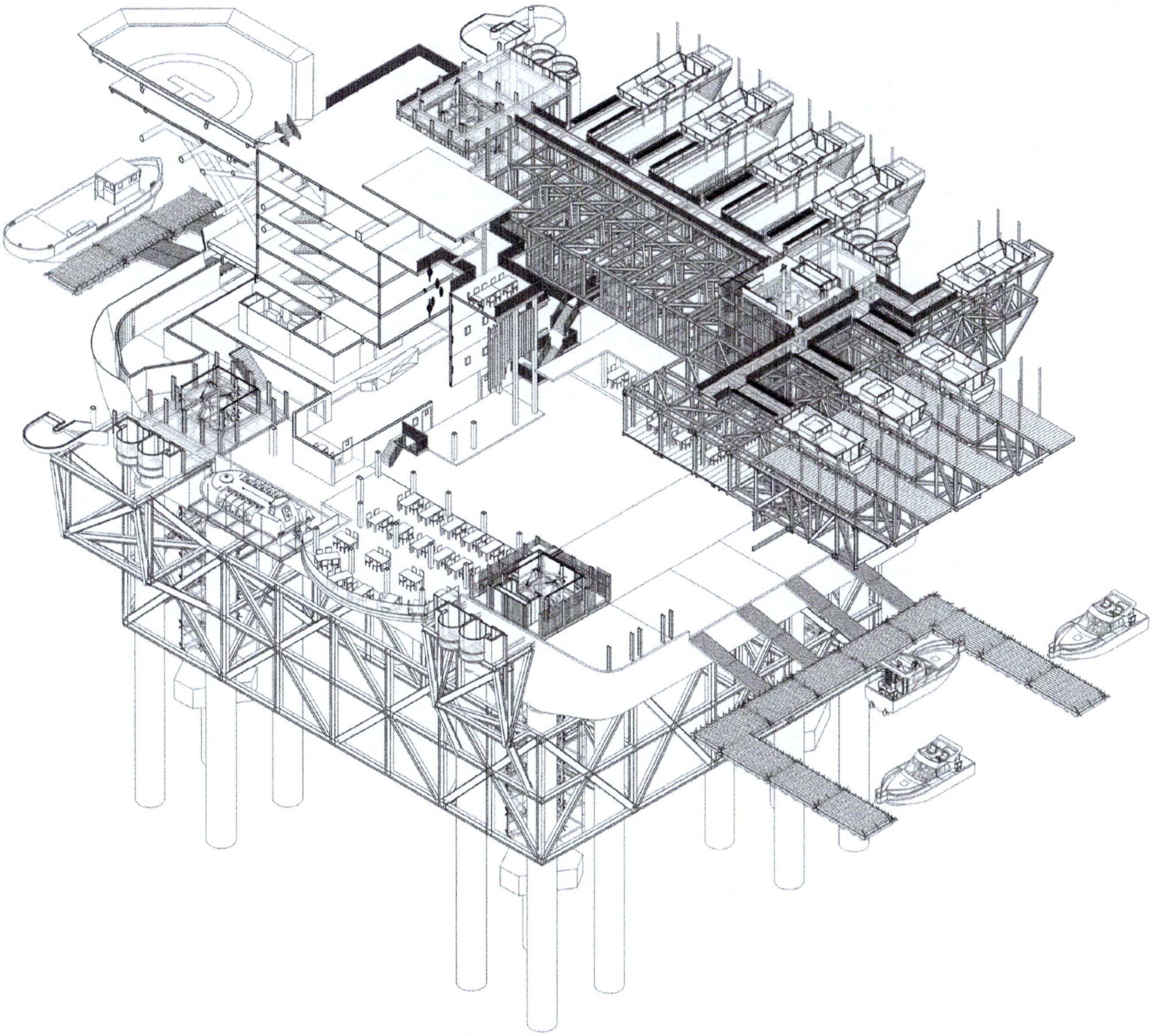

Fig. 74 (continued)

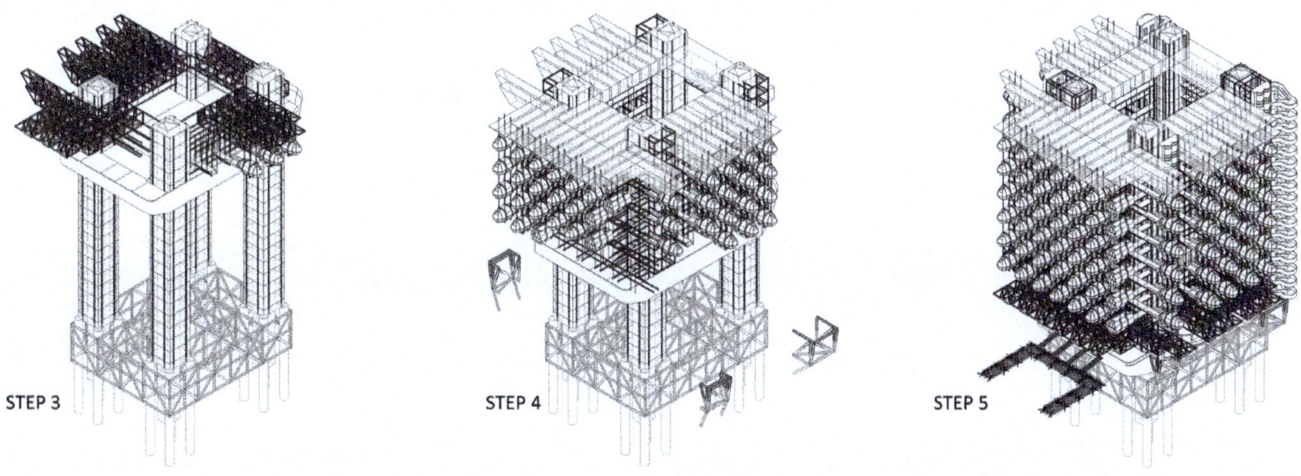

Fig. 75 (continued)

7 Semi-submersible Family Resort

The proposal for a family resort is explored in the context of repurposing a semi-submersible with a deck area comparable to 400-room Fullerton Bay Hotel, which occupies 4,200 m^2 plot on the water's edge. With 'Keppel DSS38' semi-submersible, the developed design accommodated 384 rooms in 4,830 m^2 with an efficiency of 7.95 rooms per 100 m^2. Fullerton Bay at 100 rooms on 4,200 m^2 has an efficiency of only 2.38 rooms per 100 m^2. The public areas of a semi-submersible include the area above the hull and the sheltered water body beneath the hull. This surpasses the public area of Fullerton Bay Hotel (Fig. 76).

Rooms are arranged into block configurations such as 'central tower', 'corner towers' 'pyramids' and 'inverse pyramids (Fig. 77a). The inverse pyramid has the least capacity with 196 rooms and only four rooms per 100 m^2 of hull area. The pyramid configuration has 336 rooms with 6.85 rooms per 100 m^2 of hull area and the largest capacity at 350 rooms can be accommodated in corner towers at 7 rooms per 100 m^2 of hull area. At preliminary stages, permutations are capped at twelve storeys to keep the CG of the combined structures within stability limits (Fig. 77b). A family resort has larger pool and F&B facilities. Pod rooms have occupancies of four persons capable of combining with a second pod to accommodate larger family occupancies (Fig. 77c).

The feature space is the twelve-storey atrium, which can be landscaped for communal usage. This overlooks the waterbody enclosed by the pontoons through an enlarged moonpool at deck level (Figs. 78 and 79).

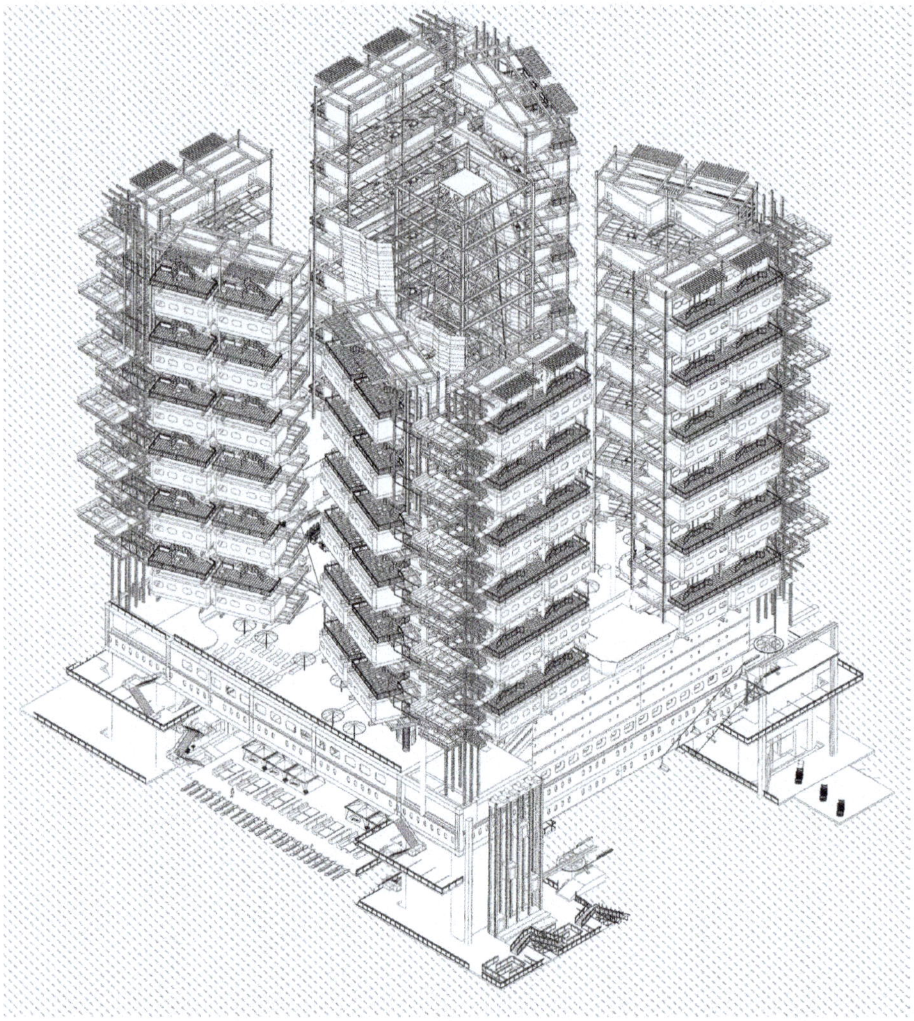

Fig. 76 A semi-submersible Family Resort with 384 rooms on a 4,830 m^2 hull/deck area

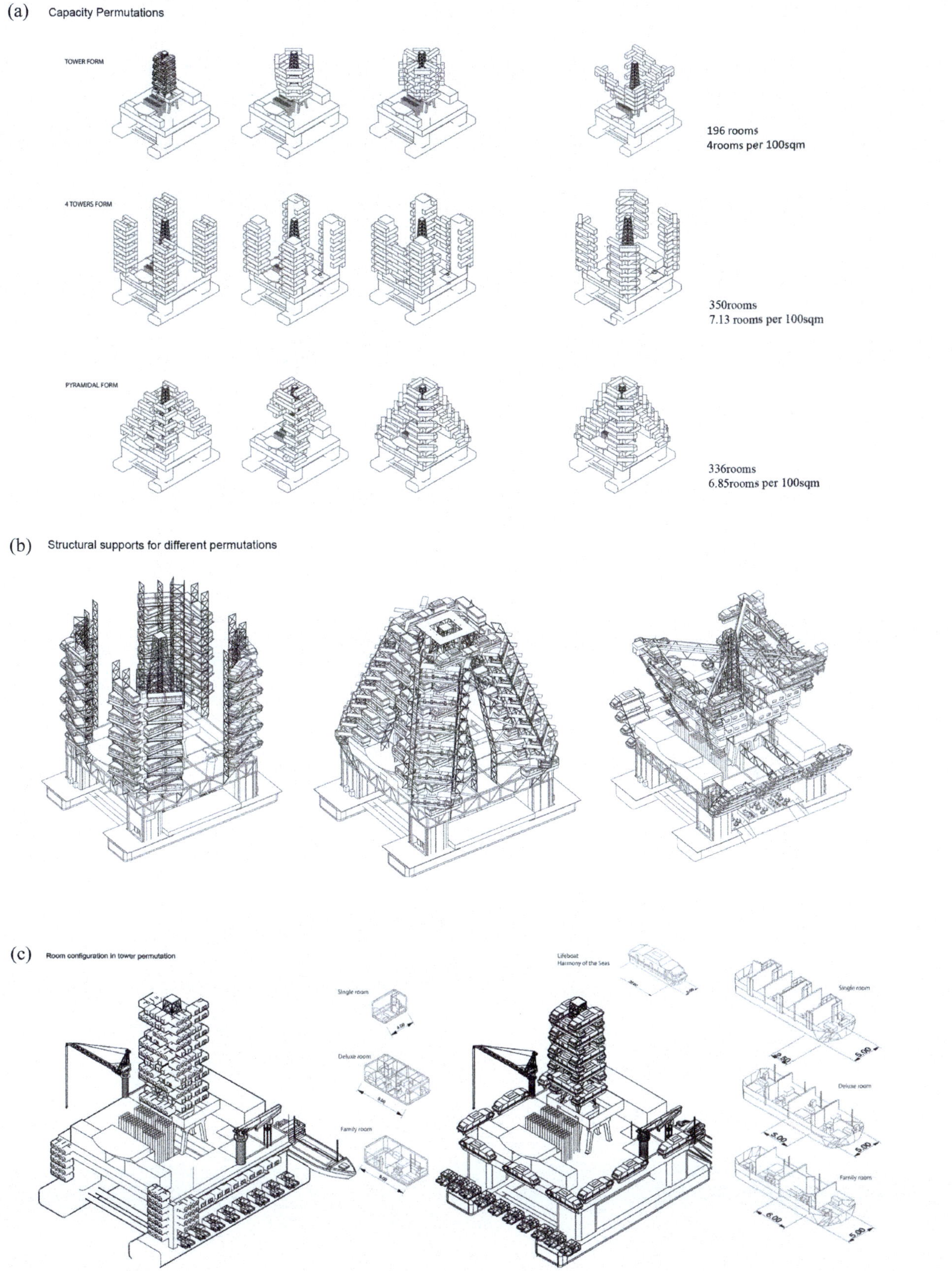

Fig. 77 **a** Capacity permutations; **b** structural supports for different permutations; and **c** optimizing room configuration to accommodate larger families

Fig. 78 To achieve the spacing for a dramatic atrium above the hull of the rig, rooms are arranged in L-shaped towers. Each L-tower has 96 rooms and panoramic views of the sea

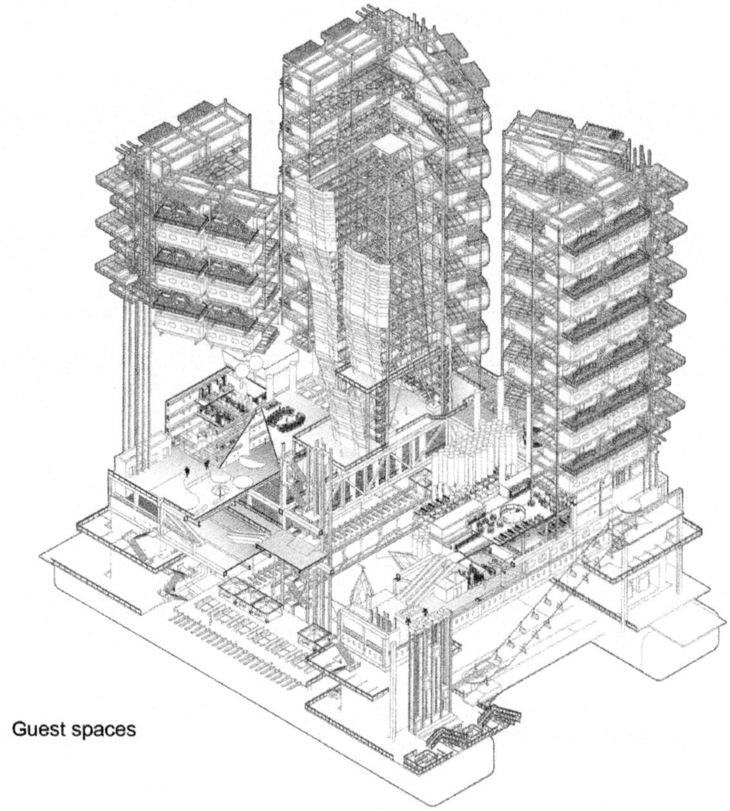

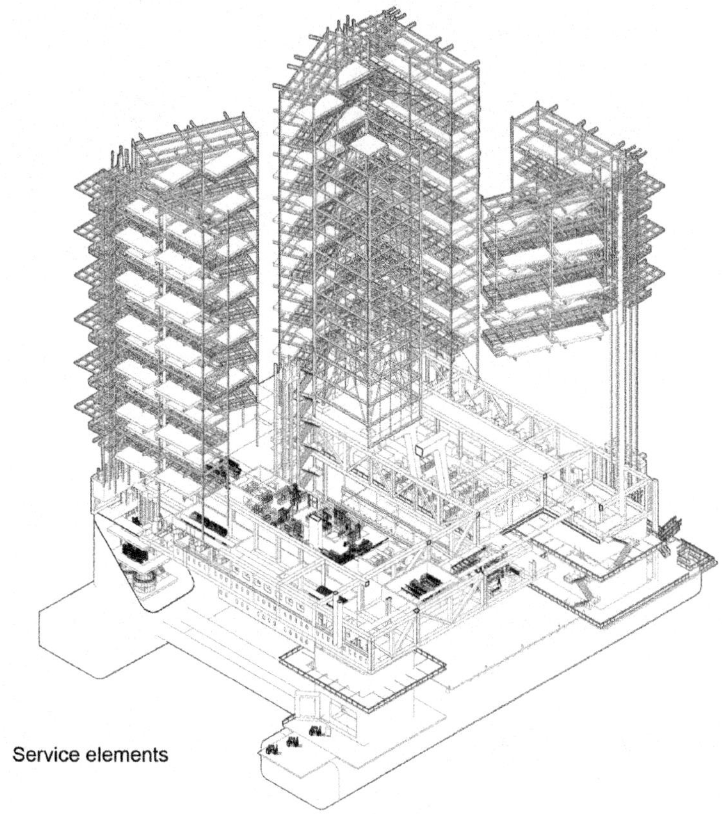

Fig. 79 Axonometric detailing the levels of the semi-submersible family resort. All indoor amenities may be accommodated at the top and middle decks of the hull. Machine rooms are accommodated at the lower deck. The pontoon level is designed for sea sports and recreation at the center, with overlooking F&B facilities. Jetties are located at the fringes for guest arrivals and departures and a separate berth is provided for service deliveries

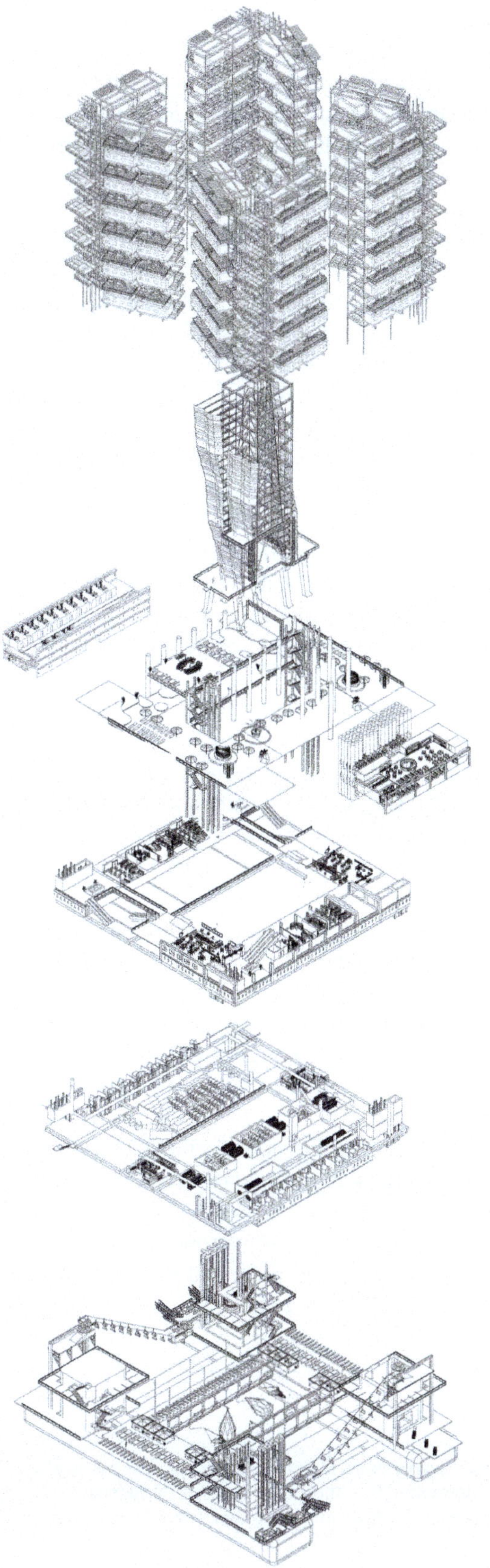

8 Summary

The standalone applications address emergent need in ways, which obviate sustained reliance on energy in a resource scarce future. Three repurposed rigs generate resources for producing water, food or energy; for example, the byproducts of cremation in the jack-up funeral facility can be used as compost for crop fertilizer. The OTEC system incorporated in the vertical waterpark obviate the reliance on fossil fuel and mitigate the extent of water consumption with solar distillation processes. The piezoelectric application for the semi-submersible prison produces excess electricity. These examples are self-sustaining and produce useful resources for settlements.

The remaining repurposed rigs introduce new business propositions with hybrids of architecture and infrastructure. The offshore transshipment LNG bunker facility allows cargo operations in the same location for bunkering of container vessels 200–400 m long, reducing the turn-around time for port berthing and refueling processes. It increases TEU capacity at anchorages and increasing harbor productivity.

The proposition of converting a semi-submersible rig into an Olympic competition pool or a FIFA soccer stadium accompanied with games villages and recreational facilities ensure its post-event viability. New business opportunities with stadia–on-lease reduce the immense investment risks of host countries and mitigate the dwindling future bids for the Olympics event.

Offshore resort hotels, which may be accessed by helicopters, hovercrafts and ferries, relieve the demand on coastal land and integrate hospitality experiences on sea, in all-weather conditions. The up-cycling of rescue craft and observation wheel capsules into room pods of varying sizes suggest erection processes which take advantage of elevating mechanism in jack-up rigs. The conversion of moonpool, derrick and pontoon structures into hotel facilities on board semi-submersible rigs results in unique architectural spaces and experiences.

Overall, all types of repurposed rigs introduce new spatial experiences while saving land and time; and generate new operational advantages with limited resources.

References

1. Bios. Why planting your ashes directly under a tree will not work and how the bios urn solves the problem. https://urnabios.com/why-directly-planting-in-cremation-ashes-directly-will-not-work/. Accessed Jan 2020
2. Bosch universal heat recovery steam boiler. Type HRSB
3. Casten S (2003) Update on U.S. steam turbine technology. In: 8th annual conference Canadian District Energy Association
4. deRooS JA (2011) Planning and programming a hotel. Cornell University, School of Hospitality Administration
5. Facultatieve Technologies. Cremators FT II–FT III. https://www.facultatieve-technologies.com/systems-products/cremators/. Accessed Dec 2019
6. Genting Singapore. Sustainable report 2014
7. Hansen EB (2008) Is water conservation at waterparks considered oxymoronic
8. International Olympic Committee (2020) Olympic marketing fact file 2020 edition
9. Keijzer E (2011) Environmental impact of funerals: life-cycle assessments of activities after life. University of Groningen, CIO
10. Kwon MY (2010) Lack of crematorium causes problem. The Korea Times
11. Lee SN (2012) An artful answer to the cremation question. Korea JoongAng Daily
12. Makai Ocean Engineering. http://www.makai.com/ocean-thermal-energy-conversion/. Accessed 2017
13. Maritime and Port Authority of Singapore (MPA). Singapore's 2019 Maritime Performance. https://www.mpa.gov.sg/web/portal/home/media-centre/news-releases/detail/38b82bb6-2f92-418c-a98d-c4c9e56f9232
14. McBride J (2018) The economics of hosting the Olympic games. Council on Foreign Relations
15. Morris S (2005) Britain's only prison ship ends up on the beach. The Guardian
16. Mytum H (2004) Mortuary monuments and burial grounds of the historic period. Springer
17. National Environment Agency. Burial, cremation and ash storage
18. Ogle D, John L (2010) Plants for saline to sodic soil conditions. Technical Note No. 43. USDA Natural Resources Conservation Service
19. OTEC Africa. http://www.otecafrica.org/Technology.aspx. Accessed 2017
20. Pang J (2019) Ash scattering in 'Garden of Peace' at Choa Chu Kang cemetery to be ready in 2020
21. Parkworld. https://www.parkworld-online.com/. Accessed 2017
22. Penal Reform International. Prison overcrowding. https://www.penalreform.org/issues/prison-conditions/key-facts/overcrowding/. Accessed Jan 2020
23. Reuters (2016) Japan's corpse hotels upset some of the neighbors
24. Shamdasani J (2018) New sea burial facility to be built in Tanah Merah. Straits Times
25. Staudt C, Ellens JH (2013) Our changing journey to the end: reshaping death, dying, and grief in America. Praeger
26. Stringam BB (2008) A comparison of vacation ownership amenities with hotel and hotel resort hotel amenities. J Retail Leis Prop 7:186–203
27. Tan A (2014) More opting for fuss free burial at sea. Straits Times
28. United Nations, Department of Economic and Social Affairs, Population Division (2019) World population ageing 2019
29. van den Bergh BJ et al (2011) Imprisonment and women's health: concerns about gender sensitivity, human rights and public health. Bulletin of the World Health Organization
30. Yun S et al (2014) Conceptual design of an offshore LNG bunkering terminal: a case study of Busan Port. J Marine Sci Technol (Springer)
31. Vettas N (2015) The impact of the 2004 Olympic games on the Greek economy. Foundation for Economic and Industrial Research, Greece

32. Walmsley R. World prison population list. Institute for Criminal Policy Research
33. Warren MW, Maples WR (1997) The anthropometry of contemporary commercial cremation. J Forensic Sci 42(3)
34. World Prison Brief. Highest to lowest prison population total. Institute for Criminal Policy Research
35. Worpole K (2003) Last landscapes: the architecture of the cemetery in the west. Reaktion Books

The manufacturer's authorised representative in the EU is Springer Nature Customer Service Centre GmbH, Europaplatz 3, 69115 Heidelberg, Germany. If you have any concerns regarding our products, please contact ProductSafety@springernature.com

Printed and bound by CPI Group (UK) Ltd, Croydon, CR0 4YY
30/09/2026
02225269-0002